G.C. Rausser GATT Negotiations and the Political Economy of Policy Reform

Gordon C. Rausser

GATT Negotiations and the Political Economy of Policy Reform

Associates

P.G. Ardeni, H. De Gorter, W.E. Foster, R. Gray, R.E. Just, B.S. Labson, D.J. Nielson, A. Schmitz, J. Vercammen, D. Zilberman

With 39 Figures

Springer

Professor Dr. Gordon C. Rausser
University of California
Department of Agricultural and
Resource Economics
207 Giannini Hall
Berkeley, CA 94720, USA

ISBN-13: 978-3-642-79286-1 e-ISBN-13: 978-3-642-79284-7
DOI: 10.1007/978-3-642-79284-7

Library of Congress Cataloging-in-Puplication Data. Rausser, Gordon C. Gatt negotiations and the political economy of policy reform / Gordon C. Rausser; associates, P. G. Ardeni ... [et al.] p. cm. Includes bibliographical references and index. (Berlin). -- (New York) 1. Uruguay Round (1987-) 2. General Agreement on Tariffs and Trade (Organization) 3. International trade. 4. Produce trade-Goverment policy. 5. Agricultural subsidies. I. Arseni, Pier Giorgio. II. Title. HF172.R38 1994 382'.92-dc20 94-33513

Softcover reprint of the hardcover 1st edition 1995

Typesetting: M. Masson-Scheurer, Homburg, Saar
31/3130-5 4 3 2 1 0 - Printed on acid-free paper

Contents

List of Authors

Pier Giorgio Ardeni
Professor, Department of Economic Sciences, University of Bologna, Bologna, Italy

Harry De Gorter
Professor, Department of Agricultural Economics, Cornell University, Ithaca, New York, USA

William E. Foster
Professor, Department of Economics and Business,
North Carolina State University, Raleigh, North Carolina, USA

Richard Gray
Professor, Department of Agricultural Economics,
University of Saskatchewan, Saskatoon, Saskatchewan, Canada

Richard E. Just
Professor, Department of Agricultural and Resource Economics,
University of Maryland, College Park, Maryland, USA

B. Stephen Labson
Economist, Australian Bureau of Agricultural and Resource Economics,
Canberra, Australia

David J. Nielson
Economist, World Bank, Washington, DC, USA

Gordon C. Rausser
Robert Gordon Sproul Distinguished Professor,
Department of Agricultural and Resource Economics,
University of California, Berkeley, California, USA

Andrew Schmitz
George W. and Elsie M. Chair Professor,
Department of Agricultural and Resource Economics,
University of California, Berkeley, California, USA

James Vercammen
Professor, Department of Agricultural Economics,
University of British Columbia., Vancouver, British Columbia, Canada

David Zilberman
Professor, Department of Agricultural and Resource Economics,
University of California, Berkeley, California, USA

Chapter 1

The Uruguay Round and the GATT Negotiations

Gordon C. Rausser

1.1 Introduction

In 1986, the Uruguay Round began in Punta del Este with much promise. In contrast to prior negotiation rounds, the US government made agricultural reform its top priority. What set this round apart from previous negotiating rounds was the explicit recognition that trade reform in agriculture requires reform of the underlying domestic agricultural policies. These special features of the Uruguay Round – namely, its focus on agriculture and on the reforms of each country's internal subsidization policies – were also largely responsible for the stalemate that plagued the Uruguay Round. Even though agriculture was 1 of 15 separate negotiation groups in the Uruguay Round, it nevertheless was a block to trade liberalization in a number of important areas such as intellectual property, services, government procurement and investment, tropical products, textiles, market access, custom evaluation, and dispute settlement.

In 1985–86, the Reagan administration, recovering from being soundly defeated in its second attempt (the first attempt occurred in 1981) to reform US agricultural policy unilaterally, turned to international negotiations as the vehicle to implement its strategy of domestic policy reform. As noted by Rob Paarlberg (1992), "what couldn't be obtained directly from congress could perhaps be achieved indirectly, through GATT." Initially, Reagan administration officials were delighted with the framework adopted in Punta del Este for the Uruguay Round negotiations. Agriculture became the fundamental core issue, a position it maintained over the seven long years of unsuccessful negotiations. Essentially, Reagan administration officials seized upon an opportunity to sidestep the domestic political-economic forces, specifically the commodity interest groups and well-established supporting institutions, that have always dominated the design of US agricultural policy. These same officials were even more delighted with the adoption of the so-called congressional "fast track" ratification procedures that require that the US Congress either accept or reject the entire GATT agreement without revision. This means that the battle lines would be drawn differently than if only agricultural issues were under examination. A coalition of interest groups that would determine whether a GATT agreement was accepted or rejected would be dominated by interests quite different from those that have prevailed in past debates over pure agricultural legislation. This all presumes that the Uruguay Round could be successfully completed and that an external agricultural code would necessarily be imposed as a constraint on future domestic US farm legislation.

Unfortunately, the Reagan administration, following 1986, largely withdrew from pursuing an active strategy of significant unilateral reforms, placing almost all their eggs in the GATT basket. This position has been largely maintained by both the Bush and the Clinton administrations. Although these officials recognized the importance of multiple issues, expanding the policy space, trading off policies, and changing the interest group landscape in the congressional forum, they seem to have dismissed the importance of these same factors in the GATT forum where agriculture has been treated as a separable negotiating issue.[1] Moreover, these same officials paid insufficient heed to the possible influence and power that could be exercised by domestic commodity interest groups on the actual GATT negotiation process and thus on the potential formation of an external agricultural code for all contracting countries.

Given this perspective, this chapter focuses on four major themes. First, contrary to the hopes and dreams of both the Reagan and Bush administrations, and more recently the Clinton administration, an external GATT code cannot be designed independently of internal political-economic forces. Second, the differences that arise between industrial and agricultural sector outcomes in the GATT are sourced largely in the relative complexity and institutional structures of internal domestic support policy for each sector. Third, sustainable reform can be achieved only by simultaneously pursuing both unilateral policy adjustments and multilateral external codes that can be imposed as constraints on the future dynamics of internal political economic forces. Under a unanimity negotiation rule, external codes cannot be formed unless countries that suffer significant consequences have found a politically robust "way out" of their internal subsidization programs. Nowhere is this more apparent than in France. Fourth, any reform package must recognize that both productive, political-economic resource transactions (PERT), and predatory, political-economic seeking transfers (PEST) policies are jointly determined and thus both unilateral and multilateral proposals must formally integrate the inherent complementarity and/or substitutability of these two types of policies. These basic themes are used to examine the historical performance of the United States in agricultural policy reform over the 7 years that have unfolded during the Uruguay Round and to examine a generic political-economic model that should prove useful in examining agricultural policy reform and trade liberalization in the current environment.

1.2 Empirical Foundation for Multilateral Trade Negotiations (MTN)

Over the last decade, numerous empirical studies on agricultural policy reform provided incentives for the Reagan, Bush, and Clinton administrations to forge ahead. The empirical foundation for a global perspective was sourced in the posi-

---

[1] Rausser and Simon (1991, 1993) show, in the context of a generic multilateral bargaining model, what disastrous consequences can arise from such a negotiating rule.

tive spillover effects that could be generated from agricultural trade liberalization. Basically, these studies simply provided formal justification for what the leaders of these administrations already knew was the right answer.

On the domestic macroeconomic front it was demonstrated that the macroeconomic economy had disastrous effects on the US agricultural sector in the early 1980s (Rausser et al. 1986) and that agriculture can cause disequilibriums in the macroeconomic economy. The latter influences emerge through (a) the general inflation rate, (b) government deficits or surpluses, and (c) the balance of trade. Each of these influences can have and have had, in turn, dramatic effects on employment, real interest rates, investment, and economic growth. A number of general equilibrium analyses have also been conducted to estimate the intersectoral effects of agricultural policies on the balance of the US economy. One study concluded that the misallocation of resources and capital to agriculture depressed the productivity of other sectors of the US economy and reduced American manufacturing exports by $7.5 billion and service exports by $3.4 billion (Hertel et al. 1989). Another study estimated that the removal of all programs which distort agricultural production or constrain input use would increase the 1991 GNP by $9.6 billion (Robinson et al. 1989).

Since the United States is a large producer of some commodities on the world market, its price supports and its accumulation of stocks can conceivably result in short-run favorable consequences for all exporters of the commodity in question. Specifically, if the internal price supports are so high as to effectively eliminate the export market as a relevant alternative, in the short run all the benefits accrue to other exporting countries. Over much of the post-World War II period, the United States has behaved as a residual supplier on world markets of many major commodities, especially cotton and food and feed grains.

To the extent that the price support programs and coupled subsidy transfers, as well as protection against import competition (such as quotas in the United States and variable levies in Europe) all induce greater production, world prices will be depressed. This is particularly evident when the US government sells unwanted stocks on the world market at less than the domestic price (through the Export Enhancement Program), makes concessional sales, or simply donates the food as aid (through PL480). These potential effects have been examined in a number of empirical analyses (Tyers and Anderson 1986; Zietz and Valdéz 1986; Roningen and Dixit 1989). For example, Roningen and Dixit estimate that eliminating US agricultural policies would increase world dairy product prices by 23.5%, sugar by 22.8%, coarse grain by 11.6%, wheat by 10.6%, rice by 2.9%, ruminant meats by 3.8%, and nonruminant meats by 3%. This would lead to corresponding costs for consumers and benefits for producers in the rest of the world.

Anderson and Tyers (1990) estimate that multilateral agricultural policy liberalization by all Organization for Economic Cooperation and Development (OECD) countries would increase the world prices of dairy products by 90%, sugar by 22%, coarse gain by 3%, wheat by 25%, rice by 18%, ruminant meats by 43%, and nonruminant meats by 10%. While these price changes would result in costs for consumers and benefits for producers in the developing world, Anderson and Tyers estimate that the net welfare of developing countries would increase by 1%. Simultaneous policy liberalization by developing countries, however, would result in a net increase in developing country welfare of up to 64%.

One justification often expressed in support of price floors and public storage programs in the United States and other industrialized countries is that they stabilize what would otherwise be an unacceptable domestic volatility in basic commodity prices, at least over the very short run.[2] Ironically, these same policies amplify rather than dampen commodity price fluctuations on international markets. One glaring example of this phenomenon is the world sugar market. The European Community (EC) and the United States both protect their domestic sugar producers – for example, in the United States through price supports, tariffs, and import quotas. These policies have been estimated to have increased price instability in the residual world market for sugar by approximately 25% (World Bank 1986 World Development Report). Moreover, because the United States have been dominant in the world sugar trade, the imposition of import quotas has lowered world sugar prices.

Not surprisingly, European and US sugar policies have also placed significant burdens of adjustment on many developing countries. The World Bank 1986 World Development Report has estimated that sugar policies of industrialized countries cost developing countries about $7.4 billion in lost export revenues during 1983 and reduced their real incomes by about $2.1 billion. Given the domestic supply response to sugar and other substitutable products, and the zero treasury provision of the US sugar program, even those developing countries who currently benefit, can expect their quota levels, and thus values, to slowly vanish. In the case of world wheat prices, Schiff (1985) has estimated that the variability could be reduced by 48% if all countries were to end their subsidization of wheat. Tyers and Anderson (1986), using a model simulating policy reform in more than a half dozen commodity markets, calculated that liberalization of agricultural policies of industrialized countries would substantially reduce the international price variability of major temperate-zone commodities: wheat by 33%, coarse grains by 10%, rice by 19%, sugar by 15%, and dairy products by 56%.

Along with the spillover costs that emerge both domestically and internationally from current agricultural policies, we must include the deadweight losses that emerge in redistributing wealth from consumers and taxpayers to agricultural producers. At the height of the subsidization of agricultural sectors in the western world (1986–87), the OECD estimated that the total consumer and taxpayer cost ran in the neighborhood of $200 billion, with the amount received by producers in the vicinity of $100 billion. Hence, the deadweight loss was estimated in these studies to be as large as $100 billion on an annualized basis (G.E. Rossmiller 1988, National Center for Food and Agricultural Policy, Washington, DC, pers. comm). These losses exclude the adverse consequences on the macroeconomies, other related sectors of an economy, and general economic growth.

---

[2] Stabilizing prices are, of course, not equivalent to stabilizing incomes. As noted by Newbery and Stiglitz (1981), stabilizing prices may actually increase income variability.

The most recent study, a 4-year effort sponsored by the World Bank and the OECD Development Center in Paris (Goldin et al. 1994) examines a partial liberalization of agricultural and manufacturing activities, neglecting liberalization in other areas such as services. For the manufacturing and agricultural sectors, a 30% reduction in all border tariffs, quotas, and other restrictions of farm and manufactured goods is specified. The authors estimate that this level of partial liberalization would result in an increase in global income of $213 billion per year. The vast majority of this increase, $190 billion, emerges from partial liberalization of agricultural trade, while $23 billion is attributed to the manufacturing sector. In the case of agricultural trade liberalization, over 60% of these gains accrue to OECD countries, with the balance allocated to developing countries. This study also found that if all trade distortions in agriculture and manufacturing stemming from subsidies and tariffs are completely removed, gains would measure $450 billion per year by the first part of the next century.

Why, in the face of such overwhelming empirical evidence, does agriculture remain such a formidable obstacle to a successful completion of the Uruguay Round? The answer is obvious: powerful interests, political-economic forces, and institutions that have emerged that legitimize and sustain those powerful interests. From the perspective of the United States, we first examine the institutions that have emerged to legitimize these powerful interests and what role these institutions have played in prior GATT negotiation rounds. This will be followed by an examination of the role of powerful interests and other forces over the last few years of the Uruguay Round.

1.3 Historical GATT Agricultural Origins

Since its inception, the GATT has been surprisingly successful in reducing barriers to trade in industrial goods. Comparable progress in agricultural trade, however, has been elusive. The most prominent fact concerning GATT actions in agriculture is its failure, time and time again, to bring trade in the sector under any internally consistent set of rules.

The genesis of this failure lies in the original GATT charter. The first paragraph of Article XI states that, "no prohibitions or restrictions other than duties, taxes, or other charges shall be instituted or maintained by any contracting party" (Dam 1970, p. 407). At the insistence of the United States, however, a second paragraph was added, permitting export and import restrictions to protect a myriad of US agricultural commodity programs. Specifically, the provisions of paragraph 1 of this Article shall not extend to the following:

"(a) Export prohibitions or restrictions temporarily applied to prevent or relieve critical shortages of foodstuffs or other products essential to the exporting contracting party;

(b) Import and export prohibitions or restrictions necessary to the application of standards or regulations for the classification, grading, or marketing of commodities in international trade;
(c) Import restrictions on any agricultural or fisheries product, imported in any form, necessary to the enforcement of government measures which operate:
   (i) to restrict the quantities of the like domestic product permitted to be marketed or produced;
   (ii) to remove a temporary surplus of the like domestic product by making the surplus available to certain groups of domestic consumers free of charge or at prices below the current market level;
   (iii) to restrict the quantities permitted to be produced of any animal product the production of which is directly dependent, wholly or mainly, on the imported commodity, if the domestic production of that commodity is relatively negligible."

Today, this US-backed exemption looks like a laundry list of the trade complaints of American producers. But for provision 2(a), the 1973 export embargo (and, perhaps, the 1974 and 1975 moratoria on sales to the Soviet Union and Poland) that grain producers still bitterly resent would have violated the GATT (for a detailed description of these measures, see US Department of Agriculture, Economic Research Service 1986). The widespread use of grading and packaging restrictions exempted in paragraph 2(b) are a prominent bone of contention for many specialty crop exporters (Wright et al. 1987); and paragraph 2(c), parts (i) and (iii) obviously legitimize the use of import restrictions to complement domestic measures that raise consumer prices above world levels, subject to only one substantial proviso:

"... any restrictions applied under (i) above shall not be such as will reduce the total of imports relative to the total of domestic production, as compared with the proportion which might reasonably be expected to rule between the two in the absence of restrictions."

The fact that quantitative restrictions were allowed if government measures operated to restrict the production or marketing of agricultural products meant that the GATT would not operate in agriculture as a counterweight to domestic producer pressures for distortionary favors (as it does in industry). This exception was clearly designed to fit the American case because the United States were the only major agricultural producer with acreage and marketing controls (in addition to price supports). Section 22 of the Agricultural Adjustment Act of 1933 already permitted tariffs and quotas on agricultural imports that interfered with domestic programs. To be sure, the Article XI loophole became the first indication that GATT would, by design, be ineffective in reforming domestic production restrictions that required trade barriers for their operation.

In 1951, Congress amended Section 22 by adding subsection (f): "No trade agreement or other international agreement heretofore or hereafter entered into by the United States shall be applied in a manner inconsistent with the requirements of this section" (Hillman 1978, p. 211). In an amendment to the Defense Production Act, Congress also authorized the Secretary of Agriculture to restrict imports of a number of products for which there were no production controls. Both were severe blows to the constraint on the exemptions in Article XI. As the Secretary-General of the GATT commented, although Article XI was "largely tailor-made to US requirements ... the tailors cut the cloth too fine" (White 1960, quoted in Dam 1970, p. 260). In the same year, the GATT Contracting Parties held that injured parties were entitled to compensation from the United States. (Thus, in 1952, the Netherlands was allowed to impede wheat flour imports from the United States in response to measures affecting their cheese exports.) Subsequently, the United States sought, and was granted in 1955, a "waiver" from the already weakened obligations in Article XI of the GATT. This waiver sanctified the full range of US interference in agricultural trade.

Having legitimized its own distortion of agricultural trade, the United States turned to using the GATT to seek market access for US exports to other countries. The United States were concerned that the unification of European agricultural policies, required by the Treaty of Rome in 1957, could greatly affect US exports. The common agricultural policy (CAP) of the EC involved the imposition of variable import levies. A variable import levy acts as a sliding tariff to ensure that no foreign goods are even imported below a certain price – an ideal and highly protective arrangement to validate domestic price-support programs in European agriculture. Because the GATT does not prohibit variable levies, there was no legal basis for the strong objections of the United States to that practice. The United States reluctantly decided to conclude the Dillon Round of trade-liberalizing negotiations (held from 1960 to 1962) with its progress on reducing industrial tariffs and with no formal agreement on trade in agriculture. It only had the EC's assurance that it would not use the variable levy to damage US exports.

The US position at the start of the Kennedy Round of trade talks (1963–1967) was that agriculture for the first time would be a major part of liberalization negotiations, equal in importance to progress on reducing trade barriers in manufactured goods. The United States were anxious to deal with the EC, but neither the United States nor the EC was amenable to fundamental compromises on their domestic policies. In fact, instead of liberalizing policies, cartelization of world trade in agriculture was broached at these talks when the Europeans proposed a *montant de soutien* that would bind the margin of government agricultural support and create a rigid world market in agricultural products. The United States flatly rejected this plan which would have, in effect, created a worldwide CAP. At the end of the Kennedy Round, the United States were again faced with the dilemma of whether or not to abandon the gains in reducing industrial tariffs in response to the lack of progress on the agricultural front; given its own historically compromised negotiating stance, its decision to set agriculture aside was not surprising.

The Tokyo Round (1973–1979) negotiations tried again to gain some ground in opening world agricultural markets. The United States did get some enlargements in Japan and EC import quotas for certain products, but attempts to reinforce the rules in the export subsidies code did not achieve much. Because world agriculture flourished during the 1970s, economic pressures did not provide any incentives to reach an agreement.

The provision dealing with subsidies is another gaping hole in the GATT framework. Although Article XVI generally prohibits the granting of export subsidies, there is an exception for primary products. While the GATT members "should seek to avoid the use of subsidies on the export of primary products," such subsidies are not forbidden. "If ... a contracting party grants directly or indirectly any form of subsidy which operates to increase the export of any primary product ... such a subsidy shall not be applied in a manner which results in that contracting party having more than an equitable share of world export trade in that product...." This exception, adopted with US support (Harris et al. 1983, p. 275) with its ambiguous restriction, has permitted excess domestic stocks to be dumped onto world markets through subsidies. This subsidization accelerated through the 1980s and is the most visible symptom of the problems in agricultural trade today. Thus, there has been a systematic failure on the part of the GATT to reform world agricultural policies, or at least it seems to contain the origin of growth in protectionism and domestic distortions.

Although the GATT Secretariat (1983) states, "it was the original purpose of the GATT ... to strengthen governments against the particular pressures emanating from national economies," agriculture was largely excluded from this process by arrangement dictated by the US Congress and trade negotiators who responded to the short-run concerns of special interests. They did not foresee that without stronger GATT constraints the EC would adopt a structure of agricultural protection in its CAP that would eventually cause severe disruption of international markets. Harris et al. (1983, p. 275) observe that "Ironically the Community's creation of the CAP in its current form, with its use of variable import levies and export refunds as its principal agricultural trade measures, was only possible as a result of earlier measures by, principally, the USA." As US agriculture has become more and more dependent on exports for its continued prosperity, the lack of discipline in agricultural policy abroad has become increasingly detrimental to the interests of the United States.

To summarize, in one round of GATT negotiations after another the major obstacle to successful negotiations has been agriculture. As Winham has said, "The main reason for the lack of progress in the Tokyo Round through mid 1977 was agriculture" (1986, p. 146). As in the Uruguay Round, the dispute between the United States and the EC over agriculture casts "a pall over the negotiation that prevented delegates from making progress in any one area" (Winham 1986, p. 137). During the Tokyo Round, political-economic forces made it indeed difficult for side payments across commodities to be arranged. US negotiators pursued a strategy of securing concessions from the EC on grains but offering concessions on milk products. Unsurprisingly, both EC grains producers and US dairy producers rejected this side payment strategy, much the same response of US oil seed and corn gluten producers to the EC "balancing" proposal during the Uruguay Round. In the earlier Kennedy Round, the United States, at one stage, attempted to link the success of the entire Round to the outcome of the negotiations on agriculture. Only when farm trade was cast aside were these negotiations completed.

Neglecting the lessons of history and prior negotiation rounds, the Uruguay Round attempted a significant structural break by linking trade reform to the design of the underlying domestic agricultural policies. The initial architects of the Uruguay Round, quite correctly, firmly held the view that agricultural trade distorting policies exist to rationalize internal policies that support and protect domestic agricultural sectors. As a result, these officials recognized that it was not possible to achieve sustainable reforms in trade policies without major reforms of each country's internal subsidization schemes. Accordingly, the ministerial declaration from Punta del Este that launched the new round stated that the negotiations would, "... aim to achieve greater liberalization of trade and agriculture and bring all measures affecting import access and export competition under strengthened and more operationally effective GATT rules and disciplines." A major objective of the current negotiations was that they were expected to enhance "discipline on the use of all direct and indirect subsidies and other measures affecting directly or indirectly agricultural trade, including the phased reduction of their negative effects and dealing with their causes." Quite obviously, this objective goes well beyond the single-minded GATT tradition of focussing on import barriers.

1.4 Uruguay Round and Commodity Interest Groups

Why have the US Executive branches of the last three administrations failed to pay heed to the Santayana insight? As Santayana said long ago, "Those who cannot remember the past are condemned to repeat it" (1905). Seven years into the Uruguay Round, we remained at a stalemate over agriculture. The lessons and insights of prior GATT negotiation rounds may well have been lost on the Reagan administration because of the Executive branch's experience in attempting to reform internal agricultural policies. In both 1981 and 1985, the Reagan administration proposals for reforming US domestic agricultural support policies were roundly and soundly defeated by the Congress. Initially, the 1985 Reagan administration proposal called for the elimination of all subsidization in agriculture. This proposal was pronounced "dead on arrival" by both influential Republicans and Democrats on Capitol Hill.

Following the defeat of the Executive branch proposal, and the subsequent passage of the 1985 Food Security Act, the Executive branch formed a working group on agricultural policy reform that encompassed both unilateral and multilateral strategies. As 1986 and 1987 unfolded, members of this working group presented various proposals to the Congress and found a welcome reception only for multilateral reform. During these early days, the issue was not so much how to minimize the pain of reform through simultaneous adjustments by several important countries but, instead, that there could well be significant gains to a number of commodity groups. Partially, because third parties could be blamed (EC, Japan), the degree of cooperation between the Congress and the Executive branch increased dramatically during this period, certainly relative to the divisive debate over the 1985 Food Security Act.

The difficulty of achieving unilateral policy reform was exacerbated during these years by the fierceness of competition for export markets and the fear that unilateral agricultural policy reform would allow competitors to gain unfair advantage on international markets while continuing to restrict access to their own markets. This led to a "prisoner's dilemma" in which many countries felt trapped in their existing policy regimes. Each country feared that unilaterally reducing its own export subsidies or eliminating its own internal level of support to farmers would result in the loss of market share. The action of one country on its own will rarely induce a significant rise in world prices. Consequently, from the standpoint of export market shares, each country finds that the potential rewards from liberalizing their agricultural policies do not warrant a unilateral move toward reform.

The view that emerged during this period was that simultaneous reform by several countries, if achieved, might provide an escape from this prisoner's dilemma. This line of reasoning characterized the mind set of participants at the outset of the Uruguay Round. In June, 1987, Clayton Yeutter, then US Trade Representative, remarked, "We are clearly not going to reduce our level of government involvement (in agriculture) unless other people move with us. We are going to go down this road together, and we are going to go down it arm in arm, and we are not going to walk 10 steps ahead of the Europeans or the Japanese or anybody else" (Rapp 1988, p. 171).

Once it became clear that all three administrations (Reagan, Bush, and Clinton) would pursue a strategy of significant policy reform through and only through the GATT, the political-economic interest groups within the United States naturally turned their attention away from obstructing unilateral reform and, instead, used the GATT negotiations as an opportunity to maintain, and in some areas enhance, their rent-seeking activities. On the other hand, the major public-interest benefit of the congressional fast-track provision was the change in the interest group landscape that would debate an external GATT agricultural code. This change in the interest group landscape would be expected to dramatically increase the probability of significant reform.

The determination of the external code, however, is not independent of the political-economic forces that exist within each country. For example, in the case of the United States, Paarlberg (1992) has documented how agricultural interest groups have been instrumental in blocking successful agricultural negotiations in the current Uruguay Round. He argues that agricultural interests supported the initial extreme proposal of the United States, which sought to eliminate, over 10 years, all agricultural subsidies that had any tendency to distort production or trade. Knowing that the probability of success was zero, a number of commodity interests actively supported this proposal. They argued that half measures will not do; no agreement is better than an agreement that fails to guarantee a "level playing field" against the heavy subsidization of foreign competitors. The resulting strategy pursued by US Trade Representatives (USTR) and US Department of Agriculture (USDA) led to one stalemate after another.

Once the two agencies, led by Clayton Yeutter at the USDA and Carla Hills at USTR, realized that a complete elimination of all subsidies had a zero chance of success, they turned toward partial but "substantial progressive reductions" in the degree of subsidization. US agricultural interests quickly adjusted to this revised goal of the Bush administration by successfully demanding that no unilateral reforms be implemented while the GATT negotiations were still underway. Farm groups successfully argued, and key members of the Congress and Executive branch agreed, that unilateral cuts would only weaken the strategic position of the US GATT negotiators. As a safety valve, US agricultural commodity interests in 1988 realized an opportunity of including in the Trade Act a position for expanded export subsidies if the Uruguay Round of GATT negotiations should fail. This provision was incorporated into the 1988 Trade Act; and to provide still further insurance in the Fall of 1990, these same interest groups were instrumental in including a "GATT trigger" provision in the final domestic budget reconciliation bill. Under the latter provision, the Secretary of Agriculture may be permitted to reverse all or any part of the $13.5 billion in domestic farm budget cuts (over 5 years) that were imposed in the final 1990 budget reconciliation bill. It can be argued that US agricultural interests might well have lost the rents generated from the Export Enhancement Program in 1989 had it not been a potential program that could be usefully traded away in the GATT negotiations.

Ironically, the Reagan and Bush administrations pursued the GATT forum in order to facilitate domestic reform of agricultural policies which, in turn, was used by commodity interest groups as an effective means to block reform. Quite obviously, the GATT negotiations are not separable from the dynamics of domestic political-economic forces. This nonseparability emerges not only for the United States but for all other major GATT participants as well. For example, COPA [Comite des Organisations Professionnelles Agricoles de la CEE (Committee of Professional Agricultural Orgnizations in the EC)] and COGECA [Comite General de la Cooperation Agricole de la CE (General Committee of Agricultural Cooperation in the EC)], the two most powerful European farm organizations, rejected proposed subsidy reductions by the EC Commission (1989/90) "without reciprocal measures taken by our GATT partners" (USDA 1989, p. 8).

In contrast to the GATT negotiation process emerging as an obstacle to recasting agricultural policies, budgetary pressures on both sides of the Atlantic have forced more traditional internal paths to reform. In the United States, the routine renewal of domestic farm programs by agricultural committees of Congress was derailed by the 1990 Budget Reconcilation Act. Due to these overriding concerns, Congress, as a whole, preempted the agricultural committees, deciding to cut US domestic farm commodity programs by a significant 25% over the ensuing 5 years. Recall, however, that US agricultural interests received as compensation a GATT trigger provision which held out the possibility of reversing the budget cuts if the Uruguay Round failed.

Internal budget pressures in the EC force a parallel internal policy reform process. A 20% increase in farm support spending in the EC helped lay the foundation for a major policy reform effort by the EC Commission. This effort has been popularized as the "MacSharry Plan," which called for approximately a 35% reduction in CAP commodity price guarantees over a 3-year period. It is interesting to note that this supply-management, subsidy-reduction proposal achieves in 3 year's time a pace of reform that is faster than the 75% cut in support levels over 10 years that was demanded by the United States, but rejected by the EC in the GATT negotiations.

1.4.1 Recent Proposals

The Dunkel plan (1991) basically took these internal budget-driven US and EC agricultural policy reforms and restructured them as the core of this proposed Uruguay Round agreement. Former GATT director, General Arthur Dunkel, simply recognized in this proposal the need to buttress MTN proposals with unilateral actions that are undertaken by various countries. In essence, this proposal simply incorporates much of what the United States and the EC have already accomplished separately and unilaterally. The proposal calls for 20–36% reductions in farm supports over a 6-year period, using a 1986–1988 (1986–1990) base level. In the case of the United States, the vast majority (14 out of 17 farm commodities) would already be in compliance; while, in the case of the EC, the only part of the Dunkel draft that appears to exceed the subsidy cuts already anticipated was a required 24% reduction, over 6 years, in the quantity of subsidized exports. This provision, along with the Dunkel requirement that export subsidies, as well as internal supports, be significantly reduced – namely, by 36% in value – turned out to be a major impediment to a resolution of the Uruguay Round.

The United States supported the Dunkel draft, while the EC objected. The former wanted to impose explicit limits on export subsidies, while the previously launched internal EC reform failed to provide any assurance that such targets would be honored. Essentially, the supply-management features of the CAP reform meant that the subsidized EC exports would eventually fall, but there were no guarantees that the constraints imposed in the Dunkel proposal could be achieved. At this juncture, a US/EC soybean trade dispute propelled the Uruguay Round forward with the Blair House discussions in November, 1992.

At the Blair House discussions, US negotiators argued for the 24% cut in subsidized export tonnage as specified in the Dunkel draft, while the EC Commission negotiators argued for 18%. Ultimately, the Blair House Accord reached a settlement of 21% over a 6-year period, which the United States have argued cannot be *reopened*, either directly or indirectly. The government of France, however, has taken violent exception, and threatened to veto the concession, arguing that the EC negotiators exceeded their assigned authority. In essence, the French are unwilling to see the EC Commission offer anything in the GATT that the EC Council has not already approved as part of the internal CAP policy reform of 1992.

Between here and the fast-track US congressional deadline of December 15, 1993, it is difficult to imagine that some compromise will not be achieved between the United States and France. This might be accomplished by not requiring commodity-by-commodity or year-by-year cuts, or perhaps by accepting the French proposal of excluding some existing stocks from the agreement and/or shifting the bulk of the cuts from the beginning of the 6-year period to near the end. The so-called peace clause, exempting agriculture from no export subsidy provision of the GATT for a period of 6 years, will likely remain a serious bone of contention; the EC seems to support an extension while the United States are adamantly opposed.

1.4.2 Potential Limitations of GATT-Motivated Internal Reform

The complexity of internal agricultural policies and their associated trade barriers, along with the nonseparability of domestic political-economic forces, has often been used to argue that the GATT is not an effective instrument for achieving significant reforms. This perspective holds that, even though the potential gains from worldwide reform of distortionary domestic and international policies in agriculture are huge, the GATT is only designed to focus on trade distortions. This perspective maintains that an attempt to use the GATT to address the internal welfare effects of policy reform is simply inappropriate.

The above criticism is misplaced for two principal reasons. First, any strategy for multilateral trade negotiations can be far more effective if complemented with the implementation of a transition path toward unilateral policy reforms. From a political-economic standpoint, it is little surprise that Japan, the EC, and the United States have been so heavily influenced by their respective agricultural commodity interest groups. There can be little doubt that the Uruguay Round of the GATT negotiations would have been far more effective if Japan, the EC, and the United States had each simultaneously pursued an active strategy of "packaging" reforms so as to achieve significant unilateral adjustments. From a political-economic standpoint, only if these countries were able to see a politically robust way out of their current subsidization schemes would it also be possible to reach a cooperative conclusion to the Uruguay Round.

None of these countries – United States, Japan, and EC or the Group of Seven – were prepared to pursue a full court press on both international and domestic fronts. Surprisingly, the reform advocates from the Executive branch expected the domestic political-economic forces to be nullified on Capitol Hill because of the fast-track provision, while agriculture was treated as a separable negotiating issue in the Uruguay Round. These naive expectations were driven by the lack of creativity in designing and packaging unilateral reforms as well as the anticipation that only agricultural and, to a lesser degree, environmental interests have emerged for each and every piece of historical legislation that has set the framework for agricultural subsidization in the United States. It now seems obvious that the design of a GATT external code for agriculture without the simultaneous active pursuit of unilateral reforms is much like expecting manna from heaven.

Second, from the standpoint of pure economics, domestic policies can and often do have large and undesirable effects on international trade flows. As shown by Schmitz and Vercammen (Chap. 2), the efficiency of domestic policies is directly linked to the trade distorting effects of these policies. Accordingly, the reduction of policy impediments to trade and the creation of more efficient domestic policies should, in fact, be part of the same agenda. In essence, these authors demonstrate that the widely used producer subsidy equivalents (PSE), nominal rates of protection, and effective rates of protection do not allow the ranking of policies according to their trade distorting effects. In contrast, the inefficiency criteria does permit such rankings. Moreover, the widely held view that the size of the trade distortion due to a particular program is positively related to the size of the producer gain from the program (as is implicit in the PSE measures) is simply fallacious.

1.4.3 The Political Economy of Red- Versus Green-Light Policies

A review of history is instructive, but transcending the current agricultural stalemate in the GATT Uruguay Round requires a serious examination of the underlying political and economic forces that justify current agricultural policies. Domestic interests are largely responsible for the emergence and evolution of the current policy environment and operate as a powerful constraint on any possible reform program, whether it be sourced in unilateral or multilateral adjustments. These domestic political-economic forces, given the specified Uruguay Round framework, are crucial in determining the extent to which trade reform will be feasible and sustainable. Thus, political-economic analysis of the *domestic* effects of reforming policies that cause *international* distortions is not only appropriate but vital to the successful implementation of any policy reform path.

The sustainability of any reform path will be enhanced by strengthening the GATT rules and disciplines in the four areas that have been the focus of the Uruguay Round: market access, export subsidies, internal support, and sanitary and physiosanitary regulations. At the end of the day, internal support represents the cornerstone of sustainable and significant reforms of agricultural policy. Here, the US proposal calls for the grouping of policies into one of three categories: (i) policies to be phased out; (ii) policies to be disciplined; and (iii) permitted policies or, equivalently, policies that can be allocated to the so-called green box. Unfortunately, the various US proposals that have been advanced fail to realize that these policies, in a political-economic sense, are jointly determined. Hence, the reform of one policy forces an adjustment in other policies and any attempt to impose binding commitment, as though these policies were separately determined, is doomed to fail. Permitted policies – including income-support not based on production, environmental and conservation programs, pesticide and fertilizer taxes, disaster assistance, market information and promotion, inspection and grading, research extension and education – cannot be reformed in the political-economic world without jointly and simultaneously reforming trade-distorting policies, such as administered price policies and income support or direct payments based on production. In the following section we will present both the theoretical and empirical support for two types of policies which are largely consistent with the green-light and red-light policy dichotomy that has frequently been used to characterize the Uruguay Round debate on reform of not only internal supports but market access for imports and export subsidies as well.

1.5 A Generic Political Economic Model (GPEM)

In the design and implementation of governmental policy in agriculture, conflicts naturally emerge between public and special interests. A conceptual formulation that attempts to explain or prescribe public policy emphasizing only one of these interests will be of little value. Frameworks that neglect political forces and the role of special interest groups will have no explanatory power. Similarly, models that presume the government has no autonomy nor any interest in the size of the economic pie will also face serious limitations as explanatory, predictive, or prescriptive frameworks.

Agricultural policy in the United States has led to both the enhancement of efficiency through *productive policies* and the transfer of wealth and income to special interests through redistributive or *predatory policies*. These two activities can be examined in terms of PESTs (predatory, or special interest) and PERTs (productive, or public interest) policies. PEST policies, or political-economic-seeking transfers, are meant to redistribute wealth from one social group to another and are not explicitly concerned with efficiency. In contrast, PERTs, or political-economic resource transactions, are intended to reduce transaction costs in the private sector by correcting market failures or providing public goods; these policies have neutral distributional effects, at least in design (Rausser 1982, 1992).

A historical review of public policy in agriculture reveals not only tension between the PERT and PEST roles of the public sector, but also some coordination between these two types of activities. As different interest groups access, influence, and pressure the political process, the government trades off PESTs and PERTs in its attempts to acquire, balance, and secure political power. At times this has led to programs that appear incoherent. Examples include conservation (which requires the retirement of vulnerable acreage) and price supports (a transfer mechanism that penalizes premature land retirement and creates incentives for overutilization of vulnerable acreage). These apparent incoherences, however, are the direct result of institutional arrangements that generate sufficient support for particular types of governmental action. Thus, a rational process generates a government portfolio of productive (PERT) and predatory (PEST) policies. In such a world, the challenge is to design and advocate policies that are both economically productive and politically sustainable.

Many examples of consumer-biased PERTs being balanced with producer-biased PESTs have been presented (Rausser 1992). Obviously, any number of combinations are possible, depending on supply-and-demand parameters, the degree of heterogeneity within various interest groups, productivity, and the relative political power of interest groups. The framework helps explain why "inefficient" policies have a resilience that would be inexplicable if the public sector paid any attention to free-trade economists. It also explains why many policies are difficult, if not impossible, to reform.

Far too frequently, PEST policies have been analyzed as though they were separate from other policies, explaining their existence by the relative organizational strengths of interest groups or the opaqueness of the policy impacts. One of the major messages of GPEM is that policies can be packaged so that consumers and producers may acquiesce to one policy setting in exchange for another. The framework applies not only to US agricultural policy, but to all types of public-sector interventions; e.g., privatization with safeguards for some social groups in formerly command economies; urban planning and the granting of zoning variances in exchange for the supply of local public goods; special worker adjustment compensation for industries facing increased international competition; and the advice and counsel to less-developed countries from the World Bank, the International Monetary Fund, and the Agency for International Development. In the final analysis, only if we can explain how policies are selected and implemented, can we hope to motivate major reforms.

1.5.1 Accounting for PERTs and PESTs

One measure of the degree of government intervention across commodity groups can be represented as a producer subsidy equivalent (PSE), the ratio of the total value of all public sector assistance to total farmer receipts. As shown in Table 1.1, for the United States, the degree of government involvement is most dramatic for products where demand is inelastic, e.g., sugar, milk, rice, and wheat. Feed grains have an intermediate level of support while sectors with more elastic demands, such as soybeans and red meats, have the lowest level of support.[3]

1.5.1.1 US PERT/PEST Decomposition

The decomposition of the public sector assistance into productive (PERT) and predatory (PEST) forms of government policy is also reported in Table 1.1. The productive category includes all expenditures by the public sector that are expected to lower transaction costs and enhance the rate of economic growth; e.g., public-good expenditures, information and marketing services, grades and standards inspections, crop insurance, public research, extension, etc. For the PEST category, all redistributive transfers from other segments of the economy to agricultural producers, e.g., deficiency payments, price supports, trade barriers, storage subsidies, input subsidies, subsidized credit, etc., are incorporated.

Note that the products with inelastic demands (sugar, milk, and rice) receive a lower proportion of their public support in the form of productive policies, while the products with elastic demand (soybeans and meats) receive a higher proportion of their support in the form of productive policies. The data is consistent with the view that coupled predatory policies are higher in sectors where demand is inelastic and where supply is very responsive to policies and lower in sectors with highly elastic demand and low supply elasticities. As noted earlier, redistribution efficiency would argue for low supply elasticities but here the joint determination

[3] For a survey of own-price elasticity estimates, see de Gorter et al. (1992). At the farm level, the most demand-inelastic commodities are sugar, milk, and rice; those commodities with intermediate degrees of demand-price inelasticity are wheat, sorghum, barley, corn, and oats; and finally, those that have the least degree of inelasticity and, in some instances, elastic demand functions, include soybeans, beef, poultry, and pork.

Table 1.1. Productive versus predatory policy interventions in US agriculture, 1982–1986 average (percentage unit values)

	Producer subsidy equivalents		
	Total	Productive (PERT)	Predatory (PEST)
Sugar	77.4	6.1	71.3
Milk	53.9	4.2	49.7
Rice	45.0	2.9	42.1
Wheat	36.5	4.9	31.6
Sorghum	31.5	4.6	26.9
Barley	28.8	6.0	22.8
Corn	27.1	4.8	22.3
Oats	7.6	4.7	2.9
Soybeans	8.5	6.3	2.2
Beef	8.7	4.8	3.9
Poultry	8.3	5.4	2.9
Pork	5.8	4.8	1.0
Average	24.6	8.75	10.2

Source: US Department of Agriculture (1988).

of PESTs and PERTs places more weight on policy-induced supply expansion. For some products that do not appear in Table 1.1, e.g., specialty perennial crops (oranges, lemons, grapes, etc.) whose demand is highly elastic, but whose productivity and supply response is low, coupled predatory policies do not generally exist. Instead, for these speciality crops, producer organizations tax their members to finance the provision of local public goods (de Gorter et al. 1992).

1.5.1.2 EC PERT/PEST Decomposition

Surprising similarities arise when we assess the historical data arising from the European Community. The CAP has subsidized agriculture to the extent that fewer people have exited farming than otherwise would have occurred. This has bestowed a positive externality to society by preserving rural landscapes, communities, and resources (Drake 1987). On the other hand, price supports and subsidies have increased the use of inputs (land, water, chemicals, and fertilizers) and production to the detriment of the environment (Mahé and Rainelli 1987). Hence, farmers generate both positive and negative externalities. Some argue that famers should be paid for the positive externality they create. For example, they should be subsidized for preserving the rural landscape (Sinclair 1985). Secondly, compensation to farmers may be required in order to induce them to use environmentally beneficial farming practices so that the negative externalities from agricultural production can be reduced.

Meanwhile, society has bestowed an important externality upon farmers themselves, namely, public research and extension (R&E) expenditures. Public R&E expenditures have played a large role in inducing technological change as a major source for agricultural output growth in European agriculture (Bouchet et al. 1989). Real prices for agricultural products in the EC, price supports notwithstanding, have shown a secular decline with few exceptions. Farmers have experienced cost reductions which, on the whole, have not been offset by the reduction in revenues due to declining prices. Hence, farmers' welfare would have declined due to R&E expenditures if price supports had not been available.

Public R&E expenditures, as a PERT that increases agricultural productivity and supply, have an important effect in increasing the cost of export restitutions of

Table 1.2. Agricultural research expenditures and manpower

	Expenditures (Constant 1980 US$ thousands)			Manpower (Workers)		
	1959	1970	1980	1959	1970	1980
Western Europe	247,984	918,634	1,489,588	6,251	12,547	19,540
Northern Europe	94,718	230,135	409,527	1,818	4,409	8,027
Central Europe	141,054	563,334	871,233	2,888	5,271	8,827
Southern Europe	39,212	125,165	208,828	1,545	2,417	2,686
Eastern Europe and Newly Independent States (NIS)	568,284	1,282,212	1,492,783	17,701	43,709	51,614
Eastern Europe	195,896	436,094	553,400	5,701	16,009	20,220
NIS	372,388	846,118	939,383	12,000	27,700	31,394
World total	2,063,553	5,358,595	7,390,043	47,163	108,510	148,039

Source: Judd et al. (1986).

the EC's CAP. This by-product effect of public R&E expenditures has become an acute problem for the EC after it became a net exporter of many commodities by the 1980s. The EC was previously a large importer, but technological advances increased domestic production such that the EC's terms of trade with the rest of the world improved. The potential deleterious effects of R&E expenditures on farmers were moderated, and the costs to consumers of price supports with technological change were unaffected. However, the situation quickly reversed itself once the EC became a major net exporter of agricultural products.

Public R&E expenditures by Europe and the Newly Independent States and the corresponding manpower figures are given in relation to world totals in Tables 1.2 and 1.3. R&E expenditures have increased substantially over time. Research expenditures have increased more rapidly in Westen Europe relative to Eastern Europe and the rest of the world (Table 1.2). The converse is true for extension expenditures as shown in Table 1.3. In terms of public R&E expenditures as a percentage of the value of agricultural product, Europe has maintained a high profile, even in relation to North America (Table 1.4).

Data from the OECD (1989) on PSEs in Table 1.5 give an indication of the breakdown between public good expenditures and transfers resulting from price supports and related subsidy programs. In terms of percent unit value, the inelastic demand sectors of sugar and milk have the highest level of support. The coarse

Table 1.3. Agricultural extension expenditures and manpower

	Expenditures (Constant 1980 US$ thousands)			Manpower (Workers)		
	1959	1970	1980	1959	1970	1980
Western Europe	234,016	457,675	514,305	15,988	24,388	27,881
Northern Europe	112,983	187,144	201,366	4,793	5,638	6,241
Central Europe	103,082	199,191	236,834	7,865	13,046	14,421
Southern Europe	17,950	71,340	76,105	3,330	5,704	7,219
Eastern Europe and Newly Independent States (NIS)	367,329	562,935	750,301	29,000	43,000	55,000
Eastern Europe	126,624	191,460	278,149	9,340	15,749	21,546
NIS	240,705	371,475	472,152	19,660	27,251	33,454
World total	1,427,913	2,722,564	3,443,489	177,521	294,483	349,337

Source: Judd et al. (1986).

Table 1.4. Research and extension expenditures as a percentage of the value of agricultural product: public sector

Subregion\ Country group	Agricultural research Expenditures 1959	1970	1980	Agricultural extension Expenditures 1959	1970	1980
Northern Europe	0.55	1.05	1.60	0.65	0.85	0.84
Central Europe	0.39	1.20	1.54	0.29	0.42	0.45
Southern Europe	0.24	0.61	0.74	0.11	0.35	0.28
Eastern Europe	0.50	0.81	0.78	0.32	0.36	0.40
Newly Independent States	0.43	0.73	0.70	0.28	0.32	0.35
North America	0.84	1.27	1.09	0.42	0.53	0.56

Source: Judd et al. (1986).

grains and soybeans have an intermediate level of support, as shown in Table 1.5, while the sectors with more elastic demand curves, such as poultry and pig meat, tend to have lower levels of support. The breakdown of support between PERTs and PESTs are given in columns (2) and (3) in Table 1.5.

Table 1.5. PERTs versus PESTs in European agriculture (1979–1985 average)

	(1) Total subsidy	(2) PERT Subsidy	(3) PEST subsidy	(4) Ratio of PERT to PEST subsidies (%)	
				EC	US
Sugar	51.0	6.4	44.6	14.2	8.6
Milk	52.0	6.8	45.2	15.1	8.4
Wheat	28.0	6.2	21.8	28.5	15.7
Coarse grains	32.0	7.6	24.4	8.4	21.6
Soybeans	48.0	8.5	39.6	21.7	291.3
Beef and veal	48.0	6.7	41.3	16.3	125.0
Poultry	21.0	8.5	12.5	67.8	285.7
Pig Meat	6.0	6.0	0	NA	380.0

Source: OECD, *PSE and CSE Tables*, October 1989.

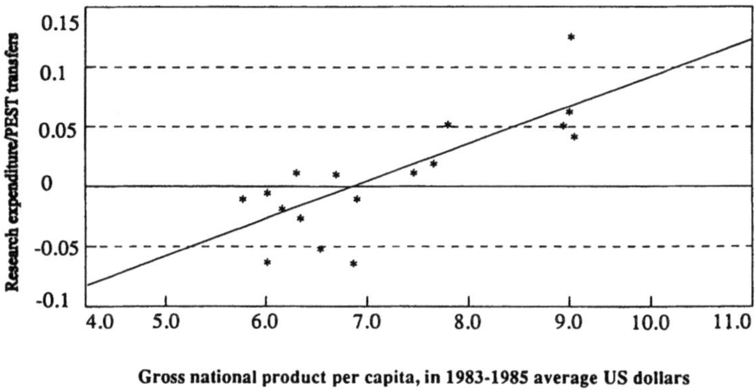

Fig. 1.1. Research expenditure/PEST transfers versu per-capital gross national product

1.5.1.3 Cross-Country Comparisons

If the PERT-PEST dichotomy is to inform policy debate in anything but a parochial sense, then it should be transportable to settings other than the United States or European Community policy arena. A data set allowing substantive inferences along these lines is available in Lee and Rausser (1991). These examine aggregate PEST transfers and PERT investments in agricultural research for 17 countries. The evidence is revealed in Fig. 1.1, which shows that the ratio of PERT to PEST transfers unambiguously increases with country income levels. In Fig. 1.2, productivity measured by value added per agricultural worker is related to the mix of PEST and PERT expenditures. This relationship suggests that one reason for the record of strong productivity growth in developed country agriculture is that, despite a strong tendency toward increased protection, there is a complementary tendency toward support for agricultural research. As Lee and Rausser note, complementary provision of PEST and PERT policies provides a consistent explanation for the otherwise paradoxical outcomes of developed countries pro-

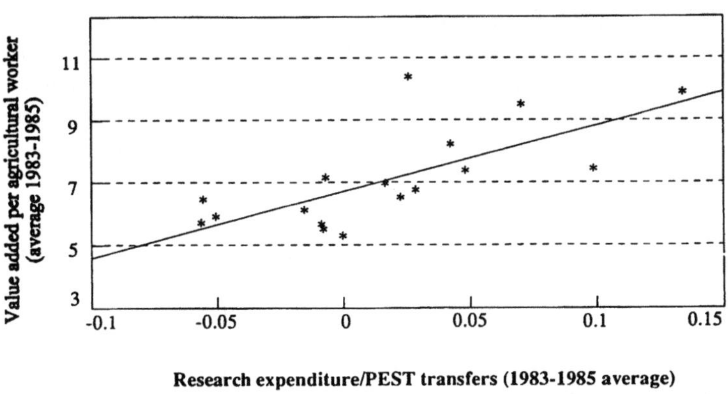

Fig 1.2. Value added per agricultural worker versus research expenditure/PEST transfers

tecting their agricultural sectors while investing in more agricultural research and generating higher levels of agricultural productivity compared to developing countries which typically tax their agricultural sectors, invest little in agricultural research, and experience generally low levels of agricultural productivity.

1.5.2 Conceptual PEST/PERT Results

For the case of public-good investments and predatory coupled transfers, it has been shown that, if a productivity-enhancing policy harms producers because of highly inelastic demand and responsive supply, but producers have more political clout than other interest groups, the amount of public-good investment will be inadequate (de Gorter et al. 1992). However, the political obstruction to public-good investments can be countered with subsidies that are tied to production, thus leading to less underinvestment in public goods than would otherwise be the case. In effect, since productive policies may harm members of special interest groups, compensation through predatory subsidies may offer a means of making the pursuit of the public interest politically viable (Rausser and Foster 1990). Transfers that seem only predatory at first glance may, in certain cases, actually be politically necessary if society is to more closely approach the optimal configuration of productive policies.[4]

Even if demand is elastic and the "representative" producer benefits from the dissemination of an advance, producers are likely to be heterogeneous in new technology adoption. Those producers who make the greatest use of the new technology will gain from its dissemination and the associated equilibrium price decrease, while others who make little or no use of the technology are likely to lose. Indeed, nonadopters always lose when the demand curve is anything less than perfectly elastic. In this setting, the key is not the elasticity of demand, but the heterogeneity of producers in their ability to take advantage of technical advances. If a sufficient number of these heterogeneous producers are harmed by the equilibrium effects of technological changes, then potential political impediments to future technical advances may arise. If so, some promise of wealth transfers from the winning consumers/taxpayers to the losing producers will be necessary to have any advance at all. As shown in Foster and Rausser (1994), wealth transfers tied to output may be a more effective means than per-firm lump-sum payments in inducing defection from the coalition of those producers least harmed by the technical change. Price-distorting payments target producers who, although harmed by the technology dissemination policy alone, expand their production the most.

[4] Political influence also operates through the public sector by frequently slowing down the redistribution of income generated by changing market conditions. Accordingly, markets with highly inelastic demand and supply conditions as well as rapidly changing technologies are those markets which generate fluctuating incomes and thus a demand for public sector "stabilization." In a world of limited knowledge of how economic systems operate, producer-interest groups have successfully argued that price stabilization programs are in the public interest. In this instance, interests who share the burden of financing such programs are led to believe that such policies are PERTs.

For many observers, it is difficult to imagine how a PERT might harm a particular segment of an industry. A number of illustrative examples, however, serve to show how structural ex ante assessments lead to PERT harming consequences. These examples include the recent dairy farmers' opposition to the market introduction of growth hormones; the obstructionist behavior of strawberry growers to Experiment Station research at the University of California; and similar behavior on the part of soybean growers to Experiment Station research at the University of Illinois. This behavior does not turn on the degree of demand elasticity but, instead, on the heterogeneous abilities of growers throughout the world to take advantage of any technological innovation that might arise from US Experiment Station research. This same phenomenon has been widely observed in much of the trade liberalization literature. In many instances, the uncertainty about whether a particular group may be a winner or loser as a result of reform can by itself lead to obstructionist behavior. From this perspective, many obstructionist coalitions include as part of their members some groups who will significantly benefit from reform.

The evaluation of whether or not farmers are harmed by a PERT turns on the effect the policy has on the value of farm assets, primarily land and human capital. Farmer wealth can be determined by the present discounted stream of returns to fixed factors and thus is a function of the discount rate and the path of the incomes of future periods. The post-PERT long-run yearly income will be greater given a long-run elastic demand, but post-PERT asset values may well be much lower. Regardless of whether the long-run elasticities of demand are elastic or inelastic, a relevant issue from a policymaking standpoint is the degree to which politicians and interest groups are myopic (Akerlof 1991). To be sure, even if farmers operate with infinite planning horizons, discounting future income streams can be expected to place more weight on short-run price and income changes, rather than long-run demand elasticities. In this respect, the overwhelming empirical evidence supports the conclusion that short-run demands for many commodities are inelastic. As a result, even if the long-run demand is elastic, farmers may have to bear a loss due to near-term short-run elasticities.

To formally demonstrate the above assertion, consider the case of a demand function that in the very short run is inelastic, but over the long run becomes perfectly elastic. In this case, whether or not the present discounted stream of farmer incomes with the PERT is less than that without the PERT depends upon the size of the supply shift, the discount rate, and the speed at which the output price adjusts to its long-run value. Let the initial supply curve be given by $P = m_0 Q$ and the post-PERT supply be given by $P = m_1 Q$, where $m_1 < m_0$. The yearly income is given by $\pi_0 = \{1/2(1/m_0)\}(P_0)^2$ for the case without the PERT and by $\pi(t) = \{1/2(1/m_1)\}[P(t)]^2$ for the case with the PERT. Price is assumed to adjust over time in accordance with $P(t) = P_0 + P_0[(m_1/m_0) - 1]e^{-\gamma t}$; that is, the very short run is perfectly inelastic but long-run price is that which prevails without the PERT (i.e., $P(t) \rightarrow P_0$ as $t = \infty$). The parameter, γ, is a measure of how fast the demand curve goes to its perfectly elastic state.

It can be verified easily that, with an instantaneous discount rate of r, the present discounted stream of income without the PERT is given by $V_0 = \{1/(2rm_0)\}(P_0)^2$ and with the PERT this stream is given by

$$V_1 = \int_0^{\infty} \frac{1}{2}\frac{1}{m_1}[P(t)]^2 e^{-rt} dt = V_0 \frac{m_0}{m_1}\left[1 + \left(\frac{m_1}{m_0} - 1\right)^2 \left(\frac{r}{r + 2\gamma}\right) + \left(\frac{m_1}{m_0} - 1\right)\frac{2r}{r + \gamma}\right].$$

The conditions under which $V_1 < V_0$ is: $r > \gamma(1/2)\{[1 + 8(m_0/m_1)]^{1/2} - 1\}$; which for small decreases in marginal cost ($m_1 \approx m_0$) implies $r > \gamma$. In other words, total farmer wealth is less with a PERT than without for sufficiently high discount rates or sufficiently low rates of adjustment to the perfectly elastic long-run demand curve.

1.6 Outline of Book

This volume is dedicated to understanding the political economy of current agricultural policies. The focus is on unilateral reform or a political robust way out for each country engaged in negotiating the GATT. Accordingly, the volume is decomposed into a group of theoretical chapters as well as empirical analyses of specific reform proposals. Specifically, the first six chapters of the book concentrate on theoretical concepts underlying policy reform. The effects of the structure of domestic policies, political and institutional constraints, the distribution of ownership of immobile resources, and heterogeneity of import-competing producers is examined. The remaining four chapters draw on the preceding results to present innovative empirical approaches to modeling the effects of alternative reform proposals.

In Chap. 2, the relationship between the structure of domestic programs and resulting international trade distortions is examined. Using agricultural support policies as a both theoretical and empirical example, Schmitz and Vercammen analyze the trade-distorting effects of public policy on internal markets. Of particular interest is the relationship between the government program characteristics and the degree of trade distortion, as well as the possibility of ranking programs according to their trade-distorting effects.

The authors challenge the widely held view that the size of the trade distortion due to a particular program is positively correlated to the size of producer gain from the program (as is implicit in PSE measures) or the degree of consumer protection [as in consumer subsidy equivalent (CSE) measures]. Instead, the authors show that the degree of trade distortion is usually negatively related to the economic efficiency of the program. This result implies that the current measures of the trade-distorting effects of government policies that rely on the size of the producer gain (such as the PSE) may seriously misrepresent the true relative and absolute distortion of these polices.

Indeed, Schmitz and Vercammen clearly demonstrate that PSEs and other related measures such as CSEs, nominal rate of protection (NRP), and effective rate of protection (ERP), do not allow the ranking of policies according to their trade-distorting effects, while the efficiency criteria does permit such rankings. The authors then present empirical evidence for several US and Canadian agricultural policies which support their theoretical findings.

In Chap. 3, Just, Rausser, and Zilberman extend the basic tools of welfare economics to incorporate political and institutional constraints. This extended welfare analysis rests on the realization that costless lump-sum (pure) transfers are rarely, if ever, feasible and the distribution of political influence and the distribution of wealth will often effect economic outcomes when the government intervenes in markets.

Absence of pure transfers is shown to imply institutional constraints while the political landscape implies political constraints. These two constraints must be considered along with the appropriate welfare function when designing feasible policy reforms. Alternative methods of balancing these concerns in the policy design process are suggested by the authors. It is also demonstrated that optimal sustainable reform policies may differ markedly from the optimal policy design when the institutional and political constraints are ignored.

The analysis of this chapter demonstrates that the political constraints will often require compensation schemes and the more costly these schemes, in terms of deadweight loss, the more binding will be the institutional constraints. Accordingly, the authors then consider the design of workable and efficient compensation schemes. In particular, design considerations arising from uncertainty, moral hazard, eligibility, and financing issues are addressed and various means for managing these concerns are suggested.

In Chap. 4, de Gorter, Nielson, and Rausser consider the joint determination of the supply of a public good and the level of producer support. In particular, the authors analyze the government's choice of the level of expenditures on agricultural research and income support for agricultural producers. If the government values the welfare of producers, and research-induced technological change affects producer welfare, then the optimal level of research and support must be determined jointly. The authors demonstrate that, when technological change lowers producer profits, the optimal level of public research is lower than in the case where the effect of research on profits is neutral or is ignored. Furthermore, as the government places greater weight on producer welfare, the optimal level of research falls.

These results have several important implications. When public research and producer income support are complements, as this model suggests, then inefficient transfer policies, such as price supports, are actually welfare enhancing. The increase in producer profits due to the transfer policy allows the government to provide more of the welfare-enhancing public good while maintaining an adequate level of producer profits. This model also gives a new perspective on the debate concerning the "under investment" in agricultural research. It has been argued that the presence of production-based producer-support programs acts as constraint on the level of public research expenditures. Increased research expenditures, it is argued, promote output-increasing technological change. The increased output augments the budgetary burden of the producer support programs, and this cost should be included as an additional element of the total public research cost. Thus, the optimal level of public research falls when the effects of producer support programs are considered. The analysis of de Gorter, Rausser, and Nielson, however, shows that concern for producer profits acts as a constraint on the optimal level of public research expenditures. Producer support programs relax this constraint and allow for a higher level of public research expenditures.

A model which explains the widespread use of seemingly inefficient redistributive mechanisms in actual government policy portfolios is presented in Chap. 5. One major result of the political-economic analysis presented is that transfers are often needed to assure the feasibility of welfare-improving policies. Given that some wasteful redistributive policies can be justified by this reasoning, however, the choice of mechanisms remains to be explained. In particular, why are some relatively distortionary mechanisms chosen when less distortionary mechanisms are available?

In Chap. 5, Rausser and Foster present a model that rationalizes the choice of relatively distortionary transfer mechanisms. The key insight of the model is that members of a potential blocking coalition to a policy are heterogeneous with respect to the harm they suffer. Thus, if the transfer mechanism can target those members of the coalition who suffer least as a consequence of the policy and can thus be made indifferent to the policy, then political opposition may be ameliorated more efficiently than by the use of a uniform transfer to all members of the coalition.

The individual harm to a potential policy opponent is usually known only to the opponent and not to the government, however; and the agent is likely to behave strategically in revealing this information, especially if it is anticipated that the size of potential compensation may depend on the revealed value. Coupled transfer schemes, however, may effectively separate the opposing coalition members and target transfers to those who require less compensation to be made indifferent. Thus, even though a transfer mechanism is distortionary and entails a deadweight loss, it may eliminate effective political opposition at a lower total cost than a less distortionary, but more uniform transfer mechanism.

Rausser and Foster use the example of agricultural producers and public policies which promote welfare-improving technical change in agriculture. This technical change causes a fall in agricultural prices and producer profits, leading to an effective coalition to oppose the policy. A nondistortionary policy would require that compensation be uncoupled from production. Note, however, that the producers' ability to benefit (or limit losses) from the technical change may be correlated to some observable measure such as output changes resulting from technology change. In this case, a coupled transfer mechanism, tied to the producers' output, will target compensation to those who are harmed least and can most efficiently be made indifferent. Thus, even though the coupled mechanism entails a greater deadweight welfare loss, the effective political opposition can be overcome at a lower total cost to consumers/taxpayers. A rigorous model of this example provides the precise conditions when a distortionary mechanism may be desirable and indicates that these conditions may often be met in practice. It is shown that distortionary transfer mechanisms will be more preferred when producers are more heterogeneous in their output response to the technological change, when fewer producers must be made indifferent in order to effectively counter opposition to the technological change, and when the supply response to the distortionary transfer is low.

In Chap. 6, Rausser and Foster extend the analysis from the preceding chapter. The model in Chap. 5 concerns the rational choice by governments (and consumers/taxpayers) for distorting transfer mechanisms. What is left to be explained, however, is the choice of the particular distorting policy. Since distorting intervention policies in both input and output markets, as well as lump-sum transfers, are potential mechanisms available to the government, Rausser and Foster seek to determine the conditions under which each of these policies would be rationally chosen.

The authors present a conceptual model where consumers/taxpayers make transfers to producers in order to overcome opposition to policy that provides a welfare-increasing public good but harms producers. They show that the optimal level of input and output subsidization and lump-sum transfers depends both on the ability of nontargeted groups to take advantage of the transfer and the social costs of the associated market distortion. Furthermore, the ability of nontargeted groups to take advantage of the transfer mechanism is shown to be a function of producer heterogeneity in both input and output markets, and the nature of the welfare-improving public good. This analysis demonstrates that forcing governments to adopt nondistorting transfer policies may actually reduce social welfare under certain conditions.

In Chap. 7, the effect of the mobility of resources and the diversification of ownership on the sustainability of trade reform is examined. A simple general equilibrium model is used to demonstrate the relationship between the gain (or harm) to an individual from trade reform and that individual's ownership portfolio of immobile resources in the import-competing sector. The results of the general equilibrium analysis are then used to derive the individual incentives for protectionist rent seeking. These incentives are shown to be a function of the individual's ownership portfolio, in particular, the ownership of immobile resources.

The authors derive three major conclusions which relate trade policy reform to the ownership and diversification of resources. First, the need for compensation will be a function of both mobility and diversification of ownership. Second, increased mobility and/or more diversified ownership will lower the incentives for protectionist rent seeking. Finally, compensation mechanisms, or other government policies, that induce greater resource mobility or ownership diversification will lead to more sustainable reform.

In Chap. 8, the focus shifts to empirical application of the theories from the preceding chapters. Just, Rausser, and Zilberman present an empirical model of the US wheat and feed grain sectors which incorporates the political and institutional constraints discussed in Chap. 3. Previous studies modeling the effects of trade reform compare the current policies to free trade with no attention given to the process, or costs, of transition. Yet, transition costs will almost surely be large and these costs will bear heavily on the political feasibility and sustainability of reform.

Political feasibility and sustainability cannot be examined with traditional economic models. This type of analysis requires models with enough temporal and sectoral complexity to permit the analysis of both distributional and transitional issues. The key requirement is that the model be able to depict both the temporal and sectoral distribution of impacts over the transition period. Chap. 8 is the demonstration of such a model.

The model focuses on the transitional effects of a 10-year program to substitute decoupled support policies (which are generally permitted under the major GATT proposals) for coupled policies (which are generally forbidden under the major proposals). The intent is to examine the feasibility and cost of using permitted policies to sufficiently compensate important special interest groups and make policy reform sustainable.

Three particular compensation policies are considered: direct payments, environmental programs that raise crop prices, and land retirement programs. Preliminary results indicate that land retirement programs are the most efficient compensation mechanism considered, but the authors caution that other policies or mixes of policies must also be considered. The model represents an important first step, however, in developing models to analyze policy reform. The principles and techniques presented in Chap. 8 offer a promising foundation on which future efforts can build.

An empirical examination of the way in which productive (welfare increasing) and predatory (welfare transferring) policies are jointly determined is presented in Chap. 9. de Gorter, Nielson, and Rausser apply the theoretical model of Chap. 4 to the US dairy sector, where price support programs transfer income to dairy producers and public research programs lower production costs and increase social welfare. Using a deterministic and partial equilibrium model of the US dairy industry, the authors model the jointly endogenous determination of research expenditures and price supports in the industry.

The model illustrates the complex ways in which both producer and consumer/taxpayer welfare is affected by the levels of these two programs. For example, increased supply (due to technological change) can lower producer profits, even with a constant level of price support. The model presents analytical expressions for the change in welfare of both groups due to changes in either program and develops the government's decision problem for the optimal joint determination of the two policy instruments. This modeling exercise allows the authors to determine the conditions under which price supports and research expenditures are complementary instruments and argue that these conditions are likely to hold.

The authors present empirical evidence to support the hypothesis that price supports and research expenditures are complementary; i.e., that research expenditures are higher with the presence of price supports than they would be otherwise. Furthermore, given this complementarity, both producers and consumers/taxpayers appear to benefit from the presence of price supports. In the absence of price support programs, a simulation model, incorporating the features of the theoretical model and empirical supply estimates, shows that government concern for producer welfare leads to drastically reduced research expenditures. This reduction would result in a decrease in consumer/taxpayer welfare, as the loss from foregone welfare-improving technological change offsets the gain from reduced expenditures on research and price supports.

These conclusions emphasize the importance of viewing government intervention programs through a wide-angle or political-economic lens. When viewed in isolation, price-support programs appear to be wasteful and welfare decreasing. On the other hand, numerous studies have found that agricultural research expenditures appear far too small. When the joint nature of these programs is recognized, however, the contradiction between underinvestment in research and overinvestment in price supports disappears.

The influence of macroeconomic conditions on the distortionary effects of agricultural policies is examined in Chap. 10. The central premise of Labson and Rausser is that macroeconomic linkages are crucial to the performance of the agriculture sector. The effects of domestic policy reform on price and output will depend on the prevailing macroeconomic environment during the reform period. Therefore, policy reform analyses which assume a static or "average" macroeconomic environment may seriously misrepresent the effects of policy reform.

Labson and Rausser present an econometric based policy simulation model of the US wheat sector and examine the short-run effects of a unilateral phased reduction on US agricultural policy instruments. The effects of the policy reform on price, output, program participation, and government payments under alternative macroeconomic conditions is simulated. The macroeconomic environments used correspond to the periods, 1975–1979 and 1981–1985, representing periods when the macroeconomy was highly favorable and unfavorable to agriculture, respectively.

The authors find that the effects of a phased reduction on agricultural policy instruments varied widely between these two scenarios. When the macroeconomic environment is favorable to agriculture, subsidies play little if any role and program participation is small. Thus, the reduction or elimination of agricultural support prices has only slight effects. Under an unfavorable macroeconomic environment, however, subsidized prices are far above the estimated free-market price and program participation is high. The reduction of agricultural policy instruments in this case leads to large effects on prices, outputs, incomes, and thus rates of return.

The importance of macroeconomic linkages to the agricultural sector and to agricultural policy reform is also examined in Chap. 11. Ardeni and Rausser briefly review past research on the influence of macroeconomic conditions on agriculture. All past research, however, has concentrated on macroeconomic linkages to agriculture – those effects that run from the macroeconomy to the agricultural sector. The agricultural feedbacks, running from the agricultural sector to the money and foreign exchange markets, have been completely ignored.

Agricultural feedbacks are potentially important, however. Even though the agricultural sector may comprise only a small proportion of total US GNP, it is a crucial export component with implications for exchange rates, it has distinct inflationary implications, and government programs in agriculture entail a large budgetary expense with implications of governmental deficits. Thus, the linkages from agriculture to the macroeconomy may be important, particularly when contemplating significant changes in these government programs.

Ardeni and Rausser carefully identify the source of potential agricultural feedbacks. A two-sector, theoretical model with all the relevant linkages is then specified and empirically estimated for the United States. The estimated model is used to simulate different scenarios for the reduction of the level of government policy instrument.

Results from the estimation and simulation indicate that linkages from the agricultural sector to the macroeconomy do indeed exist. Furthermore, these linkages are likely to have important effects during a period of agricultural policy reform. The model also indicates that government intervention in agriculture has been effective in partially insulating the sector from the negative impacts of unanticipated macroeconomic shocks.

Taken together, the research in this volume represents a significant advance in both the theoretical and applied analysis of government policy reform. A wide variety of domestic forces are shown to influence the selection and effects of government agricultural policies. These forces range from political and institutional factors to factor mobility to broad economywide policies and conditions such as exchange rates and money supply.

While the initiative for agricultural policy reform may stem from international trade issues, it is these domestic forces that have shaped the current policy mix. Analysis of trade reform, either positive or normative, must explicitly recognize these domestic linkages if the predictions and/or prescriptions are to be useful. Moreover, while the concentration in this volume is on agricultural issues, the theoretical foundations, as well as many of the modeling techniques, are equally relevant to trade issues in other sectors.

1.7 Concluding Remarks

The challenge is to identify policy reforms that make economic sense and are politically robust. It is now being recognized that political and economic forces must be jointly considered in matters of both design and implementation of public policies (Rausser 1982). Much recent investigation has been devoted to theoretical and empirical models of public sector decision making in an attempt to accomplish this task. Nevertheless, we have only begun to scratch the surface in the development of operational frameworks for blending productive and rent-seeking policies. In general, work in this area must recognize that these two types of policies go hand in hand; frequently, predatory policies are offered as compensation to those that are harmed as a result of the implementation of productive policies.

Opportunities for restructuring the tradeoff between *public* and *special* interests have often appeared greatest during times of economic crises. However, the sustainability of the restructured tradeoffs and the new mix of productive and predatory policies has been shown time and time again to depend critically upon changes in the underlying institutional configuration. In the case of US agricultural policy, an example of institutional changes that could alter the level and distribution of political power might arise from the current GATT negotiations. As previously noted, in the Uruguay Round of the GATT negotiations, it was accepted early by all parties that distortionary trade policies in agriculture exist to rationalize internal country policies; thus, both sets of policies should be included in the negotiations. Accordingly, a number of proposals have been tabled in Geneva for reducing internal country coupled PEST policies and substituting PERT policies (Rausser 1992).

For many US interests, the GATT is indeed a high stakes game. It is ironic that the pathway to a potential GATT agricultural accord, the Uruguay Round, became a favorite excuse for at least maintaining (e.g., 1989–90 EC Commission proposed internal commodity price cuts) or increasing subsidies (e.g., 1993 US Export Enhancement Program subsidy directed against Canada). Short of an over-riding external influence (budgetary forces), interest groups have been able to effectively argue that any and all reforms would undermine bargaining creditibility in the international GATT negotiations. This may prove to be a small price to pay if we arrive at a conclusion to the Uruguay Round. Some might even characterize a successful conclusion as a set of rules less aggressive than the Blair House Accord; at least, the process would be initialized for future GATT rounds and the probability of backsliding would reduced significantly. In other words, setting the terms of internal reforms into an international agreement can enhance the durability of reform and possibly set in motion further reform steps. Regardless, the interest-group configuration in the United States, if a GATT accord emerges, could be dramatically different than the political landscape that has existed over much of the prior 60 or so years. Agriculture will no longer be compartmentalized. Agricultural sector issues will be linked with other trade issues, thus widening the vested interests that will determine whether a GATT agricultural code is accepted or rejected. If a GATT agreement is accepted by the US Congress, the executive branch will lean on the external code as a basis for credible commitments to achieving more PERTs in exchange for fewer PESTs. It is, however, difficult if not impossible to achieve an external GATT code without simultaneously pursuing a strategy of unilateral reform, or a political robust way-out for each country engaged in negotiating the GATT.

References

Akerlof GA (1991) Procrastination and obedience. Richard T Ely Lecture, Am Econ Rev 81 (2):1–20

Anderson K, Tyers R (1990) How developing countries would gain from agricultural trade liberalization in the Uruguay round. In: Goldin I, Knudsen O (eds) Agricultural trade liberalization: implications for developing countries. The World Bank, Washington, DC, pp 41–47

Bouchet FD, Orden D, Norton G (1989) Sources of growth in French agriculture. Am J Agric Econ 71:280–293

Dam KW (1970) The GATT: law and economic organization. University of Chicago Press, Chicago

de Gorter H, Nielson D, Rausser GC (1992) Productive and predatory public policies: research expenditures and producer subsidies in agriculture. Am J Agric Econ, 74 (1):27–37

Drake L (1987) The value of preserving the agricultural landscape. 5th Eur Congr of Agricultural Economists. Resource Adjustment and European Agriculture, Working Group A5 Papers, Balatonszeplak, Hungary, pp 11–18

Dunkel A (1991) Draft Final Act. Prepared by GATT Director General Arthur Dunkel and submitted to the General Agreement on Tariffs and Trade, Geneva

Foster WE, Rausser GC (1994) Price-distorting compensation serving the consumer and taxpayer interests. Public Choice, 77 (2):275–291

GATT Secretariat. (1983) International Trade, 1982–83. Geneva

Goldin I, Knudsen O, van der Mensbrugghe D (1994) Trade liberalization: global economic indications. World Bank, Washington, DC (in press)

Harris S, Swinbank A, Wilkinson G (1983) The food and farm policies of the European Community. Wiley, New York

Hertel TW, Thompson RL, Tsigas ME (1989) Economywide effects of unilateral trade and policy liberalization in US agriculture. In: Stoeckel AB, Vincent D, Cuthbertson S (eds) Macroeconomic consequences of farm support policies. Duke University Press, Durham, pp 269–292

Hillman JS (1978) Nontariff agricultural trade barriers. University of Nebraska Press, Lincoln

Judd MA, Boyce JK, Evenson RE (1986) Investing in agricultural supply: the determinants of agricultural research and extension investment. Econ Dev Cult change 35:78–11

Lee DR, Rausser GC (1991) The structure of research and transfer policies in international agriculture: evidence and implications. A paper presented to the International Association of Agricultural Economists, Tokyo, Japan, August

Mahé LP, Rainelli P (1987) Impact des pratiques et des politiques agricoles sur environment. In: Bonnieux F, Rainelli P (eds) Environnement et ressources naturelles, Cahiers d'economie et socologie rurales, 4:9–31

Mueller DC (1989) Public choice II. Cambridge University Press, New York

Newbery DMG, Stiglitz JE (1981) The theory of commodity price stabilization: a study in the economics of risk. Clarendon Press, Oxford and Oxford University Press, New York

OECD (various years) National policies and agricultural trade. OECD, Paris

OECD (1989) PSI and CSE Tables. Paris, October

Paarlberg R (1992) How agriculture blocked the Uruguay Round. SIS Rev 2 (1):27–42

Rapp D (1988) How the US got into agriculture and why it can't get out. Congr Q, Washington, DC

Rausser GC (1982) Political economic markets: PESTs and PERTs in food and agriculture. Am J Agric Econ 64 (5):821–823

Rausser GC (1992) Predatory versus productive government: The case of US agricultural policies. J Econ Perspect 6 (3):133–157

Rausser GC, Foster WE (1990) Political preference functions and public policy reform. Am J Agric Econ 72 (3):641–652

Rausser GC, Simon LK (1991) A noncooperative model of collective decision making: a multilateral bargaining approach. Working Paper No 620, Department of Agricultural and Resource Economics, University of California, Berkeley

Rausser GC, Simon LK (1993) The political economy of alliances: structure and performance. Working Paper No 665, Department of Agricultural and Resource Economics, University of California, Berkeley

Rausser GC, Zusman P (1992) Public policy: explanation and constitutional prescription. Am J Agric Econ 74 (2):247–257

Rausser GC, Chalfant JA, Love HA, Stamoulis KG (1986) Macroeconomic linkages, taxes, and subsidies in the US agricultural sector. Am J Agric Econ 68:399–412

Robinson S, Kilkenny M, Adelman I (1989) The effect of agricultural trade liberalization on the US economy: projections to 1991. In: Stoeckel AB, Vincent D, Cuthbertson S (eds) Macroeconomic consequences of farm support policies. Duke University Press, Durham, pp 222–259

Roningen VO, Dixit PM (1989) Economic implications of agricultural policy reform in industrial market economics. US Department of Agriculture Staff Report, AGES89-36, Washington, DC

Santayana G (1905) Life of reason, vol 1, Chap 12

Sinclair G (1985) How to help farmers and keep England beautiful. Research report commissioned by CPRE, Council for National Parks and World Wildlife Fund, United Kingdom

Schiff M (1985) An econometric analysis of the world wheat market and assimilation of alternative policies, 1960–80. Staff Report AGES-850827, US Department of Agriculture, ERS, International Economics Division, Washington, DC

Tyers R, Anderson K (1986) Distortions in world food markets: a quantitative assessment. Paper prepared for the World Bank, World Development Report 1986. Washington, DC

USDA (1989) Western agriculture and trade report, ERS, RS-89-2, Economic Research Service, Washington, DC

US Department of Agriculture (1986) Embargoes, surplus, disposal, and US agriculture. Economic Research Service. United States Government Printing Office, Washington, DC

US Department of Agriculture (1988) Estimates of producer and consumer equivalents: government intervention in agriculture. Economic Research Service, ATAD Staff Rep No AGES 880127, April

Wicksell K (1967) A new principle of just taxation. Finanztheoretische Untersuchungen, Jena. Reprinted in: Musgrave RA, Peacock AT (eds) Classics in the theory of public finance, St. Martin's Press, New York, p 108

White EW (1960) Europe in the GATT. Address given at Europe House, London

Winham GR (1986) International Trade and the Tokyo Round Negotiation. Princeton University Press, Princeton

World Bank (1986) World development report 1986. Oxford University Press, New York

Wright BD, Stamoulis KG, Nishiyama Y, Pick D, Carter C, McCalla A (1987) Trade barriers and macroeconomic factors affecting exports of California specialty crops. Agricultural Issues Center Symposium on Marketing California Speciality Crops: Worldwide Competition and Constraints, Sacramento. Mimeograph

Zietz J, Valdéz A (1986) The costs of protectionism to developing countries. World Bank Staff Working Papers No 769, The World Bank, Washington, DC

Chapter 2

Efficiency of Farm Programs and Their Trade-Distorting Effects

Andrew Schmitz and James Vercammen

2.1 Introduction

In many countries, governments intervene in agricultural markets in order to increase the welfare of producers at the expense of consumers and taxpayers. Not surprisingly, a debate has emerged regarding the extent to which government intervention distorts the trade of agricultural commodities. A widely held belief is that the size of the transfer to producers or the degree of protection for producers is positively related to the size of the trade distortion associated with the government program. Given this belief, it has been argued that the problem of measuring trade distortions effectively reduces to one of measuring levels of producer support and/or protection. In fact, much of the background research for the GATT negotiations has been steered in this direction. It has generally been found that farm programs have a high level of support as measured by the Producer Subsidy Equivalent (PSE).

In this chapter, we show that the size of a trade distortion is often not well linked to the level of producer support or protection but is closely related to a more fundamental variable – the economic efficiency of a government program. Hence, when evaluating farm programs in terms of their effect on trade, it is generally not sufficient to examine only the level of support or the degree of protection. Rather, it is necessary to examine the determinants of program efficiency and the relationship between program efficiency and the associated trade distortion. We show that, if a given program is efficient, it cannot be trade distorting even though the transfer may be large. This has important implications because, as Gardner (1987) shows in the context of US agriculture, many programs are relatively efficient. Because efficient programs do not distort trade, it follows, by definition, that the gains from the removal of efficient programs are zero.

Specifically, the purpose of this chapter is to provide a new look at how farm programs can be ranked according to their trade-distorting effects. We propose a general criterion. Trade-distorting policies can be ranked according to their degree of economic efficiency. For most classes of farm programs, the more efficient a program, the lower its trade-distorting effect. Thus, a necessary and sufficient condition for a policy to be non-trade-distorting is that the program transfer be 100% efficient. This occurs when the size of the producer subsidy is equal to the size of government plus consumer transfer. Although our approach provides a number of new and useful insights into the problem of trade distortion measurement, it should be viewed as complementary to, and not a substitute for, the traditional measures.

The criterion we propose virtually always holds when the world price of a commodity is not affected by an individual country's programs – i.e., the small-country case. In the large-country case, the criterion breaks down in a number of instances; thus, several caveats must be added to strengthen the results. For this reason, we will consider small- and large-country policies separately.

We will examine traditional measures of trade distortions, such as the PSE, throughout this chapter. In addition to the theoretical section, we will provide an empirical counterpart by examining select farm programs in Canada and the United States. When one does the theoretical and empirical analysis on the efficiency of farm programs, it becomes apparent that the PSE and other related measures of trade distortions have major shortcomings.

2.2 A General Classification Scheme

Bhagwati and Srinivasan (1984) cite four principal types of distortions that can arise in the international trade arena. The distortions caused by government intervention can be conveniently classified using this taxonomy. First, domestic producers and consumers may face the same internal price (once the appropriate marketing margins are accounted for), but this price differs from the world price of the commodity. This type of distortion will arise when border policies such as quotas, tariffs, or export taxes are imposed.

A second type of distortion arises when the domestic consumer price equals the border price but differs from the domestic producer supply price. Policies which cause this type of distortion include a producer tax or subsidy that may take the form of a deficiency payment, floor price, or stabilization payment. Alternatively, a third type of distortion results when the domestic producer price equals the border price but differs from the domestic consumer price. Such a distortion would typically be due to a consumer tax or subsidy. Finally, a fourth type of distortion arises when a policy creates a wedge between the price that producers pay for their factors of production and the undistorted factor price.

It should be clear from the above discussion that a necessary condition for a trade distortion is that a price wedge exists between one or more sectors of the world economy. It appears, therefore, that an index capable of measuring the trade-distorting effects of policy should be based on some type of aggregation of the prevailing price wedges. In fact, a number of such indices have been developed, including the PSE, the consumer subsidy equivalent (CSE), the trade distortion equivalent (TDE), and the nominal rate of protection (NRP) and the effective rate of protection (ERP).

Roughly speaking, the PSE is a measure of the percentage of producer income attributable to government intervention. The CSE is analogously defined. The TDE, a refinement of the PSE, nets out the components of the PSE that do not directly distort trade. The NRP is based on a comparison of domestic and world market prices. The ERP is similar to the NRP except that it also accounts for differences in the level of protection among inputs and the final product. For a more detailed description of these indices, see Zietz and Valdés (1988).

Ideally, an index of trade distortion should be able to accurately measure the size of the trade loss stemming from a particular policy. Or, at the very least, an index should be capable of ranking policies both within and across countries in terms of increasing trade distortion. Unfortunately, the indices based on price wedges are not capable of doing either with reasonable consistency. These type of indices often fail because they are based on ex post prices, which may be decoupled from ex ante incentive prices. Another reason for their failure is that these indices generally do not take into account the relative supply and demand elasticity differences when making across-country and across-commodity comparisons. These and other weakness will be discussed in greater detail later in the chapter.

2.3 The Concept of Economic Efficiency

Before discussing our approach, which relates the efficiency of a policy to its effect on trade, it is useful to define a number of concepts that will be used throughout the analysis and to develop an important qualification of our results. Figure 2.1 illustrates a closed economy with demand given by D and supply given by S. Before any policy is introduced, the market clearing price and quantity are P_f and Q_f, respectively. Now, suppose the government supports the producer price to P_s. Because producers will respond to the higher price and produce Q_s rather than Q_f, it is necessary for the consumer price to fall to P_c in order to induce them to consume the extra production. Since the government must pay producers the difference between the consumer price and the support price, it spends the amount, a + b + c + d + e.

Since producers gain a + b and consumers gain c + d, it is obvious that producers and consumers do not receive the full amount of the government transfer. Indeed, the government transfer exceeds the sum of producer and consumer gain by area e. Area e is the social welfare cost (i.e., the efficiency loss) of the farm program.

Formally, the degree of efficiency (E) of this program is defined as the ratio of producer gain and the sum of government expenditures and consumer costs. In Fig. 2.1, E = (a + b)/[(a + b + c + d + e) – (c + d)] = (a + b)/(a + b + e). Because area *e* grows proportionately faster than government expenditures, the efficiency of the program decreases as the price support increases.[1] It is important to note that if the supply curve is price inelastic (i.e., vertical), then the price support program is 100% efficient. This is because all of the government transfer is actually received by the producers. In such a case, output is not affected by the price support program.

[1] For example, Fig. 2.1 illustrates that, if the price support $P_s - P_f$ doubles, the size of area e increases by a factor of 4, while the size of area a + b increases by a factor of less than 3.

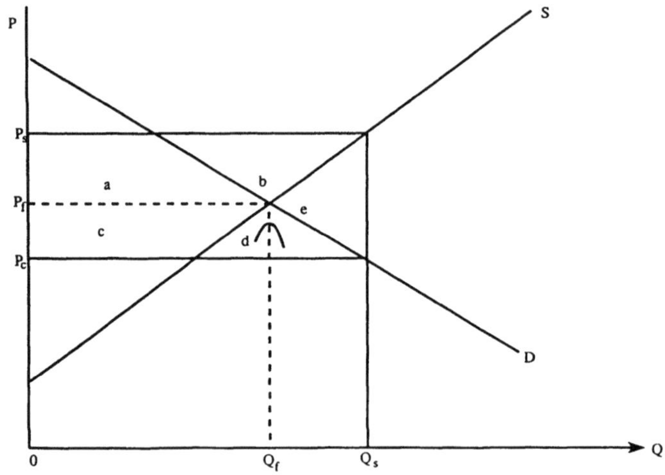

Fig. 2.1. Program efficiency in a closed economy

In the above context, the PSE measure can be specified. The US Department of Agriculture (1988) defines the PSE as the ratio of producer support from a farm program to producer revenue, including government support. Chattin (1987) provides a good description of how the PSE has typically been used as a measure of government subsidies. In Fig. 2.1, the PSE is equal to $[(P_S - P_f)Q_S]/P_SQ_S = (P_S - P_f)/P_S$. Notice that, given the support price, P_S, the PSE is not affected by the nature of the supply curve even though the degree of efficiency of the farm program is affected. Thus, there is an inconsistency since the producer benefit from the government transfer is related to the nature of the supply curve.

Because we assumed that the commodity in question is not traded (closed economy), it necessarily follows that there is no relationship between the efficiency of the policy and the level of trade distortion. This result also holds when a country insulates its producers and consumers by using an import or export quota. (For example, a subsidy provided to producers protected by an import quota is generally inefficient but it may not be trade distorting.) In such cases, the trade distortion is directly determined by the trade restriction and is no longer linked to other domestic policies. Consequently, for the remainder of the analysis we shall focus on open economies which do not place quantitative restrictions on trade.

Finally, it is important to note that the underlying framework of our analysis is inherently static in nature and must, therefore, be interpreted carefully in a dynamic context. For example, the US target price and acreage set-aside policy may be both efficient and non-trade-distorting in the short run because price and output are decoupled. However, over a period of several years farmers may attempt to have the program yields adjusted upward. If this should happen, the target price/acreage set-aside policy will no longer be efficient and will, consequently, distort trade. These dynamic effects are not explicitly captured by the criteria proposed in this chapter; therefore, our results must be interpreted accordingly.

2.4 Small-Country Policies

2.4.1 Efficiency-Distortion Relationship

The notion of program efficiency is perhaps most relevant in the context of international trade. Fig. 2.2 illustrates the case where the government of a net exporting country introduces a deficiency payment to raise its producer price to P_s, which is above the world price, P_f (with corresponding consumption, D_f). From the supply schedule, S, it follows that producers will increase production from Q_f to Q_s. This increase in production represents a trade distortion.

The cost to the government of providing the deficiency payment is given by area a + b in Fig. 2.2. Notice, however, that producer welfare has only increased by area a. Consequently, the efficiency of the program is given by a/(a + b). There are two important points to note. The first is that, as the deficiency payment increases, area b grows proportionately faster than area a, implying that the efficiency of a policy is inversely related to the size of the deficiency payment.[2] Since the trade distortion necessarily increases as the deficiency payment increases, it therefore follows that the efficiency of a policy and the level of trade distortion are inversely related. The second point is that, the steeper the supply curve, the greater the efficiency of the transfer and the smaller the trade distortion for a given per-unit subsidy.

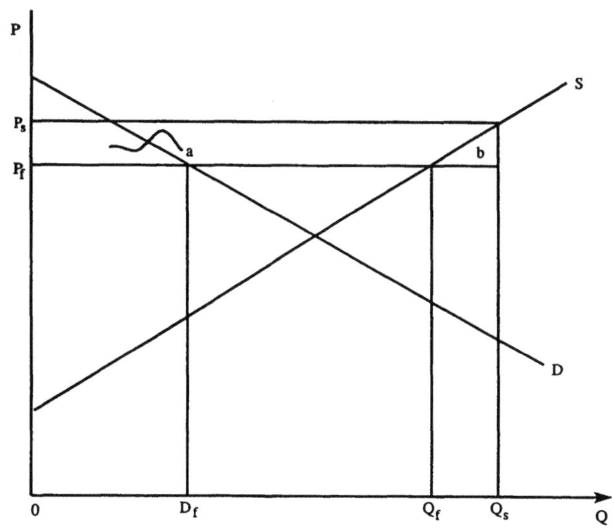

Fig. 2.2. Program efficiency and trade distortion

---

[2] To see this result in Fig. 2.2, notice that if the size of area b increases by a factor of 4, for example, then the size of area will increase by a factor of less than 3 when the price support is doubled.

2.4.2 Rankings Across Commodities

The above analysis must be extended to rank the programs affecting different commodity groups. For example, suppose an equal price support is applied to two commodities, one with a relatively low export volume, commodity 1, and one with a relatively high export volume, commodity 2. Figure 2.3 illustrates the excess (export) supply curves of these two commodities. For simplicity, assume that both commodities are exported at a world price, P_f. If the price of both of these commodities is supported to P_s, then the exports of commodity 1 will increase from $Q^1{}_f$ to $Q^1{}_s$ and exports of commodity 2 will increase by an identical amount, $Q^2{}_f$ to $Q^2{}_s$. Based on our previous discussion, it is clear that the program affecting commodity 11 is less efficient than program 2, even though the trade effects are identical.[3] However, measuring trade distortions in an absolute sense is not correct. Clearly, a trade distortion must be viewed in a relative sense. In Fig. 2.3, the relative trade distortion for commodity 1 is $(Q^1{}_s - Q^1{}_f)/Q^1{}_s$ and, for commodity 2, is $(Q^2{}_s - Q^2{}_f)/Q^2{}_s$. Because $Q^1{}_s - Q^1{}_f = Q^2{}_s - Q^2{}_f$ and $Q^2{}_s > Q^1{}_s$, it follows that the program affecting commodity 2 (the relatively efficient program) causes a lower relative trade distortion than the program affecting commodity 1. Thus, our result – a relatively more efficient program will cause a relatively smaller trade distortion.

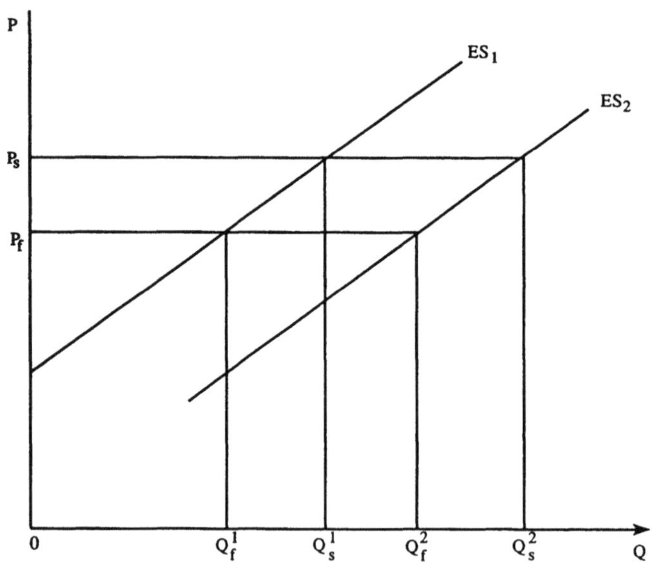

Fig. 2.3. Efficiency and trade distortion under alternative outputs

[3] This is because the dead-weight loss associated with the price support is the same for commodities 1 and 2, but the transfer to the producers of commodity 2 is larger than the transfer to the producers of commodity 1. Thus, producers of commodity 2 are receiving a relatively higher transfer.

2.4.3 Supply-Side Polices Versus Border Policies

In the previous section we demonstrated that more efficient programs imply lower relative trade distortions regardless of the export volume of the commodity. In this section we show that certain types of policies are intrinsically more efficient than others, even though both types of policies may have similar effects on trade. In particular, for equal levels of trade distortion, programs that affect the producer price only (i.e., pure supply-side policies) are always more efficient than programs that affect both producer price and domestic consumer price (i.e., combined policies).

Figure 2.4 illustrates this point. Domestic demand and supply are given by D and S, respectively. At the world price, P_f, domestic production is $0Q_f$, domestic consumption is $0D_f$, and imports are D_fD_f. If a per-unit tariff of $P_s - P_f$ is imposed on imports, then both domestic producer and consumer prices rise to P_s and imports are reduced to Q_sD_s. As a result of the tariff, domestic producers gain area c, the government gains area e + f in tariff revenue, and domestic consumers lose area c + d + e + f + g. The efficiency of the program is, therefore, given by c/(c + d + g) and the absolute trade distortion is given by $Q_fQ_s + D_sD_f$.

Suppose that a support price of P'_s rather than a tariff had been used to support producer income in Fig. 2.4. In this case, producers would gain area a + c and the government would spend area a + b + c + d + e, implying that the program efficiency is given by (a + c)/(a + b + c + d + e). Notice that the absolute trade distortion caused by the price support, $Q_fQ'_s$, is identical to the absolute distortion caused by the tariff (i.e., $Q_fQ'_s = Q_fQ_s + D_sD_f$). However, the price support, which affects the producer price only, is more efficient than the tariff, which affects both the producer and consumer prices; i.e., c/(c + d + g) < (a + c)/(a + b + c + d + e).[4]

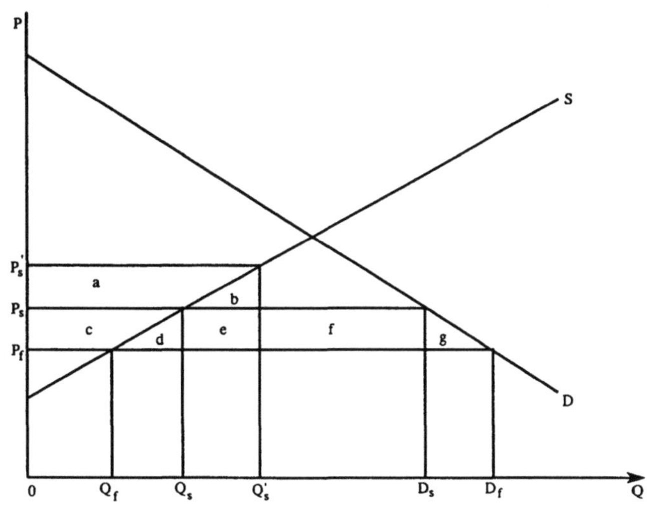

Fig. 2.4. Comparison of tariff efficiency to subsidy efficiency

[4] This result can easily be proven using simple geometry.

As a result of the intrinsic differences in program efficiencies, it is necessary to qualify the general relationship between program efficiency and trade distortion. For all pure supply-side policies, the original result holds: The more efficient a particular policy, the lower the relative trade distortion will be. Also, if a pure supply-side policy is less efficient than a combined supply-side/demand-side policy, then it necessarily must be more trade distorting. However, if a pure supply-side policy is more efficient than a combined policy, then it is not possible to tell which distorts trade more (unless the specific elasticities are known), since supply-side policies are intrinsically more efficient.

2.4.4 Ex Ante Versus Ex Post Policies

Some farm programs are ex ante in that the price support is known at the time of planting. In other programs, however, the policy is announced after production decisions have been made (e.g., the 1988 drought payments in Canada and the United States). The distinction between which programs are ex ante in nature and which are ex post can be important when examining trade impacts. It is also important to consider whether producers anticipate ex post programs prior to production. Indeed, at the extreme, the distinction between ex ante and ex post programs becomes irrelevant when programs are perfectly anticipated by producers. When this is the case, a dynamic expectations model is required to capture the true effects of farm programs on trade flows. The following illustrates our point.

Consider Fig. 2.5, where D is domestic demand (assumed perfectly inelastic) and S is domestic supply. Initially, producers expect world price, P_S, so they produce Q_S and expect to export D_0Q_S. However, after planting, suppose that the price drops to P_W and the government introduces a deficiency payment of size $(P_S - P_W)Q_S$. Notice that this is a 100% efficient program since the total transfer is received by producers or, alternatively, output has not been affected. Also notice

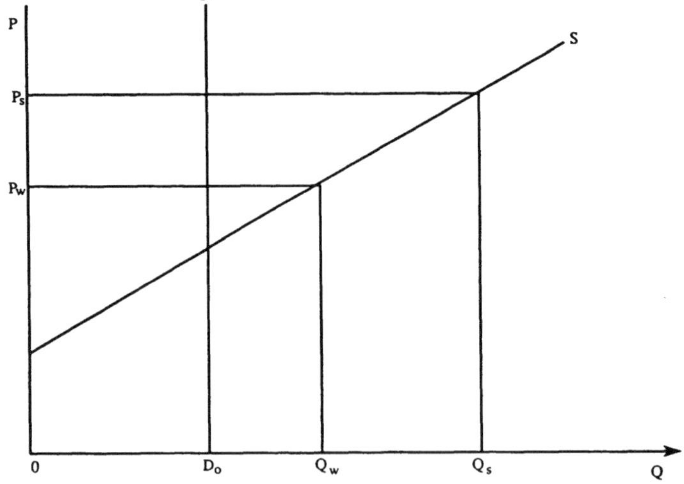

Fig. 2.5. Comparison of ex ante and ex post policies

that the volume of exports has not been affected by the deficiency payment, implying that the trade distortion is zero even though the PSE would take on a positive value.

In the following year, if the deficiency payment is totally capitalized into expectations (i.e., producers expect the support price P_s) but the world price remains at P_w, then a trade distortion of Q_wQ_s will result. This expectation has the same effect on efficiency and trade, as does an ex ante price support of P_s. On the other hand, if the transfer does not influence price expectations (i.e., producers expect P_w), then the deficiency payment from the previous year is not trade distorting. Interestingly, with a vertical supply curve, it makes no difference how a deficiency payment affects price expectations since, in such a case, neither a deficiency payment nor a price support can be trade distorting. In general, when the supply curve is less than perfectly inelastic, ex ante price supports will be less efficient and more trade distorting than will ad hoc type programs such as drought payments or deficiency type payments that are not linked to support payments.

2.4.5 Input Subsidies

Finally, it is useful to examine the efficiency and trade effects of an input subsidy. In Fig. 2.6, domestic demand is given by D and domestic supply is initially given by S_0, implying that, at the world price, P_f, consumption is D_f, production is Q_f, and exports are D_fQ_f. Suppose that the government subsidizes the inputs used by producers by $P_s - P_f$ per unit of output. This effectively shifts the supply curve down to S_1 and induces producers to supply Q_s; consequently, exports increase by Q_fQ_s.

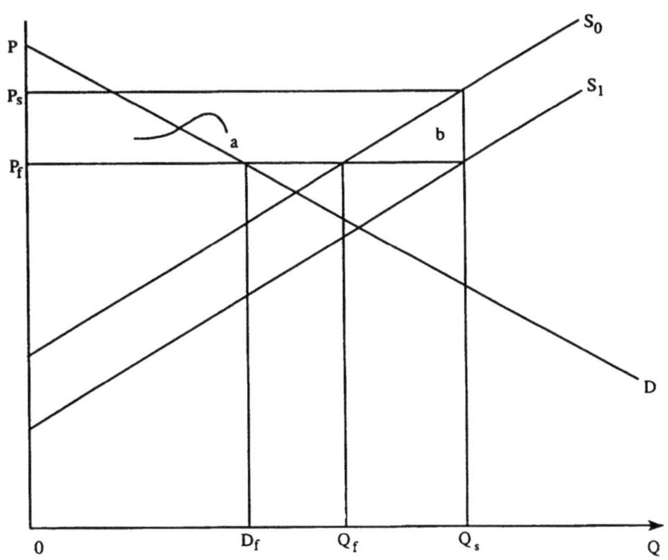

Fig. 2.6. Efficiency and trade effects of an input subsidy

The cost to the government of providing the input subsidy is given by area a + b. (Note that area a + b equals the area between the two supply curves up to Q_s.) However, because producers only benefit by area a, it follows that the efficiency of the subsidy is a/(a + b).

Note that the input subsidy has the same effect on producer welfare, government expenditures, and trade as does a producer price support of P_s. However, the PSE measure differs. The PSE associated with the input subsidy is given by $(P_s - P_f)/P_f$, and the PSE associated with the price support is given by $(P_s - P_f)/P_s$. Consequently, farm programs which include input subsidies (e.g., Canadian freight subsidy) will register a higher PSE than will a price support policy, even if both policies have the same effect on trade.

2.5 Large-Country Policies

In many circumstances, a country's policies will have a non-negligible effect on the world price, implying that the small-country assumption is no longer valid (for example, the US wheat trade). In this section the relationship between the efficiency of a program and the program's effect on trade is developed for a large country. The results are not nearly so concise as they were in the small-country case, however. Nevertheless, with sufficient caveats, a number of weaker results can still be set forth.

2.5.1 Price Supports

Figure 2.7 depicts a model for an export good where price supports are used in the absence of quantitative restrictions. With a price support of P_s, trade is disorted by

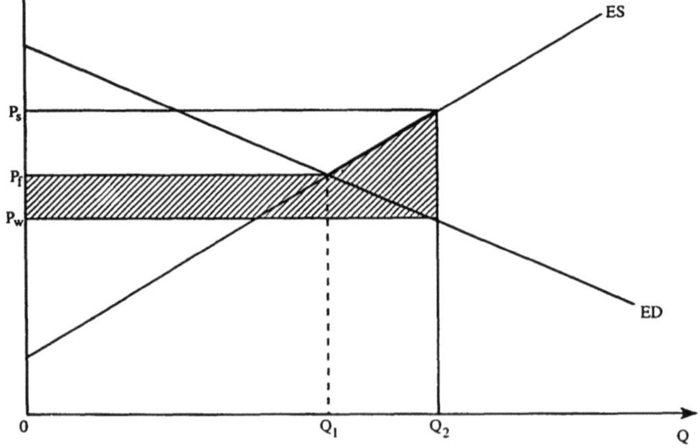

Fig. 2.7. Price supports in the large-country case

Q_1Q_2. The cost of the program, measured in absolute terms, is the cross-hatched area. It therefore follows that, the more inefficient a program, the more its trade-distorting effect and, for a given price support, the more elastic the excess demand schedule, the smaller the loss in efficiency; likewise, the more inelastic the excess supply curve, the smaller the loss in efficiency.

In comparing programs of different commodities, the following apply: (1) if the goods have identical supply and demand elasticities, the program having the higher price support is more inefficient and more trade distorting; (2) if two goods have the same price support and demand elasticities, the program corresponding to the good with the more elastic supply is less efficient and more trade distorting; and (3) if the commodities have the same price support and supply elasticities, the program corresponding to the good with the more inelastic excess demand is the less efficient and more trade distorting.

Figure 2.8 shows how both the efficiency and PSE criteria can break down in the large-country case. Two goods are considered with different supply and de-

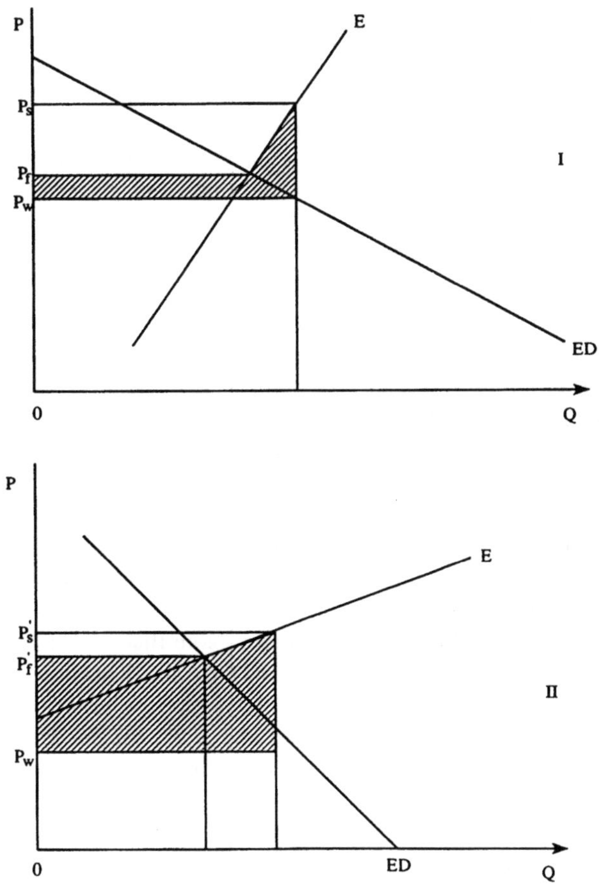

Fig. 2.8. Efficiency under alternative supply and demand elasticities

mand elasticities; both have price supports (P_s and P'_s). Note that program II is much more inefficient than program I, even though both affect trade equally. The PSE is larger for program II than for program I, if measured at prices P_w and P'_w rather than at free-trade prices P_f and P'_f. The opposite is true if the PSE is measured at free-trade prices. In either case, the PSE has the same problem as the efficiency measure in that it cannot rank programs according to their trade-distorting effects.

In Fig. 2.8, note the difference in the effect of price supports on the total value of exports. With program I, the total value of exports increases, while in program II it decreases. Thus, it appears that the lower the efficiency of a program, the greater the relative fall in the value of exports, even though the effects of the programs on physical trade do not vary. This result follows because, the more inelastic the excess demand, the lower the program efficiency; in addition, as the price falls to importers due to domestic price supports, the total value of exports drops if the excess demand curve is price inelastic.

2.5.2 Price Supports and Acreage Controls

Programs in the United States for commodities such as corn and wheat combine price supports and acreage set-asides. These cases are depicted in Fig. 2.9. The free-trade price is P_f, and free-trade exports are Q_f. Suppose that, at support price P_1, production is controlled to the free-trade level. This program is efficient and not trade distorting. However, the government has transferred the area, P_1abP_f, to producers. Note that, if the support is at P_2, the program is still efficient, even though government transfers increase to P_2cbP_f.

A problem with PSE measures is apparent. Large PSE measures are attached to the program, even though it is efficient and, hence, not trade distorting. Understanding this is crucial since, in the United States, major commodities such as corn and wheat have these types of programs. As always, an empirical question is: how far is production to the right of Q_f (due to the Farm Bill). For a given commodity, the further production is to the right of Q_f when combined with price supports, the greater the trade-distorting effects and the degree of inefficiency.

Suppose, in Fig. 2.9, a price support of P_1 generated an output of Q_1 and, hence, a trade-distorting effect of Q_fQ_1. The efficiency loss of the program is the cross-hatched area. It follows that, for a given price support and acreage set-asides, the greater the elasticity of the excess demand curve and the more inelastic the free-trade supply curve, the less inefficient the program. Note that, while the free-trade supply curve elasticity does not change the size of the trade distortion, it does affect the program's degree of inefficiency. Figure 2.10 depicts programs for goods I and II. The price support is set at P_s for goods I and at P'_s for goods II. The PSE is the largest for goods I, which is produced under the most efficient program; i.e., the cross-hatched area for goods II is far greater than for goods I. This model demonstrates that, given the production controls, the PSE can change without affecting either trade distortions or the farm program's degree of inefficiency. For example, if the price support for goods II were increased to P*, the PSE would

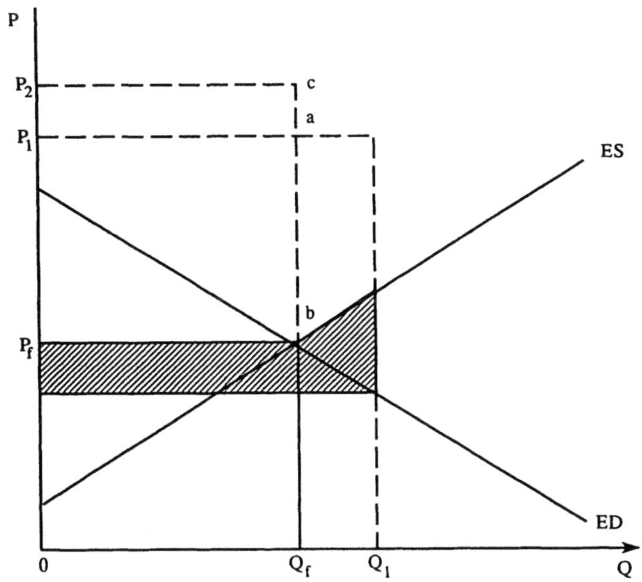

Fig. 2.9. Price supports, acreage set-asides, and efficiency

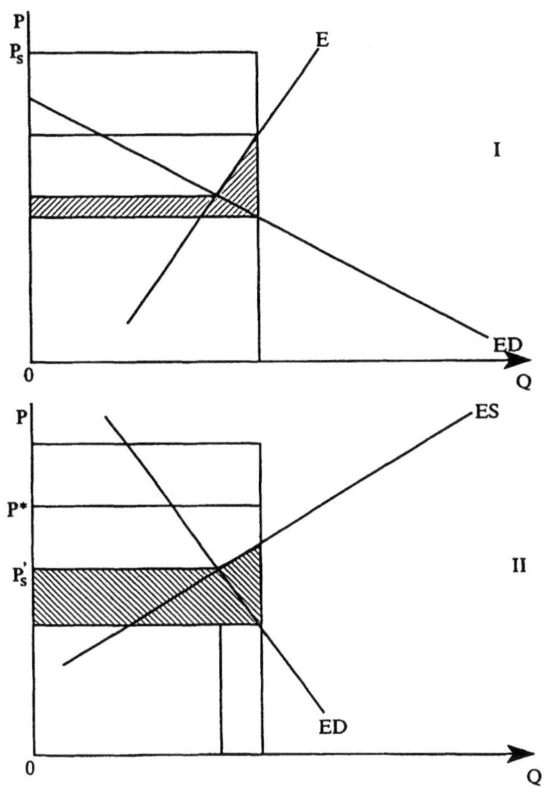

Fig. 2.10. Price supports, acreage set-asides, and the producer subsidy equivalent

increase for goods II but nothing else would change except that the government would increase its payments to producers.

2.6 Empirical Results

In this section, a number of the relationships derived earlier are reexamined in the context of Canadian and US agricultural policy. The ultimate objective is to demonstrate that ranking policies according to their effects on trade is equivalent to ranking them by their degree of economic efficiency. To contrast this approach, the same policies will be ranked according to their associated PSE measure. Not surprisingly, the PSE ranking scheme does not consistently rank policies in terms of their effect on trade; the TDE, however, does.

2.6.1 Canadian Agricultural Programs

The programs included in the empirical exercise cover a wide range of Canadian policies. Fulton et al. (1989) provide a complete description of the most important of these policies. A summary of their work is provided below:

- *Western Grain Transportation Act (WGTA).* Under this act, the Canadian government provides rail companies with an annual payment of up to $658.6 million (plus an inflationary index) to cover the transportation of eligible grain shipments to select shipping terminals. The program, enacted in 1983, is generally perceived by farmers as being a relatively stable policy.
- *Western Grain Stabilization Act (WGSA).* This voluntary program allows farmers to contribute a percentage of their gross sales to a stabilization fund. The Canadian government also contributes to the fund by adding approximately 50% of the producer contribution. The program is designed to pay out when net cash receipts from the seven major grains produced in the prairie region falls below 90% of the 5-year average net cash flow. Payouts to the prairie region equalled $223 million, $522 million, $859 million, and $1,398 million for the crop years 1983/84 to 1986/87, respectively. In 1987, it was necessary for the federal government to inject $750 million into the fund to offset the program's large deficit.
- *Crop Insurance.* This is essentially a voluntary production-guarantee program. Premiums paid by producers are matched by the federal government; administrative costs are paid by the provincial governments. If necessary, the crop insurance program is supplemented with disaster relief payments to offset the effects of drought (such as in 1988/89) or flooding.
- *Special Canada Grains Program (SCGP).* The SCGP was introduced in 1986 to facilitate a $1 billion transfer from the federal government to Canadian grain and oilseed producers. The purpose of the payout was to offset the loss incurred by producers resulting from the subsidy war between the United States and the European Community (EC). An additional $1.1 billion was paid out in 1987 as part of the same program.

- *Canadian Wheat Board (CWB) Initial Price Guarantee.* As part of the price-pooling scheme implemented by the CWB, western Canadian wheat producers are guaranteed an initial payment for their commodity when delivery is made. The CWB, upon consultation with the federal government, sets the initial price at a level somewhat below the expected export price of the commodity. After all grain deliveries have been made, the residual balance in the CWB accounts is paid out to producers. If the residual balance is negative (due to an unexpected fall in the export price), then the federal government pays the deficit.
- *Agriculture Stabilization Act (ASA).* The federal and provincial ASA programs provide price stabilization to a large number of commodities, including beef and pork. The federal ASA program requires no producer levies and makes payouts based on a trigger mechanism related to historical market prices and the indexed cost of production.
- In the prairie region, wheat is the predominant agricultural commodity produced. Of the agricultural programs described above, all but the ASA have a direct affect on the revenues farmers derive from wheat production. In recent years, transfers from the Canadian government to wheat producers have been relatively large (Table 2.1). The efficiency of these transfers, as indicated in the previous section, depends on the elasticity of wheat supply and the extent to which ex post transfers affect ex ante production.

Incorporating data from Table 2.1, Table 2.2 details the efficiency of aggregate transfers to wheat producers in 1986 as well as the PSE associated with the transfers and the relative trade effects. The measures are derived under alternative supply elasticity assumptions and alternative assumptions regarding the proportion of the transfer affecting the ex ante supply price. Not surprisingly (but probably unrealistically), if none of the transfers influence production, then the policies are 100% efficient and result in a zero trade distortion, regardless of the elasticity of supply. On the other hand, if all of the transfers are assumed to affect production, then transfer efficiency is 96.2% for the inelastic supply curve and 84.3% for the elastic supply curve. The efficient transfer is associated with a low trade distortion, while the inefficient transfer is associated with a relatively large trade distortion. If the supply elasticity equaled 0.4, and half of the transfers in 1986 influenced production, then a 13.3% distortion in trade (implying increased exports of roughly 3.7 million tonnes) would have resulted.

Interestingly, the level of trade distortion and the PSE measure are negatively correlated in Table 2.2. Note that when the transfers are 100% efficient (implying a zero trade distortion), the PSE takes on its maximum value of 46.9%. Conversely, the least efficient program, which causes the highest relative trade distortion, has the lowest PSE (equal to 37.8%). This is because the transfer per unit of post-policy production is lower for more highly elastic supply curves, or when a higher fraction of the transfer is assumed to affect production. Clearly, the PSE would serve poorly as a indicator of trade distortions in this case. It is also useful to rank a number of other Canadian agricultural policies using the methodology developed in this chapter. To contrast how the efficiency of many of these policies has evolved, we will consider policies in place in both 1976 and 1986. Table 2.3

Table 2.1. Transfers from the Canadian government to wheat producers and producer subsidy equivalent (PSE) measures: 1982–1986

	1982	1983	1984	1985	1986	1982–1986 (average)
Level of production (millions of tonnes)	27	27	27	27	27	27
Producer price (Canadian dollars/tonne)	180	181	172	147	117	159
Value to producers (millions of Canadian dollars)	4,835	4,992	4,140	4,289	5,162	4,684
Total policy transfers to producers (millions of Canadian dollars)	681.7	1,007.9	1,413.9	1,452.3	2,575.5	1,426.3
Total income/price supports	24.6	189.8	484.9	733.0	1,640.6	614.6
Agricultural Stabilization Act[a]	0.0	10.8	0.0	0.0	0.0	2.2
Western Grain Stabilization Act[a]	–38.8	108.1	327.8	516.3	850.5	352.8
Crop insurance[a]	61.9	67.5	174.7	211.7	131.4	129.4
Special Canada Grain Program[a]	0.0	0.0	0.0	0.0	506.7	101.3
Canadian Wheat Board pool deficit	0.0	0.0	0.0	23.0	0.0	4.6
Export credits	11.1	6.5	4.6	1.2	0.0	4.7
Corn competitive	–8.6	0.1	–0.8	0.0	0.0	–1.9
Two-price wheat	–1.0	–3.2	–21.4	–19.2	152.0	21.4

Total input subsidies	25.0	27.4	39.0	87.3	59.1	47.6
Fuel subsidy	13.1	13.5	20.4	39.8	40.5	25.5
Other financial assistance	11.9	13.9	18.6	47.5	18.6	22.1
Total marketing subsidies	454.9	609.3	683.5	463.2	704.6	583.1
Transportation	422.4	574.2	645.9	433.9	679.8	551.2
Inspection services	29.6	31.7	32.1	26.7	22.3	28.5
Marketing/promotion	2.9	3.4	5.5	2.6	2.5	3.4
Total long-term structure	76.7	88.9	96.6	61.9	58.4	76.5
Research and advisory	51.5	61.3	59.6	46.7	43.6	52.5
Development, structure	25.2	27.6	37.0	15.2	14.8	24.0
Provincial programs	100.5	92.5	109.9	106.9	112.8	104.5
Producer Subsidy Equivalent	14.2%	20.2%	34.2%	33.9%	49.9%	30.4%
Total Producer Subsidy Equivalent (Canadian dollars/tonne)	25.5	38.0	66.7	59.8	82.0	54.4
Total Producer Subsidy Equivalent (US dollars/tonne)	20.7	30.9	51.5	43.8	59.0	41.2

[a]Payment to producers in addition to price received.
Source: US Department of Agriculture (1988).

Table 2.2. Policy induced trade effects and efficiency of Canadian wheat policy: 1986

Supply elasticity[a]	Fraction of transfer coupled to production: 0	0.5	1	Results
	percent			
0.1	0.00	4.10	7.60	Trade Distortion
0.1	100.00	98.00	96.20	Efficiency
0.1	46.90	45.90	45.00	Producer Subsidy Equivalent
0.4	0.00	13.30	21.70	Trade Distortion
0.4	100.00	93.40	89.20	Efficiency
0.4	46.90	43.40	40.90	Producer Subsidy Equivalent
0.75	0.00	20.80	31.30	Trade Distortion
0.75	100.00	89.60	84.30	Efficiency
0.75	46.90	41.20	37.80	Producer Subsidy Equivalent

[a] Elasticities based on 1982–1986 production and 1986 world price.
Note: Total transfers in 1986 were $2.5755 billion Canadian (see Table 2.1).
Source: Computed using data from United States Department of Agriculture (1988).

details Canadian domestic production and consumption levels, as well as the price wedges which existed for the selected commodities in 1976 [estimated by Harling and Thomas (1983)] and 1986 [estimated by the US Department of Agriculture (1988)]. With no Canadian government intervention, all of the price wedges would necessarily be equal to zero. The price wedges were calculated using the following price definitions: Supply price is the price which producers are assumed to use when making their ex ante supply decisions. Incentive price is supply price plus any input subsidies measured in terms of value per unit output. Market price is the total price received by producers, including decoupled ex post transfers. The price wedges for wheat and barley in 1986 were calculated using the assumption that the value of all programs other than WGSA and SCGP payments, divided by total production, equals the difference between the producer incentive price and the export price.

Table 2.4 details the assumed demand and supply elasticities and the derived estimates of consumer, producer, treasury, and social gains and losses from these Canadian programs. Table 2.5 presents these programs (both the 1976 and 1986 groupings) ranked in terms of their economic efficiency, CSE, PSE, and TDE measures. It also lists the calculated trade distortions, expressed in terms of absolute volume and value (measured at border prices), and as a percentage of post-policy production.

Table 2.5 demonstrates a number of noteworthy points. First, ranking policies according to increasing relative trade distortion is equivalent to ranking them according to decreasing economic efficiency. The 1976 policies tend to be relatively more efficient than those in 1986; they also tend to be relatively less trade distorting. The relationship between policy efficiency and trade distortion breaks down if policies are ranked in terms of volume of trade distorted as opposed to volume of trade distorted relative to post-policy production. For example, the Canadian sugar policy is the most inefficient but does not result in the highest absolute trade distortion.

Not surprisingly, policies ranked according to increasing PSE measures are not consistent with the original policy ranking. For example, sugar production in 1986 has the third highest PSE (49.38%), yet causes the highest trade distortion. Also, potatoes have a higher PSE than pork but the relative trade loss is lower for potatoes. Note that 1976 barley policies and 1986 wheat policies have roughly the same proportional effect on trade, yet the PSE for wheat is nearly four times that of barley. Finally, the TDE, developed by Rausser and Wright (1987), provides a consistent ranking of policies because it nets out ex post transfers which, by assumption, do not affect the producer incentive price.

2.6.2 A Brief Look at US Agricultural Policy

In the previous section, a number of Canadian policies were examined in detail. The general relationship between program efficiency and trade distortion stood up well in the Canadian case since, in many of the agricultural commodities in which it trades, Canada is classified as a small country. In this section, we examine certain US programs, but in less detail. To facilitate this discussion, it is useful to

Table 2.3. Canadian post-policy production and consumption levels, and policy-induced price effects

	Post-policy			Price differential between			
Commodity	Production	Consumption	Border price	Consumer and border	Supply and border	Incentive and border	Market and supply
	1,000 tonnes		Canadian dollars/tonne				
1976							
Potatoes	2,314	1,595	132.27	0.00	2.65	3.51	0.00
Pork[a]	538	571	1,403.00	11.20	11.20	8.55	0.00
Beef	1,081	1,143	2,161.66	66.14	66.22	71.81	0.00
Wheat	20,175	5,166	139.19	–18.79	–4.81	10.84	0.00
Oats	4,533	4,053	93.04	0.00	0.00	12.23	0.00
Barley	10,610	6,377	90.27	0.00	0.00	12.73	0.00
1986							
Wheat	28,118	4,535	123.59	33.52	11.73	44.98	48.27
Barley	13,605	6,531	49.08	0.00	18.29	36.54	17.04
Sugar	816	945	23.52	4.66	15.69	19.36	0.00

[a] Date for 1976 based on 1975–1977 averages.
Source: Harling and Thomas (1983); US Department of Agriculture (1988).

Table 2.4. Canadian policy-induced consumer and producer transfers and social cost

					Producers			
	Elasticities		Consumers		Coupled		Decoupled[b]	Treasury
Commodity	Demand[a]	Supply[a]	Gain	Social cost	Gain	Social cost	Gain	Net cost
			millions of Canadian dollars					
1976								
Potatoes	NR[c]	0.08	0.00	0.000	8.11	0.009	0.00	8.12
Pork	–0.23	0.64	–6.40	0.006	4.59	0.009	0.00	–1.80
Beef	–0.39	0.56	–76.04	0.438	76.93	0.701	0.00	2.03
Wheat	–0.10	0.52	96.31	0.757	214.27	4.428	0.00	315.01
Oats	NR[c]	0.50	0.00	0.000	53.62	1.822	0.00	55.44
Barley	NR[c]	0.50	0.00	0.000	130.30	4.762	0.00	135.07
1986								
Wheat	–0.10	0.30	–153.63	1.622	1,210.43	54.314	1,357.26	2,468.37
Barley	NR[c]	0.30	0.00	0.000	441.61	55.517	367.88	865.01
Sugar	–0.13	1.09	–4.45	0.047	9.88	5.915	0.00	11.39

[a] Demand elasticities are measured at consumer prices. Pork and beef supply elastitices are also measured at consumer prices; all other supply elasticities are measured at border prices.
[b] By definition, decoupled transfers do not have a social cost.
[c] Not required.
Source: Harling and Thomas (1983); Clark (1980).

Table 2.5. Transfer efficiency, price wedges, and trade loss for selected Canadian commodities

Commodity	Transfer efficiency[a]	Consumer subsidy equivalent	Producer subsidy equivalent	Trade distortion equivalent	Trade loss		
					Volume	Value[b]	Percentage
	Percent				1,000 tonnes	$1,000 Canadian	
1976							
Potatoes	99.89	0.00	2.60	2.59	4.91	650	0.21
Pork	99.68	0.79	0.60	0.61	3.12	4,380	0.58
Beef	98.54	2.97	3.22	3.22	32.75	70,788	3.03
Wheat	97.98	–15.61	8.07	7.23	736.41	102,501	3.65
Oats	96.71	0.00	13.14	11.62	297.93	27,719	6.57
Barley	96.47	0.00	14.10	12.36	748.12	67,533	7.05
1986							
Wheat	95.71	21.34	50.79	26.68	2,511.78	310,431	8.93
Barley	88.83	0.00	67.34	42.68	3,038.67	149,138	22.33
Sugar	62.37	16.54	49.38	45.15	631.37	14,850	77.37

[a] Efficiency is calculated as the ratio of producer gain (basis incentive price) to the sum of treasury cost (less decoupled transfers) and consumer loss.
[b] The trade loss value is the trade loss volume at the border price.
Computed using data from Tables 2.3 and 2.4.

contrast the effects of US policy on the production of rice (where the United States have very little influence on world price) to policy effects on the production of wheat (where the United States has significant influence on the world price.

- *Rice versus Wheat.* Consider ranking a relatively low volume export good, such as rice, and a relatively high volume export good, such as wheat, in terms of the degree of efficiency and effect on trade of their respective policies. In Fig. 2.11, the excess supply curve for rice is S_r, and the excess supply and demand curves for wheat are *ES* and *ED*, respectively, giving a free trade price of P'_W. A price support of P_s is used in rice (which is above the work price, P_W). Wheat supported a price of P'_s, and because four sets of indicators production restricted to Q. Notice that the wheat program is more inefficient than the rice program. This would be the case even if a higher price support of P^* was used for rice and production was restricted to Q_2. This result holds because, as already stated, the degree of inefficiency of a program depends on

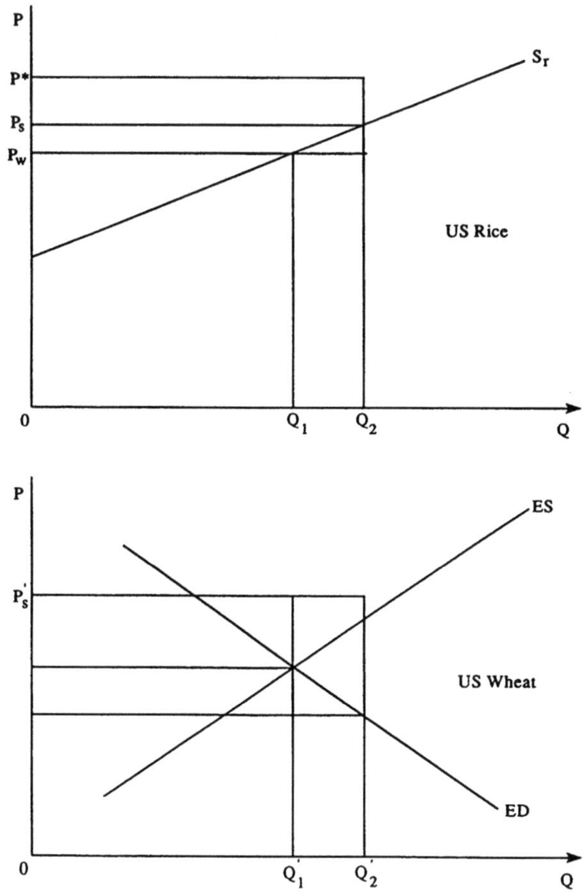

Fig. 2.11. Comparison of small-country and large-country price supports

the elasticity of the excess demand curve. In the small-country case, the excess production generated by price supports can be sold without affecting price, since the excess demand curve is essentially horizontal.

- *Imported Commodities.* Consider the case of an import good such as sugar. Historically, US sugar has been protected by import quotas. As a result of the wedge between the US domestic price and the world price, the US sugar program is believed to be inefficient. In Fig. 2.12(a), S is the supply curve of US sugar and D is demand. Given the "distorted" world price of P_w, a quota which restricts imports to Q_1Q_2 has an efficiency loss of the cross-hatched

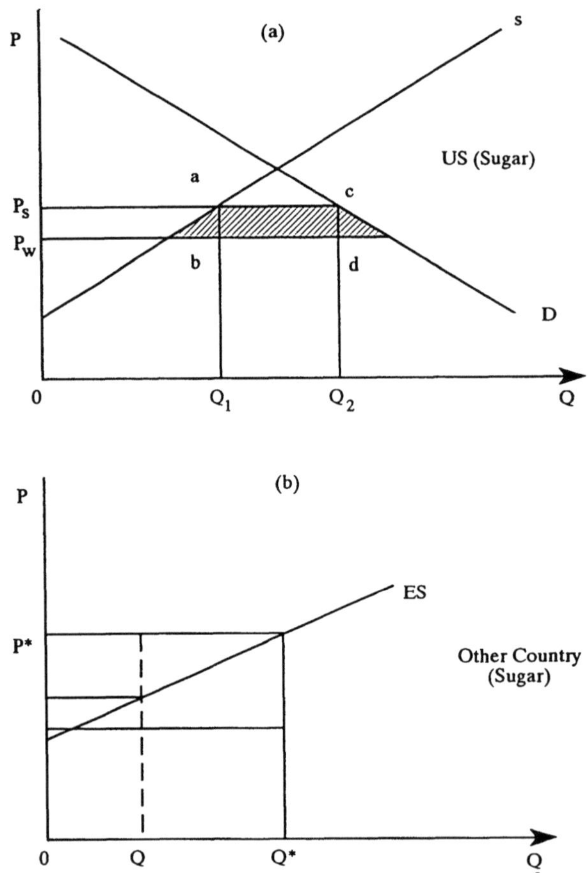

Fig. 2.12. Trade distortions with two large countries

area. (Note that the abcd quota rents go to foreign exporters.) Thus, the sugar program appears to be both inefficient and trade distorting.[5]

These conclusions can be misleading if used as a basis for freeing up international sugar markets. The world price, if distorted by other countries' policies, is not the appropriate price to use as a base upon which to measure efficiency. Recall that, in earlier models, the free-trade price was always used as the point of reference. However, in Fig. 2.12a, the reference price, P, need not equal the free-trade price; in fact, for sugar, it is well below the free-trade price (Schmitz and Vercammen, 1990).

Table 2.6 illustrates that the measure of a program's efficiency critically depends on the external reference price (which is typically the world price). Due to the increase in the world sugar price between 1988 and 1989, for example, the US gain from import quota removal decreased from $242 million in 1988 to $150 million in 1989. Presumably, these gains would decrease still further (and perhaps be eliminated) if world prices continue to rise. The point is that efficiency is a relative rather than an absolute concept. As we show below, the relationship between the efficiency of a program and its trade-distorting effects will generally be misleading unless the correct reference price is used.

The major problem with using the world price, P_w (Fig. 2.12a), as a reference point when making trade distortioning and efficiency calculations can be shown in

Table 2.6. Effects of removing sugar quotas in the United States: 1988–89

	1988	1989
	$ Million (US)	
Consumer benefit		
Purchases in the domestic and import markets	997	894
Quota rents recovered	157	137
Deadweight loss recovered	126	74
Total consumer benefit	1,280	1,105
Producer subsidy cost	–1,038	–955
Net welfare gain	242	150

Estimated by the staff of the US Trade Commission.

[5] As indicated earlier, one must be careful when applying the efficiency criteria to policies that involve quantitative restrictions. If the import quota is the only policy in place, then the quota is equivalent to an import tariff and our earlier results hold. However, if other domestic policies are used in conjunction with import quotas, then the efficiency criteria generally break down. Indeed, controlling the production of a commodity – as is done with supply management in Canada, for example – will generally be inefficient but not trade distorting if imports are restricted (via a quota) to the free-trade level.

a dynamic context. Referring to Fig. 2.12a and b, suppose the true free-trade price is P_s but exporters (i.e., the "other countries") subsidize production by using a support price, P^*, thereby driving the world price from P_s to P_w as output expanded from Q to Q*. (Note that P_w is no longer the free trade price.) The United States may impose an import quota of Q_1Q_2 in reaction to foreign exporter subsidies. However, this quota would merely reestablish the free-trade level of imports for the United States and exporters would have to restrict production to Q. Clearly, if the world returned to free trade in sugar, there would be no impact on the United States in terms of trade volumes or producer prices. In this example, import quotas, given a distorted price of P_w, need not be trade distorting if measured relative to the free-trade solution.

To support this assertion, Schmitz and Vercammen (1990) found, in a separate study using reasonable elasticity estimates, that the free-trade sugar price may indeed approximate the current US sugar price of $0.20 per pound. Alternatively stated, Schmitz and Vercammen show that unilateral liberalization by the EC may result in the world price rising to the level of the current US price, since the US import quota would not be binding under a complete EC phase-out.

Thus, there is a problem with both the PSE and efficiency measures when they are based on the distorted price, P_w, rather than a free-trade price. It is true that current quotas may be inefficient, given existing world prices. However, it does not automatically follow that a free-trade solution (which is efficient) would alter US sugar production, prices, or imports.

In recent years, it has been argued that sugar receives relatively more protection than corn or wheat, for example, and that the sugar program is relatively inefficient. However, even if the sugar program is relatively less efficient than programs governing corn or wheat (when measured using the current world price), it does not follow that trade liberalization will affect the trade volumes of sugar relatively more than corn or wheat. Indeed, as we have shown, removing sugar programs worldwide may well affect changes in trade volumes less than analogous removals of the corn or wheat programs.

- *Implications.* Some of the major US programs are considered in Table 2.7. Alston and Carter (1989), using an efficiency criteria (i.e., the fifth column of values), found the dairy program to be the most inefficient. Cotton, wheat, and sugar programs ranked behind dairy; even though sugar, followed by dairy and then rice, had the highest PSE. The inconsistency of the PSE as a measure in ranking commodity programs according to decreasing efficiency (or equivalently, increasing trade distortion) is similar to the findings detailed earlier in this study for the Canadian programs.

A second important implication is that estimates of the efficiency of an import quota, and the trade distortion resulting from it, will generally be biased unless an adjusted reference price is used. In particular, if a depressed world price rather than the free-trade price is used as a reference, the efficiency of the import quota will be understated and the size of the trade distortion will be overstated. As discussed earlier, it does not follow that a multilateral liberalization of the world

Table 2.7. US Income transfer costs and market elasticities

Commodity	Demand elasticity	Supply elasticity	Degree of protection (100 = World price)	Average cost per dollar transferred (1987)[a]		Producer subsidy equivalent[b]
				(a)	(b)	
Feed Grains	0.75	0.40	127	$1.16	$1.38	27.06
Wheat	3.00	0.50	137	$1.54	$1.85	36.49
Rice	2.20	0.60	145	$1.20	$1.44	45.20
Cotton	0.50	0.50	140	$1.67	$2.00	[c]
Sugar	0.20	0.40	178	$1.82	$1.82	77.37
Dairy	0.65	0.50	154	$2.00	$2.22	53.93
Tobacco, peanuts, and wool	1.50	0.40		$0.80	$0.92	

[a] Average transfer costs from Gardner (1987) reported on column (a). Costs in column (b) reflect the assumption that the deadweight cost of raising taxes is 20%.
[b] Producer Subsidy Equivalent figures reflect 1982–1986 average.
[c] Blanks indicate data not available.
Source: Alston and Carter (1989); Gardner (1987).

sugar trade will bring about major changes in US sugar prices and levels of imports.

Finally, we have shown that the relationship between farm programs and their corresponding affect on trade relies heavily on demand and supply elasticities. In particular, the more inelastic the excess supply curve and the more elastic the excess demand schedule, the more efficient a specified policy will be. The elasticities in Table 2.7 (computed by Gardner, 1987) tend to be consistent with relatively efficient programs. Comparable estimates used by the OECD, however, are more inelastic with respect to price elasticities of demand. This, therefore, gives rise to some disagreement over the extent to which US programs are efficient and distort trade.

2.7 Conclusions

A great deal of effort has been devoted to computing PSE measures for the GATT negotiations. We have shown that additional research should have been carried out on the efficiency of farm programs. With further research, it would have been possible to determine which programs are trade distorting and which are not. Generally, traditional measures are not able to distinguish income distribution aspects of a program from trade-distorting components. We show that not all farm programs are trade distorting, even though there may be a large associated transfer.

It is still open to debate, and has yet to be empirically shown, whether or not US farm programs are actually trade distorting, given that many incorporate production controls and, thus, are at least partially decoupled. Our results show that the size of trade distortions cannot be deduced from the many PSE estimates available. If they could, it would necessarily follow that US farm programs do distort trade.

As a society we are ultimately interested in trade and other associated gains from changes in farm programs, both at home and abroad. Supporters of the GATT believe that "the gains from trade" are large numbers. In this chapter, we have identified the relationship between the efficiency of a farm program and the program's effect on trade. The next research endeavor, therefore, should examine how program efficiency relates to the gains from a change in farm policy. The analysis of this latter relationship follows directly from the results established here, since the gains from a program change are closely related to the reduction in the corresponding trade distortions because both depend upon net social cost.

The framework discussed here can also be used to assess alternative policy prescriptions for the EC. For example, it has been argued that EC policies are inefficient since they result in a sizeable net loss in welfare. Movement toward more efficient policies, such as a production quotas, could allow the EC to maintain farm incomes at current levels while significantly reducing the impact of their policies on world trade.

References

Alston J, Carter C (1989) Causes and consequences of farm policy. Paper prepared for the 64th Annual Western Economics Association International Conference, June 18–22, Lake Tahoe, CA

Bhagwati JN, Srinivasan TN (1984) Lectures on international trade. MIT Press, Cambridge, MA

Chattin BL (1987) Measures of government intervention in US wheat, corn, sugar and dairy markets. Paper presented at American Agricultural Economics Association Annual Meeting, Aug 2–5, East Lansing, MI

Clark S (1980) An economic model of soil conservation – depletion of dryland farming in Saskatchewan. PhD Dissertation, Department of Agricultural Economics, University of Saskatchewan, Saskatoon

Fulton M, Rosaasen K and Schmitz A (1989) Canadian agricultural policy and prairie agriculture, Economic Council of Canada, Ottawa

Harling KF, and Thomas RL (1983) The economic effects of intervention in Canadian agriculture, Can J Agric Econ 31:153–176

Gardner BL (1987) Causes of US farm commodity programs. J Polit Econ 95:290–310

Rausser GC, Wright B (1987) Alternative strategies for trade policy reform, Working Paper No 441, Department of Agricultural and Resource Economics, University of California, Berkeley

Schmitz A, Vercammen J (1990) Trade liberalization in the world sugar market: how level is the playing field? Working Paper No 563, Department of Agricultural and Resource Economics, University of California, Berkeley

United States Department of Agriculture (1988) Estimates of producer and consumer subsidy equivalents: government intervention in agriculture, 1982–86. ERS Staff Report No AGE880127. Economic Research Service, Washington, DC

Zietz J, Valdés A (1988) Agriculture in the GATT: an analysis of alternative approaches to reform. International Food Policy Research Institute, Research Rep 70, Washington, DC

Chapter 3

Compensation and Political Feasibility: Facilitating Welfare Improving Coalitions

Richard E. Just, Gordon C. Rausser, and David Zilberman

3.1 Introduction

There is a broad consensus among economists, policymakers, and informed citizens that free-market economies do not preclude many inefficiencies in resource allocation. There are many circumstances where policy interventions have the potential to achieve a "Pareto improvement" in the sense of making some people better off without making others worse off. The challenge to any particular market-oriented society is to design those mechanisms, organizations, and institutions that identify the limitations of free markets and result in those reforms that seek Pareto-improved outcomes.

Several phenomena have been identified that lead to suboptimal performance of competitive markets. Traditional welfare economics has identified externalities and public goods as sources of "market failure." Without adequate intervention, competitive markets may lead to the excessive generation of those goods responsible for negative externalities and the underprovision of goods resulting in positive externalities. An incomplete set of risk markets is another potential reason for suboptimal behavior.

In general, limited or incomplete information is a major reason for imperfect market performance. Moreover, Simon (1957) identified "bounded rationality" in order to formally recognize that humans have limited ability to store and process information. Traditional welfare economics has also isolated monopolistic behavior as sources of imperfection, calling for antitrust regulation in some instances and public sector regulation where a "natural monopoly" is warranted. To these possible sources of market imperfections, other phenomena including ill-defined property rights, principal-agent distinctions, asymmetric information, moral hazard, and adverse selection can be added to the list, under various circumstances.

In a seminal piece, Coase (1960) elaborated on the role of property rights in explaining suboptimal performance of market economies. He argued that ill-defined property rights and liability rules lead to misallocations of resources. In a world of zero transaction costs, a well-functioning legal system and a clear definition of entitlement allow economic agents to achieve efficient resource allocation by private negotiations. Of course, we do not live in a world of zero transaction costs. Stiglitz (1985) has used imperfect informational frameworks to argue that, even when economic agents are rational and property rights are well defined, resource allocation may be inefficient. In this framework, scarcity and unequal distribution of information explains various inefficiencies in resource allocation as well as many of the observed institutional and contractual arrangements that have emerged. In many cases, information deficiencies lead to market failure; therefore, the design of policy interventions must consider screening and monitoring possibilities to obtain desired outcomes.

The above approaches view the market as the dominant mechanism for resource allocation. Inefficiencies that arise are the result of limitation in the *economic man* and the undesirable properties of informational, legal, and technological structures. Recent theories of political economy introduce a new source for inefficiency, *political failure*, and a new factor affecting resource allocation, *the government*. As Aaron noted in his 1989 Ely lecture, we must design and implement policies that are not only robust economically but also politically.

In the neoclassical Arrow-Debreu model, there is one weighting scheme that affects prices and quantities; and this is the distribution of wealth. Political economy models introduce another element – distribution of political power – that affects resource allocation through legislation and executive branch policies. Wealth and political power are not unrelated; wealth actually may be used to obtain political power. Still, the distribution of economic and political power may be sufficiently different for the political system to substantially modify the outcomes of the pure economic system.

Political and economic considerations often lead to policy choices that are suboptimal from the standpoint of first-best outcomes but are, nevertheless, implemented to enhance the welfare of groups with substantial political power. In this second-best world, how can conventional welfare analysis be modified to isolate political-economic coalitions that will sustain welfare-improving policy reform? Can policy reforms be sustained without the simultaneous institution of compensation schemes that effectively counter the obstructionist activities of powerful interest groups? Both of these questions will be investigated in this chapter by modifying and extending the standard tools of welfare economics employed in Chap. 2. In contrast to other studies in this volume, a complete political-economic framework which admits both political and economic markets will not be pursued. No attempt will be made to explain government behavior. Instead, we will recognize that policy reforms that seek to improve efficiency of resource use can only be implemented if some account is taken of the distribution of political power and institutional feasibility. In essence, the proposed framework will modify standard efficiency analysis by identifying policy options that improve efficiency and are politically feasible. It will extend standard analyses by formally incorporating institutional mechanisms that sustain policy reform through the implementation of compensatory transfer schemes that assist in the formation of supporting political coalitions.

3.2 Foundations of Modified and Extended Welfare Analysis

Traditionally, the analysis of economic welfare has focused on measurement of the welfare of producers and consumers. Producers' and consumers' welfare is usually measured by producers' and consumers' surplus (Currie et al. 1971). In traditional economic welfare analysis, if a policy makes the sum of the two surpluses increase, it is pronounced a Pareto-preferred policy even though one group suffers a loss because the other group's position is improved more than the first group loses. In this sense, traditional welfare economics has focused primarily on economic efficiency and only secondarily on equity. The validity of such a practice assumes that lump-sum transfers among groups are both politically and institutionally feasible as well as costless.

In reality, costless lump-sum transfers are rarely, if ever, possible. Agencies incur costs in administering transfers; and, in many circumstances, the behavior of individuals is affected by their wealth which, in turn, is affected by transfers. If so, economic efficiency is altered by a transfer scheme and the sum of producer and consumer surplus is no longer a valid measure of the associated change in efficiency. What this means is that policy analysis cannot separate the problems of maximizing the size of the pie and distributing the pie. The two must be considered simultaneously.

Problems of political feasibility and/or institutional bottlenecks must be basic components of any operational analysis of economic policies. Political considerations imply that many alternative economic policies are not feasible or cannot be sustained without the simultaneous institution of compensation schemes that meet the distributional concerns of various special interest groups. With reforms and the implementation of compensation schemes comes the need for new institutions to ensure that the reforms are sustained and the compensation schemes are effectively administered. Accordingly, institutional feasibility and associated costs must also be integrated into the modified and extended analysis of economic efficiency.

To incorporate political and institutional considerations, conventional economic welfare analysis of public policy must be extended. In this section, we first consider how institutional feasibility modifies the tradeoffs that would otherwise be possible with lump-sum transfers. Political feasibility is then incorporated and several alternative criteria are considered to complete the framework for analyzing policy reforms.

3.2.1 Institutional Feasibility in Economic Welfare Analysis

The problem raised by nonexistence of true lump-sum transfers where redistribution must be accomplished by compensation schemes that alter economic efficiency is represented in Fig. 3.1. Here, the economic welfare of two groups are depicted – group X on the horizontal axis and group Y on the vertical axis. Suppose that the current state of the economy resulting from current policies and resource endowments is at point A. Now consider a change in policy – for example, a reduction of producer subsidies in agriculture – that would result in a move to point B. By this change, group X (producers) becomes worse off and group Y (consumers) becomes better off.

According to the criteria of traditional welfare economic practice, the improvement in economic efficiency of this change would be evaluated according to the difference in the broken lines with slope –1 that run through points A and B – evaluated on either axis. For example, the change in the sum of welfare across

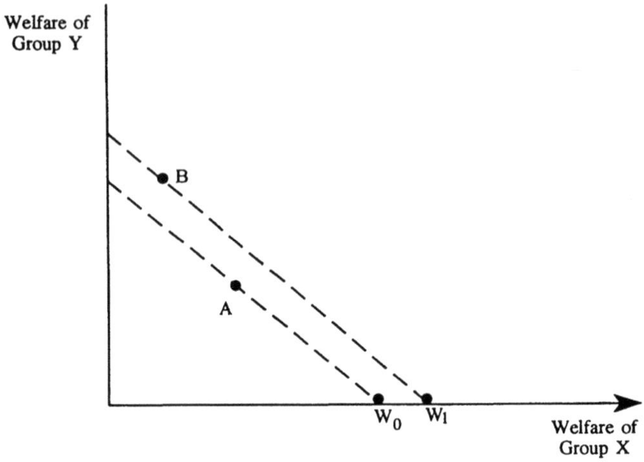

Fig. 3.1. Comparison of policies with lump-sum redistribution

both groups is from W_0 to W_1. Movement along the broken lines assumes costless lump-sum redistribution. Thus, the most efficient policy can be selected and then the welfare can be redistributed according to other criteria.

Figure 3.2 represents the same two policy alternatives but assumes that redistribution must be accomplished through specific mechanisms that fail to admit lump-sum transfers. For example, points along curve B_1 may represent the welfare

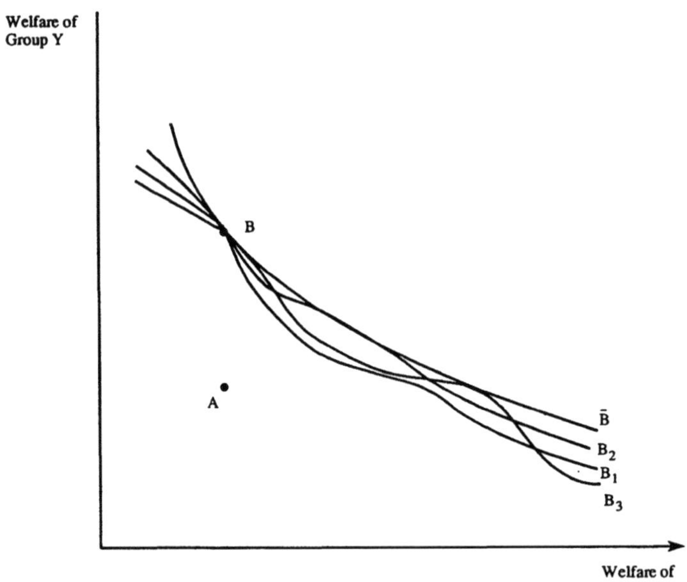

Fig. 3.2. Comparison of policies with institutionally feasible redistribution

tradeoffs that are feasible between the two groups by paying producers an initial once-and-for-all compensation payment for the loss of subsidies. In the case of the United States, the necessary funds for compensation and the costs of administering the scheme could be financed from general income tax revenues. In this context, it is important that welfare tradeoffs incorporate the institutions and their associated costs through which the compensation is to be achieved. Movement along curve B_1 represents changing the level of compensation. These tradeoff curves can bend upward because income taxation becomes increasingly distorting as the amount increases, because producers receive decreasing benefits due to the necessary intertemporal reallocation of income, etc.

In reality, there may be many alternative mechanisms or compensation schemes whereby the distribution of benefits from policy B can be altered. Each would have a tradeoff curve such as B_2, B_3, ..., which would reflect the change in welfare for each group given the particular institution's set up or use, its costs, and the way those costs are financed (e.g., income taxes, producer levies, etc.). Once all such compensations schemes are identified and evaluated, one can find the envelope curve, $\overline{B}$, which represents the best redistributions that can be attained under policy B given all compensation possibilities. Policy B would be preferred to point A if the envelope curve, $\overline{B}$, passes above and to the right of point A.[1]

3.2.2 Political Feasibility in Economic Welfare Analysis

Political forces play a major role in comparing types of policies, compensation schemes, and their outcomes. To incorporate these forces, first note that certain types of policies may not be politically acceptable. For example, an export tax is illegal under the US Constitution, so such a policy could not be considered as a policy B for improving upon the current state A. Second, certain types of compensation schemes may be politically unacceptable because of their implications for certain special interest groups. For example, financing compensation for farmers (even partially) by direct reduction in social security payments may be politically unacceptable even though other schemes that indirectly reduce the real income of social security recipients may be politically acceptable. For purposes of later politically feasible considerations, all such politically unacceptable compensation mechanisms should be ignored in constructing the envelope curve, $\overline{B}$ in Fig. 3.2.

Aside from these considerations, a criterion function must be superimposed on the analysis. This is demonstrated in Fig. 3.3. Again, point B represents a policy alternative to the current policy, A. The envelope of institutionally feasible redistributions under policy B is represented by $\overline{B}$. Now consider indifference curves associated with criterion functions such as the curve, P. This curve runs through

[1] To simplify the discussion here, we only consider comparison of policy B to the single point, A, represented in the figures. This assumes that policy A is defined by the various redistribution schemes that are currently in place. In reality, one could also consider institutionally feasible redistributions of the point A just as for policy B. In this context, policy B would be clearly preferred to policy A only if its envelope curve of institutionally feasible redistributions were everywhere above that for policy A. This concept is similar to Samuelson's (1956) analysis of utility possibilities frontiers.

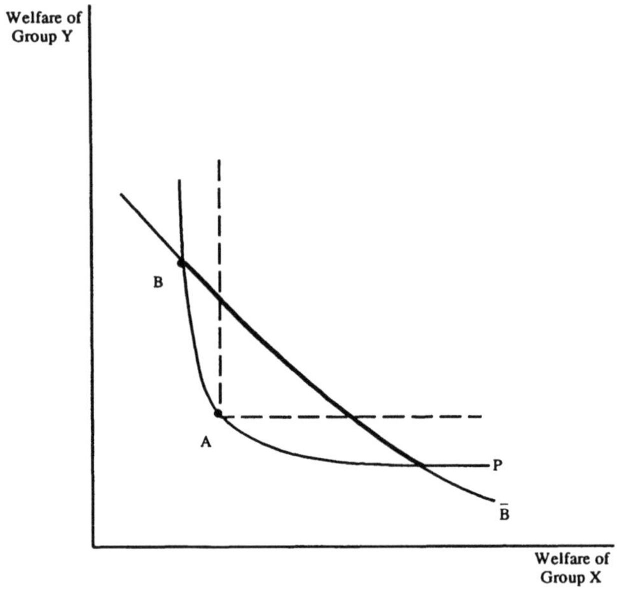

Fig. 3.3. Joint consideration of political and institutional feasibility of alternative policies

point A and is defined such that all points above it are politically preferred to point A and all points below are not. To be politically preferred means that policymakers as a group would not only choose outcomes above the curve over the current situation at point A but also that outcomes above the curve would be politically sustainable; i.e., opposition from certain interest groups would not develop such that the policy would be reversed or altered once instituted.

In Fig. 3.3, policy B would be both institutionally feasible and politically preferred to point A if and only if the envelope curve, $\bar{B}$, of institutionally feasible redistributions has some segment above and right of the political indifference curve, P, which passes through point A. This segment is represented by the heavy segment of the curve, $\bar{B}$, in Fig. 3.3. Only points on this segment (and points both above P and below $\bar{B}$ associated with inefficient compensation schemes) merit substantive consideration. Points along this segment do not necessarily satisfy the criteria for a Pareto welfare improvement. Only the points on this segment that are both above and right of point A (the points between the two broken lines extending from point A in Fig. 3.3) would constitute a Pareto improvement, i.e., would have one group better off without the other being worse off. Of course, some portions of the heavy segment represent alternatives where one group is hurt but by a small enough amount relative to the other's gain that the resulting outcome is politically stable and preferred.

3.2.3 Optimization in Policy Selection

In traditional economic welfare analysis of public policy, the criterion for policy selection is maximization of social welfare. Social welfare is represented mathematically by a social welfare function that is a function of each individual's or group's welfare. The exact form of this function has been the subject of considerable debate, and agreement has not been reached. The traditional criteria associated with potential compensation based on producer and consumer surplus, in effect, assume that social welfare is the simple sum of all individuals' welfare. From society's point of view, one dollar is just as valuable in one person's hands as another's. Some (e.g., Rawls 1971) have purposed functions on moral and philosophical grounds that are heavily oriented toward income equality; e.g., the welfare of society is defined by the welfare of the least well-off individual. Others view the public-sector criterion function as determined by the political institutions or constitution currently in place which makes this criterion function essentially synonymous with the political preference function (Rausser and Freebairn 1974; Buchanan and Tullock 1975; Zusman 1976).

For purposes of discussion, we assume here that a social welfare function as well as a political preference function can be used to evaluate alternative policy reforms. This suggests that several alternative criteria could be used to evaluate policy reforms. One possibility would be to maximize social welfare, *S*, subject to institutional and political feasibility. Another would be simply to maximize political preferences, *P*, subject to institutional feasibility and then examine the implications for social welfare. Yet another possibility would be to maximize political preferences subject to institutional feasibility and a constraint that social welfare not violate some prespecified value.

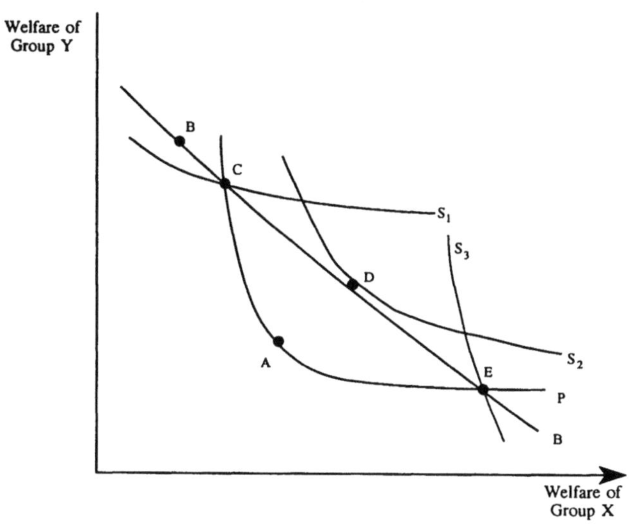

Fig. 3.4. Maximization of social welfare subject to political and institutional feasibility

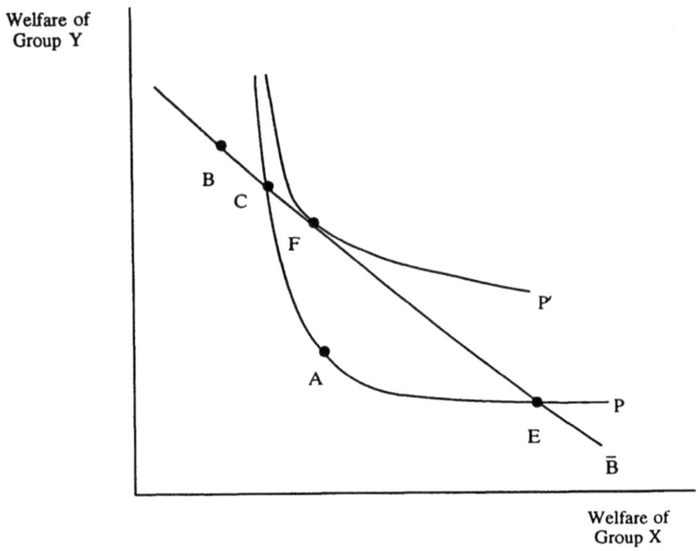

Fig. 3.5. Maximization of political preverences subject to institutional feasibility

These three alternatives are represented in Figs. 3.4, 3.5, and 3.6. In all three figures, the current state is represented by point A and the outcome under an alternative policy is represented by point B. Institutionally feasible redistributions of point B are represented by the curve, $\overline{B}$; and the indifference curve associated with political preferences passing through point A is represented by the curve, P.

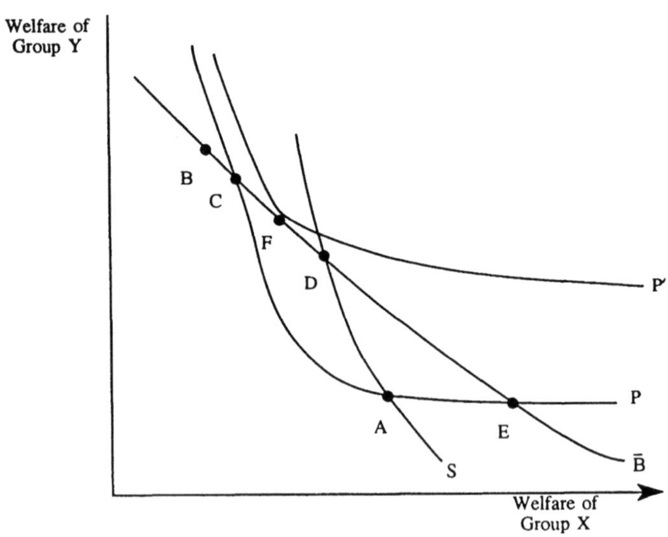

Fig. 3.6. Conflict of social and political values

Figure 3.4 considers the maximization of social welfare subject to political and institutional feasibility. Note that only points both above P and below $\bar{B}$ or on are both politically and institutionally feasible. Thus, the object is to choose the point among this feasible set that attains the highest social indifference curve. Following standard assumptions where social indifference curves do not cross and higher curves lie upward and to the right, point C will be the optimum if the social indifference map contains an indifference curve, such as S_1, where the social indifference curves are less steep than $\bar{B}$ and thus tend to favor group Y. Point D will be the optimum if one of the social indifference curves, such as S_2, has a tangency to the segment of CE of $\bar{B}$. Point E will be chosen if the social indifference map contains an indifference curve such as S_3 which runs through point E with steeper slope than $\bar{B}$.

Figure 3.5 represents the case of maximizing political preferences subject to institutional feasibility. The curve, P', represents another political indifference curve defined similarly to P that is just tangent to $\bar{B}$ at point F. By the standard properties of an indifference map, such a curve must exist and have a tangency point, F, between points C and E on $\bar{B}$.

In many circumstances, the improvement of social welfare would also be an improvement according to political preferences and vice versa. However, this need not be the case. For example, improvements in social welfare do not correspond to an improvement according to political preferences in the cases of social indifference curves, S_1 and S_3, in Fig. 3.4. In fact, if social welfare were maximized only with respect to feasible institutional redistributions and without regard to political feasibility, then politically inferior outcomes would result in these cases. This demonstrates one reason for the political inadequacy of traditional welfare economics.

To demonstrate the potential incompatibility, Fig. 3.6 shows how social welfare can be lost if political preferences are pursued only with regard to institutional feasibility and without regard to social welfare consequences. This happens when social values are in conflict with political preferences. In Fig. 3.6, political preferences are heavily oriented toward group Y, i.e., an improvement for group Y is considerably more valuable than an equal improvement for group X. This orientation is somewhat extreme in the sense that the tangency or political optimality point tends to move left on higher indifference curves (such as from point A to point F). This might occur when group Y is heavily concentrated and organized and, thus, can bring considerable political pressure to the policy process as opportunities arise.

On the other hand, the social welfare function is relatively more heavily oriented toward group X (the indifference curve, S, running through point A is more steeply sloped than the political indifference curves). With one of the more egalitarian welfare functions, this could occur if group X were a relatively poor group compared to group Y. The result is that the point of political preference optimality under policy B, which occurs at point F, is at a lower level of social welfare than is point A. Thus, maximizing political preferences can actually reduce social welfare.

To avoid this possibility, one criterion is to maximize political preferences subject to institutional feasibility and a constraint that social welfare not be reduced. In the case of Fig. 3.6, this results in choosing point D. This point is on the institutionally and politically feasible segment CE of $\bar{\bar{B}}$ and attains the same social welfare as the initial point A but maximizes political preferences subject to the stated constraints.

The above treatment does not attempt to resolve which of these three criteria to use in policy formulation but, rather, to point out that significant differences may exist among choices under the three criteria. Because the concept of a social welfare function is illusive and difficult to measure, it is easy to disregard. Moreover, it has little, if any, predictive power in explaining what a particular government might do in the context of policy reform. Nevertheless, in a prescriptive sense, greater awareness and sensitivity to situations where social welfare and political interests may be in conflict can help to identify cases where social welfare can be improved while satisfying political and institutional concerns.

3.3 Endogenous Preferences, Dynamic Adjustment, and Uncertainty

Several generalizations must be recognized before any of the above criteria can be employed to assess policy reforms. One such generalization is the dynamic implication of adjustment and change over time. Dynamic considerations in the measurement of conventional economic welfare for individual groups has been investigated and the associated methodology has been developed (Just et al. 1982). However, dynamic generalizations can also amplify the differences between political and social preferences. Consider, for example, the case where policies and compensation schemes are selected to maximize political preferences only with respect to institutional feasibility, then policy changes will tend to improve the well-being of strong political groups. Eventually this will tend to lead to further inequality and greater failure to satisfy social welfare criteria if the two are in conflict. In other words, political and social preferences will tend to diverge. On the other hand, if policy choices are made to maximize social welfare subject to political and institutional feasibility, then *less powerful* groups can be served. If this economic strengthening of the weaker groups also strengthens them politically, then the eventual dynamic outcome is to cause political and social preferences to converge.

Some observers have argued that rent seeking in a capitalist economy can eventually cause social and political preferences to diverge to the point of collapse of the political structure. Accordingly, a large and discrete correction may be required from time to time when the divergence between political and social preferences gets too large. Because of the discrete nature of such adjustments, the costs and associated inefficiencies can be substantial so the long-run consequences should be kept in mind in formulating short-run policies. Yet another, and perhaps more realistic, alternative is that certain *watchdog* special interest groups can be successful in limiting the divergence between political and social preferences. In any case, the dynamic effects of the criteria of policy selection on the distribution of political power and the ultimate conflict of social and political values is an important issue for policymakers to bear in mind. Moreover, any successful compensation schemes must take into account these dynamic effects of policy selection criteria.

3.3.1 Uncertainty

Another important generalization needed in our framework is uncertainty. In reality, policymakers and economic analysts may have a limited understanding of the tradeoffs. Many unknown factors may cause actual outcomes to be considerably different than anticipated at the time of policy formulation. This can be true both with respect to the impacts on welfare of various interest groups and by unanticipated reactions by special interest groups.

The problem of uncertainty is demonstrated in Fig. 3.7. The welfare of the two individual groups is measured on the respective axes by certainty equivalents (the expected real income discounted for risk). Rather than representing institutionally feasible tradeoffs from the expected new policy outcome at point B with a single curve, this figure represents, say, a 90% confidence band around the expected tradeoff curve. In other words, given uncertainties about how the economy operates and how individuals are affected, policymakers can only be 90% sure that the actual effects of a compensation scheme will fall between the two curves, $\overline{B}_L$ and $\overline{B}_U$. There is a difference between the two curves at point B because the initial outcome of the new policy is uncertain. The difference between the two curves likely widens in moving away from point B because the effects of various compensation schemes introduce further uncertainties.

In addition to uncertainty about the institutional feasibilities, policymakers also have uncertainty about political reactions and the resulting political sustainability of policies and compensation alternatives. This uncertainty is also represented in Fig 3.7 by a confidence band. Instead of a single curve representing political indifference associated with point A, a 90% confidence band is represented by the area between the curves, P_L and P_U These curves may be tangent at point A representing known political acceptability of the status quo but diverge in moving farther away from the current situation because greater uncertainty about political fallout is associated with larger changes.

For the uncertainties represented in Fig. 3.7, the range of feasible choices must be less than in the case of certainty if some minimal level of confidence in the outcome is to be maintained. For example, suppose a specific compensation scheme operated at a particular level produces a 90% confidence outcome as represented by the segment, $C_L C_U$. Then policy B with this particular compensation scheme would be both institutionally and politically feasible with 90% confidence if and only if the segment $C_L C_U$ lies everywhere above P_U. Noting that the confidence band on institutionally feasible tradeoffs through compensation diverges in moving away from point B and that the confidence band on political indifference diverges in moving away from point A, it is clear that greater uncer-

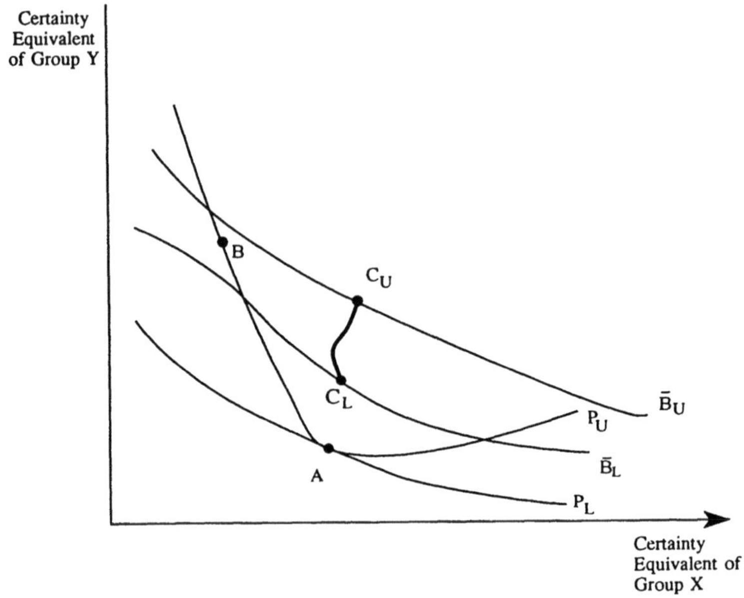

Fig. 3.7. Uncertainty in policy formulation and compensation

tainty tends to cause acceptable policy/compensation schemes to favor the status quo (be close to point A) and involve smaller magnitudes of compensation (be close to point B). In other words, when policymakers recognize their limited understanding of the underlying economy, they will tend to change policies less than otherwise.

3.4 Alternative Compensation Schemes

The "possibility frontiers" of the two preceding sections are partially based on the design of compensation schemes. Accordingly, a critical component of the extended welfare analysis is the structuring of compensation mechanisms. Historically, compensation schemes have been used frequently as elements of programs and policies intended to provide public goods or advance public objectives. For example, establishment of transportation facilities or military bases have involved voluntary or involuntary (confiscation) land transfers with compensation to the owners. Tax incentives are used to induce enterprises to invest in certain activities (e.g., oil drilling) or locations (e.g., ghettos, low-income rural areas). Land grants were given in exchange for construction work on railroads. Severance pay and special early retirement arrangements are used as mechanisms to trim labor forces of public and private organizations. A developer who is granted an exemption from zoning or building codes may, in exchange, be required to construct a park or low-income housing.

Compensations are expenditures or foregone revenues that are made to facilitate a project and are not spent on goods purchased through normal market transactions. On occasion, compensation is given in situations where owners of property rights are forced to transfer these rights (for example, when land is confiscated for construction of public facilities). In other situations, compensation recipients do not give up property rights but may be losing income, amenities, or asset values as a result of a policy or project. Such compensation may occur when a project or policy causes pecuniary externalities (reducing a price of output or increasing prices of inputs) that negatively affect the compensated party. The compensations are not necessarily equal to the actual loss; they are not determined according to welfare economic formulae but, rather, reflect political forces. In these situations, the receipts of compensation do not have a property right but have other types of entitlement.

The entitlements that are behind many compensation arrangements are sometimes vaguely defined and are protected by the current political landscape. For example, many of the existing farm programs are based on the notion that farmers are entitled to *fair prices, fair income,* or some notion of *fair returns.* Recent frameworks have been advanced that view commodity policies as compensation for government activities (research and development, extension) that have increased welfare overall but have adversely affected farmers (see Chaps. 4 and 5). Exact measures of *fair income* and *fair prices* are not well defined or agreed upon; hence, the design and implementation of a compensation scheme is not an easy task.

There are, of course, many other problems that must be faced in the practical implementation of compensation. Determining fair compensation under uncertainty (ex post versus ex ante), reducing moral hazard concerns, specifying eligibility, articulating less politically powerful interest groups, financing compensation, and designing credible threats all make the implementation of compensation schemes difficult to structure. In this section, each of these issues is considered in turn.

3.4.1 Ex Ante Versus Ex Post Compensation

One effective way to reduce the uncertainty borne by policymakers in evaluating institutional and political feasibility is to condition compensation on the occurrence of events that are unknown at the time of policy formulation. For example, a transfer payment to farmers in lieu of future government program subsidies could be formulated such that the payment would be small if future free-market prices turn out to be high and vice versa with low future prices. Such a scheme constitutes ex post compensation because the amount of payment is determined after random events are realized. It reduces political uncertainty in the above example in the following way: If future prices turn out to be high, then government transfers to farmers are reduced; this situation tends to generate political support from farmers. On the other hand, this situation would tend to generate political opposition from consumers if the large subsidies were paid to farmers at a time when consumers are suffering from high food prices. Similarly, conditioning large transfers on the event of low prices would tend to placate farmers' interests; this situation is also less likely to aggravate consumers.

A useful consideration with respect to ex post compensation relative to ex ante compensation is that it enables a spreading of risk between government and the private sector that can reduce the overall cost of risk for society. In the above example, the risk faced by both farmers and consumers is reduced because the ex post compensation rule makes them better off when they are in an unfavorable situation and worse off when they are in a favorable situation. At the same time, government attains the many benefits of flexible policy discussed by Just and Rausser (1984).

In the case of ex ante compensation, transfers are made before the states of nature are realized and the value of some key variables are known. With this approach, the entitlement is to the expected levels of some key measures of performance, perhaps adjusted for risk. Use of ex ante compensation arrangements prevents moral hazard problems and reduces the monitoring and administration costs of implementation. The reduced transaction cost of ex ante compensation, vis-à-vis ex post compensation, is one reason that this approach is frequently employed by the public sector.

The ex ante compensation approach has, however, several drawbacks relative to ex post schemes. First, with ex ante compensation the randomness facing the compensated agent does not necessarily decline. The mean income (or return) of the agent is increased by the compensation even though the distribution around the mean may not change. Ex post compensation tends to reduce and even eliminate the randomness and variability of the income or return of the compensated agents. It is also likely to increase the income mean. When the mean income effect of the two compensation schemes is equal, economic agents that are risk averse favor the ex post rule that reduces their risk. In other words, the use of ex post compensation may offer risk-bearing cost saving.

A second advantage of ex post rules is that, with more information, they facilitate decision making. The decisions on whether to and how much to compensate are determined after the state of nature is revealed, not before. One basic tenet of the economics of information is that, even under risk neutrality, efficiency is increased as more information is used by decision makers. The difference in the use of information under the two decision rules leads to different choices; and they make the resource allocation and choice under ex post rules superior.

3.4.2 Moral Hazard and Monitoring

Entitlements to compensation are often affected by the actions of the recipients. This can, of course, give rise to *moral hazard* problems. When producers know that they will be compensated for low market prices, they will fail to adjust their production. Similarly, the literature on pollution control (Baumal and Oates 1974) has shown that, when firms are paid (subsidized) for reducing their pollution generation, some may actually increase their pollution to be entitled to larger payments. To prevent these moral hazards, compensations should not be based on activities that can be manipulated by the recipients of the compensation.

Second, when entitlements are based on incomes or returns, determination of actual compensation levels requires *monitoring*. Recipients of compensation have a strong incentive to overstate their losses; accordingly, effective compensation mechanisms must be designed around easily verifiable and observable variables. Monitoring is a resource-consuming activity that sometimes requires a staff of claims adjusters and formal surveillance mechanisms. Improvements in computerized data management has reduced the cost of monitoring and allows the expansion of verifiable compensation schemes. For example, there is growing evidence that the availability of aerial surveillance data on land-use patterns and growth status of crops over time has improved the effectiveness of crop insurance and drought-protection schemes.

3.4.3 Eligibility

A third issue in the design of actual compensation relates to *eligibility*. In a policy-reform context, compensation is based on hypothetical losses of groups who benefit from current programs. The determination of the hypothetical loss is guided by economic principles but may require specific details about the individuals and their activities, especially when the compensated population is heterogeneous.

There is an obvious conflict between eligibility stated in terms of actual activity levels and reducing moral hazard. To reduce moral hazard problems, past behavior may be used to determine eligibility. But, again, individuals may realize that present behavior may partially influence eligibility in the future and behave accordingly. The main challenge in designing a compensation scheme relates to discriminating among individuals so as to yield a good approximation of the hypothetical loss while minimizing moral hazard problems.

The compensation of groups with considerable political clout may be greater than the loss they incur as a result of a policy reform while other groups who incur losses may not be compensated at all. Changes in the distribution of political power and preferences are likely to influence the weight given to losses of different groups and thereby alter the level of compensation. Affirmative action presents an obvious example of a policy that attempts to compensate for past discrimination. Farm workers in California have attempted and, to some extent, succeeded in using their political power to alter agricultural labor legislation and public research agenda in ways that serve their interests. Some of the strict environmental regulations of the present may be in response to the inattention to the environmental consequences of policy measures and agricultural practices in the past.

3.4.4 Formation and Articulation of Political Interest Groups

The extent to which social welfare interests can be served depends on the possibility of organizing new political groups that can support and sustain outcomes consistent with social welfare improvement. When a policy that is socially preferable is not politically feasible, it is because the groups helped thereby are not organized sufficiently to represent their political interests relative to the groups that are harmed. If this organization can be provided, then the socially preferable change can be instituted and the groups helped thereby can be organized to make the change politically sustainable.

The problem that must be considered, however, is that if such an organization is feasible and economical then why has is not occurred previously? A common reason is that the transactions costs of organization are too large. This tends to be the case with large groups and small individual effects. On the other hand, these conditions also tend to make such organizations have public good characteristics (some of the individuals helped by the organization can receive the benefits even if they do not participate). In any case, when transactions costs are important, they must be weighed against the social benefits of the changes under consideration. If the transactions costs of organization are greater than the increase in social benefits that could be attained without regard to political feasibility, then a socially suboptimal policy that is politically feasible may be the best alternative.

3.4.5 Financing Compensation and Credible Threats

For most policy reforms there will be winners and losers. Given an initial estimate of the distribution of gains and losses, the government could present to the different groups a proposed compensation and associated financing burden. If sufficient support arises, legislation could be introduced that executes the proposed reform and compensation scheme. If support does not arise, the government could conduct a *willingness-to-pay* exercise by requesting that groups proposing the suggested compensation scheme say how much they are prepared to pay (winners) and receive (losers) for cooperation. The key question at this juncture is whether the amount available for compensation from winners is sufficient to tempt enough losers such that adequate political support exists to implement reform. If the political power of groups who support the reform plan is sufficient, the process is completed. If this is not the case, a search must begin for an alternative supporting coalition.

For the process to generate credible threats, groups must be subject to exclusion from any supporting coalition. In particular, the compensation scheme should be designed to encourage groups to accept the distribution of burden and transfers in the first round of the negotiation process. Losers who do not agree may be harmed in subsequent stages by not being compensated, while gainers who decline to support the scheme may suffer by being forced to pay the assigned compensation scheme if legislation is passed without their support. Formally, suppose there are K losers and J winners. Let L_k denote the loss of group k (expected loss under this neutrality or the certainty equivalent under risk aversion) and G_j denote the gain of group j, where k = 1 ... K; j = 1, ... J. If $\Sigma_K L_K < \Sigma_j G_j$, group k will receive L_k and group j will pay $(G_j/\Sigma_j G_j)\, \Sigma_K L_K$.

In the above structure, if total gains exceed total losses, a compensation scheme can be internally financed. Suppose, however, that all parties do not accept this compensation scheme. Instead, suppose

$j = 1, J_1$ accept the plan;
$j = J_{1+1}, J$ reject the plan;
$k = 1, K_1$ accept the plan;

and

$k = K_{1+1}, K$ reject the plan.

If those who accept the plan, $J = 1,J_1$ and $k = 1,K_1$, have sufficient political power, the proposed compensation scheme can be implemented. If not, one approach is to ask each group who rejected the proposed compensation scheme to submit their respective willingness to pay. If these amounts are W_j and R_k, the funds available for financing compensation would be G_j from $j = 1,J_1$ and W_j from $j = J_1,J$ and the demands for compensation would be L_k from $k = 1,K_1$ and R_k from $k = K_{1+1},K$.

Based on the willingness-to-pay responses, the next stage could consist of formulating a compensation plan where all the groups from $k = 1,K_{1+1}$ will be compensated and where some groups from $k = K_{1+1}, K$ will be compensated such that the total compensation does not exceed the total willingness to pay by gainers and the political power of the supporting group is sufficient to implement the proposed reform.

If the willingness to pay does not exceed the minimum cost required to finance the compensation scheme, a third round could be conducted with the purpose of raising the contribution of winners. Negotiation could continue until a solution is obtained or the process is terminated because political support of the actual compensation cannot be achieved. This process assures that losers accepting the proposed compensation in the first round actually receive the agreed-upon levels while other groups that bargain risk losing compensation. This design provides a credible threat and increases the incentive for losers to support the proposed reform.

The above algorithm presumes that the benefits of any proposed reform are sufficiently concentrated such that specific interest groups can be identified. If benefits are widely dispersed, however, the burden of financing any compensation scheme could be placed upon the general population. The actual financing could occur through general tax revenues or could take the form of long-term governmental debt to mitigate short-run, potentially undesirable macroeconomic effects or temporally concentrated tax burdens.

Many alternative designs also exist for the losers of any particular policy reform. Compensation will be incurred only when these losers are sufficiently concentrated and have some political influence. Of course, compensation to these groups does not have to be equivalent to the capitalized loss under reform. In many instances, these capitalized losses will be difficult to estimate and subject to error.[2] Compensation at these levels should effectively counter the opposition to

---

[2] Only when market valuations of assets reflecting the effects of policy reform are continuously available is it possible to get precise estimates of the amount of "capitalized losses."

proposed reforms. However, expenditure minimizing compensation schemes can be constructed which result in payments to losers that are just sufficient to mitigate any politically obstructionist strategy they might pursue. This leads to partial compensation schemes that reflect the distribution of political influence among both concentrated losers and concentrated winners. Finally, compensation schemes can be "in kind" transfers, as demonstrated in Chaps. 5 and 6.

3.5 Concluding Remarks

The purpose of this chapter has been to develop operational formulations that can be used to assess policy reforms. Three alternative formulations were specified:

- Maximization of social welfare subject to institutional and political feasibility.
- Maximization of political preferences subject to institutional feasibility.
- Maximization of political preferences subject to institutional feasibility and minimal levels of social welfare.

We have argued that political and institutional considerations must be explicitly recognized in the design and evaluation of sustainable policy reforms. In particular, the focus of this chapter has been on *politically feasible* compensation mechanisms for groups within a particular country that may be harmed by policy reform.

A particular country's policy reforms, motivated by trade liberalization or otherwise, can entail substantial adjustment cost, especially in highly protected sectors. In many Organization for Economic Cooperation and Development (OECD) countries, one of the most highly protected sectors is agriculture. In the case of the United States, much of agriculture is highly capitalized with equipment and land; and, thus, reduced protection will cause the value of these assets to decline (Rausser 1992). Reform proposals can include policies for facilitating adjustment and for providing compensation to those disadvantaged by reductions in the level of protection.

Compensation of losers must be considered, in part, to prevent obstructionist political tactics. Since policy reform may generate benefits to other groups, it would, indeed, be desirable if these beneficiary groups have, or could acquire, some influence in the policy reform process. To the extent that these winners are concentrated and can be identified, they can share in financing the burden of compensation. Regardless, to sustain any policy reforms that might be implemented, means must be found to placate obstructionistic coalitions.

For each of the three alternative formulations considered in this chapter, there are many operational problems that must be faced in the implementation of actual compensation. Determining eligibility and interest group representation, fair compensation under uncertainty (ex post versus ex ante), financing compensation, designing credible threats, and reducing moral hazard concerns, all make the design and implementation of compensation schemes difficult. Various means for managing these concerns have been outlined in this chapter. In subsequent chapters (especially Chaps. 4, 5, and 6), conceptual formulations are constructed for explaining mixes of policies, some of which involve creative compensation schemes. In these subsequent conceptual formulations, the focus will be on endogenous political-economic determination of policy mixes. In contrast, the current chapter has been largely concerned with prescriptive criteria that can be used to evaluate specific discrete policy reforms.

References

Aaron HJ (1989) Politics and the professors revisited. Am Econ Rev 79:1–15

Baumal WJ, Oates WE (1974) The theory of environmental policy. Prentice-Hall, Englewood Cliffs

Buchanan JM, Tullock G (1975) Polluters' profits and political response: direct control versus taxes. Am Econ Rev 65:139–147

Coase RH (1960) The problem of social cost. J Law Econ 1:1–56

Currie JM, Murphy JA, Schmitz A (1971) The concept of economic surplus and its use on economic analysis. Econ J 81:741–799

Just RE, Hueth DL, Schmitz A (1982) Applied welfare economics and public policy. Prentice-Hall, New York

Just RE, Rausser GC (1984) Uncertain economic environments and conditional policies. In: Rausser GC, Farrell KR (eds) Alternative agricultural and food policies and the 1985 farm bill. Giannini Foundation of Agricultural Economics, University of California, Berkeley, pp 101–132

Rausser GC (1992) Predatory versus productive governments: the case of US agricultural policy. J Econ Perspec 6; No. 3:133–157

Rausser GC, Freebairn JW (1974) Estimation of policy preference functions: an application to US beef import policy. Rev Econ Stat 6:37–449

Rawls JA (1971) Theory of justice. Oxford University Press, New York

Samuelson PA (1956) Social indifference curves. Q J Econ LXX; 1:1–22

Simon H (1957) Administrative Behavior, 2nd edn. Free Press, New York

Stiglitz J (1985) Information and economic analysis: a perspective. Econ J 95 (Suppl):21–41

Zusman P (1976) The incorporation and measurement of social power in economic models. Int Econ Rev 17:447–462

Chapter 4

The Political Economy of Redistributive Policies and the Provision of Public Good in Agriculture

Harry de Gorter, David J. Nielson, and Gordon C. Rausser

4.1 Introduction

Within the literature on public choice, much attention has been directed toward understanding how the interests of various constituents and groups of constituents are reflected in the process of policy determination. When the interests and the resources of the various constituents differ, policies whose aims are redistributive frequently emerge. Redistributive policies emerge in some cases as a result of competition and rent seeking among the various interest groups (PESTs). In other cases, redistributive policy initiatives are generated in response to altruistic or ideological motivations. Magee et al. (1989) have recently characterized the processes which determine the existence of such policies as processes whose inner logic guides societies into political equilibrium. In so doing, such policies can be thought of as establishing a state of *political efficiency*. Although policies of market intervention which are motivated by redistributive goals are not economically efficient, they exist because they are politically efficient.

A second broad class of governmental policy is designed to improve allocative efficiency through collective action (PERTs). Such policies become necessary as collective responses to situations in which competitive market forces do not result in Pareto-optimal outcomes. The literature which has analyzed such situations typically focuses on how incentive structures can prevent individuals from achieving outcomes which serve their own interest. These policies exist because they are economically efficient.

Wicksell (1967) was early to recognize the distinction between these two types of policies and argued the necessity of organizing government so that the provision of the two types of policies would be decided upon in separate and qualitatively different processes. Separate processes, he argued, are required in order to reach socially optimal outcomes because of differences in the nature of the motivations for the two types of policies. Mueller (1989), in his survey of the public choice literature, indicated the conceptual and practical advantages of considering the two types of policies separately. In essence, for both Mueller and Wicksel, when the objective is to design democratic decision-making frameworks which will ensure Pareto-optimal outcomes, the best the literature has been able to offer requires separate decision-making institutions and processes for each of the two types of policies.

Although no general framework is known for optimally organizing the joint determination of the two types of policies, it is nevertheless frequently the case that they are not separable. Public policies whose intended purpose is to establish allocative efficiency generally have distributional consequences. Similarly, policies designed to be redistributive will also, insofar as they alter incentives and market behavior, impact upon the outcome of policies which are designed to establish allocative efficiency. While not dealt with extensively, such interactions between the two types of policies have not been lost upon agricultural economists. Rausser (1982), Rausser and de Gorter (1990), and Gardner (1989) have all commented on the existence of interactions between agricultural policies which are oriented toward redistributive goals, on the one hand, and toward allocative efficiencies, on the other. Their work has emphasized the potential importance of explicitly recognizing that these two types of policies are, in practice, jointly determined.

Rather than designing optimal institutional frameworks within which to jointly determine the two policies, we analyze governments as jointly determining the two policies, optimal or not, within the existing political framework. We examine the effect that the joint determination of the two types of policies has upon the functional relationship between the policies, given the existing institutional framework. It is suggested that the joint determination of the two types of policy instruments, if not explicitly recognized, can lead to results which are both puzzling and subject to misinterpretation.

We examine the joint determination of public policies that provide agricultural research and subsidies to farmers. The public provision of agricultural research is one of the most important ways in which governments in industrial countries intervene in agricultural markets in pursuit of greater allocative efficiency. Policies designed to subsidize agricultural production processes are important examples of public policies which are essentially redistributive in their objectives and lead to political efficiency.

In examining the factors which affect how these two types of agricultural policy instruments are determined, we will attempt to uncover a possible explanation for the puzzlingly consistent finding that governments *underinvest* in agricultural research. Governments are said to have underinvested in agricultural research on the basis of evidence that public expenditures on agricultural research have historically generated higher estimated returns per dollar expended than could have been gained as a return to alternative investment opportunities. Numerous research efforts have been in agreement in concluding from the available evidence that governments have underinvested in agricultural research. Several compelling explanations for the phenomenon have been provided (Fishel 1971; Arndt et al. 1977; Hertford and Schmitz 1977; Ruttan 1982, 1983). We suggest that the explanation may be linked to the simultaneous provision of agricultural research and income-supporting commodity market intervention and the interaction between the effects of the two types of policy.

We will also critically examine the notion that public expenditures on agricultural research are limited by the expansionary effect which they have upon the social costs of production subsidy programs. This notion is found to be potentially misleading because it does not fully account for the jointness of the two policies.

A complete analysis of the interactions between the motivations for and the consequences of these two particular types of policies are shown to result in a relationship of functional complementarity. This functional complementarity means that the existence of commodity market intervention leads to more, rather than less, publically funded research than would otherwise be the case. The potential functional complementarity between the two types of policy also creates the possibility that, counter to traditional wisdom, government intervention in the form of a production subsidy may actually be Pareto-improving relative to the outcome which would obtain in its absence.

4.2 A Stylized Model of an Agricultural Market

It is assumed that governments behave in their orientation toward agriculture as if they were choosing levels of policy instruments so as to maximize an objective function (Rausser and Freebairn 1974; Peltzman 1976; Becker 1983). The arguments of the government's objective function are taken to be weighted income levels of the members of the society. The weights reflect the government's relative level of concern with the welfare of farmers and consumers (taxpayers). As shown in Rausser and Foster (1990), the weights are indicative of a politically efficient outcome of the various actions taken by both producers and consumers in their efforts to influence policy.

The government is assumed to have control over only two instruments that impact on the commodity market of interest. The two instruments available to the government are: a fixed production subsidy per unit of agricultural output, denoted as r, and the level of public expenditures upon agricultural research, denoted as E. In attempting to achieve a maximum for its objective function, the government is allowed to choose levels of both instruments simultaneously. Each of the instruments available to the government will affect the level of welfare for both consumers and producers. In general, the effects of either of the instruments upon the welfare of each group will not be separable from the effects of the other instrument. Thus, in order to maximize its objectives, the government will elect to exercise its freedom to choose jointly the preferred combination of the two instruments.

The market for the hypothetical agricultural commodity of interest is assumed to be free from distortions other than those imposed by the government in the form of the two policy instruments, r and E. It is further assumed that a large number of identical producers, m, are price takers and produce a product which is consumed by an even larger number, n, of identical consumers. The price and quantity which characterize equilibrium in the market for the commodity are assumed to be determined competitively.

The n identical consumers are each assumed to behave as if they were choosing a level of consumption of the agricultural commodity so as to maximize a concave and twice differentiable utility function which is additively separable in the consumption of the agricultural good. Consumers as taxpayers are coerced by the government, through the collection of tax revenues, to pay the entire budgetary cost of both governmental policy instruments. Consumers act so as to solve the following maximization problem:

$$\max_{q^d} U = U(q^d) + \left[m_0 - pq^d - \frac{E}{n} - r\left(\frac{1}{n}\right)S(p + r)\right]\lambda, \qquad (1)$$

where
q^d = the quantity of the agricultural good demanded;
p = the price of the agricultural good;
E = the government's expenditures on agricultural research;
r = the per unit subsidy to producers of the agricultural commodity;
S = the total supply of the agricultural commodity;
n = the number of consumers (assumed to be identical);
m_0 = the endowment of each individual consumer;
$U(q^d)$ = each individual consumer's utility as a function of the quantity of the commodity consumed;
λ = the marginal utility of income.

When n is large, the necessary condition which characterizes a utility maximizing level of consumption of the commodity for each individual consumer is described by the expression, $U_q(q^d) - \lambda p = 0$. This first-order condition can be inverted to yield the individual's Marshallian demand schedule, $q^d = q(p) = U^{-1}{}_q(p)$. The summation of the demands of individuals yields the industry demand schedule,

$$D(p) = {}^n{}_{i=1}\Sigma\, q^d{}_i(p).$$

Assume that m identical producers each act so as to maximize profits (π_i):

$$\max_{q^s{}_i} \pi_i = (p + r)q^s{}_i - C_i(q^s{}_i, E), \qquad (2)$$

where $q^s{}_i$ represents the level of production for individual i and $C_i(q^s{}_i,E)$ represents the cost function of individual i. Profits are defined here to reflect the returns to owned assets such as land, unhired labor, capital, and management. Accordingly, costs are defined to represent only the cost incurred in employing inputs which are purchased or rented as flow inputs. Given this specification of profits and costs, the necessary condition for profit maximization by each producer is $(p + r) - C^s{}_q(q^s{}_i,E) = 0$. This condition can be used to determine the individual's supply schedule as a function of $p + r$ and E:

$$q^s{}_i\,(p + r, E).$$

Summation over the supply schedules of individuals yields the industry supply schedule, $S(p + r, E)$.

The market clears at the combination of quantity and price levels at which $D(p) = S(p + r, E)$. The market-clearing level of price is denoted as $p* = p*(r, E)$ and the market-clearing level of output is denoted as $Q* = Q*(r, E) = D*(p*) = S(p* + r, E)$ where the superscript star denotes the market-clearing levels. Total differentiation of the necessary conditions for maximum levels of profit and utility reveal that, given identical consumers and identical producers,

$$\frac{dQ^*}{dr} = \frac{\eta^d \eta^s Q^*}{(\eta^d - \eta^s)p^*} \geq 0 \tag{3}$$

and

$$\frac{dQ^*}{dE} = -\left(\frac{dQ^*}{dr}\right) C_{QE} = -\left(\frac{\eta^d \eta^s Q^*}{(\eta^d - \eta^s)p^*}\right) C_{QE} \geq 0 \tag{4}$$

where η^s = price elasticity of supply
η^d = price elasticity of demand
C = aggregate cost function
C_{QE} = derivative of aggregate marginal cost with respect to E.

Likewise,

$$\frac{dp^*}{dr} = -\left(\frac{\eta^s}{\eta^s - \eta^d}\right) \leq 0 \tag{5}$$

and

$$\frac{dp^*}{dE} = -\left(\frac{dp^*}{dr}\right) C_{QE} = \left(\frac{C_{QE}\, \eta^s}{\eta^s - \eta^d}\right) \leq 0. \tag{6}$$

Expressions (3) through (6) reveal that both policy instruments have positive impacts upon the market-clearing level of quantity. Both r and E have negative impacts upon the market-clearing price level. The exact way in which the subsidy instrument affects output and price is dependent upon the elasticities of supply and demand with respect to price. The more elastic the supply and demand schedules, the greater the impact the subsidy has on output. The effect of a subsidy on the market-clearing price is greater, the more elastic the supply schedule. The effect of a subsidy on the market-clearing price is less, the more elastic the demand schedule. The impact of the research expenditure instrument upon market clearing prices and quantities is dependent upon the price elasticities of supply and demand schedules as well as upon C_{QE}. The magnitude of C_{QE} reflects the particular way in which the effects of research affect marginal costs.

4.3 The Government's Policy Decisions

Government's choice problem is to jointly select the politically optimal level of r and E. The arguments of government's objective function are assumed to be the politically weighted welfare of both consumers and producers. We assume the weights are given and invariant to changes in each group's welfare. Suppose that the government's objectives can be characterized by the preference function $V(V_1,V_2)$ which takes the form of a weighted sum of perceived consumer (V_1) and producer welfare (V_2). The government selects the level of the two policy instruments which are available to it (r and E) so as to maximize $V(V_1,V_2)$. This maximization problem may be represented as

$$\max_{r,E} V = w_1V_1(p^*,M^*) + w_2V_2(\pi^*) \tag{7}$$

where
w_1 = preference weight assigned to consumers;
w_2 = preference weight assigned to producers;
$M^* = n[m_0 - E/n - r(1/n)S^*]$;
$\pi^* = (p^* + r)S^* - C$;
$S^* = S(p^* + r, E)$;
$C = C(S, E)$.

In order to solve its optimization problem, the government chooses the instruments, r and E, so as to satisfy the following necessary conditions for a maximum

$$w_1V_p P_r + V_Y[w_1M_r^* + w_2\pi_r^*) = 0 \tag{8}$$

and

$$w_1V_pP_E + V_Y[w_1M_E^* + w_2\pi_E^*] = 0 \tag{9}$$

where
$V_{1M^*} = V_{\pi^*} = V_Y$ = marginal utility of income
$M^*_E = -r(S_pP_E + S_E) - 1 < 0$
$M^*_R = -S - r[S_p(1 + P_r)] < 0$
$\pi^*_E = SP_E - C_E \gtreqless 0$
$\pi^*_R = S(1 + P_r) \geq 0.$

Because $S_E = C_{QE}S_p$ and recall that $P_E = -C_{QE}P_r$, then M_E can be written as $-rS_PC_{QE}(1 - P_r) - 1$ where $-1 \leq P_r \leq 0$. Hence, the maximum value of M_E is $rS_E - 1$ and the minimum value is -1. Subsidies have no effect on producer profits if the demand curve is perfectly inelastic or the supply curve is perfectly elastic; i.e., $p_r = 0$ only when $P_r = -1$. These expressions emphasize that each policy affects each group differently, with farmers always gaining from production subsidies but possibly losing from publicly funded research expenditures. Consumers lose as taxpayers, but the loss due to research expenditures can easily be offset by the gains to consumers generated by price reductions resulting from the cost-reducing effects of research.

Utilizing the definitions of M^* and π^* from above and employing Roy's Identity to the indirect consumer welfare function $V_1(p,M)$, the necessary conditions can be expressed as

$$-w_1\left[DP_r + S + r\frac{dS}{dp}(1 + P_r)\right] + w_2[S(1 + P_r)] = 0 \tag{10a}$$

and

$$-w_1\left[DP_E + 1 + r\frac{dS}{dE}\right] + w_2[SP_E - C_E] = 0. \tag{11a}$$

Expressions (10a) and (11a) characterize the way in which the welfare of consumers and producers are balanced against each other in the government's choice of r and E. Expression (10a) indicates that, if the government objective function is to be maximized with respect to r, the level of the subsidy must be chosen such that the weighted marginal cost to consumers of increasing the subsidy, $w_1[DP_r + S + r(dS/dp)(1 + P_r)]$, is just equivalent to the weighted marginal benefit of the subsidy to producers, $w_2S(1 + P_r)$. Similarly, (11a) indicates that government should, in the interest of maximizing its objective function, choose E such that the weighted marginal cost to consumers of additional E (i.e., $w_1[DP_E + 1 + r(dS/dE)]$ is just equated with the weighted marginal benefit to producers, $w_2(SP_E - C_E)$.

Given that the market is assumed to clear in response to market forces at the level of output at which $S = D$, conditions (10a) and (11a) can be simplified and rearranged to read as follows:

$$r = \left(\frac{w_2}{w_1} - 1\right)\frac{p}{\eta_s} \tag{10b}$$

and

$$-C_E = \left(\frac{w_1}{w_2}\right)\left[1 + r\frac{dS}{dE}\right] - \left(1 - \frac{w_1}{w_2}\right)QP_E. \tag{11b}$$

Expression (10b) can be interpreted to indicate that a government which acts so as to maximize the objective function $V(V_1, V_2)$ will offer the positively (negatively) valued per-unit subsidy r to producers if it has assigned a larger (smaller) welfare weight to producers (w_2) than it has assigned to consumers (w_1). Further, the chosen level of subsidy r will have larger absolute value, the less elastic the industry supply schedule with respect to the output price. These are standard results which hold whether research expenditures are treated as exogenously determined or are determined endogenously and jointly with the selection of r.

4.4 Underinvestment in Agricultural Research

Expression (11b) is instructive in explaining the phenomenon of underinvestment in agricultural research. This expression indicates that if the costs and benefits of public investment in agricultural research are evaluated under an implicit assumption of equality between the preference weights assigned to consumers and producers (i.e., as if $w_2/w_1 = 1$) and without neglecting the effect of existing subsidies on the social costs of research results at the margin – i.e., implicitly setting

$r(dS/dE) = 0$, the condition which will characterize the appropriate choice of E will be

$$-C_E = 1. \tag{11c}$$

This is the benchmark condition used in much of the underinvestment literature to characterize an efficient allocation of resources to research.

In the case of the agricultural policies of the industrialized countries, it is clear that subsidies to agricultural producers are, in general, positive. It is also clear that the results of research have the effect of shifting the supply function downward and outward. This combination ensures that the results of agricultural research efforts will increase the cost to taxpayers of any existing commodity subsidization scheme. If the effect of $r(dS/dE)$ is properly incorporated into the measurement of the social costs associated with expenditures upon research, still ignoring for the moment the implicit differential between w_2 and w_1, the resulting first-order condition characterizing the optimal choice of E is

$$-C_E = 1 + r(dS/dE). \tag{11d}$$

Other costs which have not been explicitly accounted for here (for example, the deadweight losses incurred in the imposition of the taxes necessary to generate revenues to cover the full amount of government expenditures) would also appear as positively valued terms on the right-hand side of expression (11d). They would have the effect of pushing the value of the right-hand side even further above the value of 1.

Expression (11d) describes a condition characterized by less investment in research, i.e., by a lower E, than is prescribed by expression (11c). Analyses which have incorporated the positive effect of E upon the cost of existing subsidy programs will expect a lower level of E to be chosen than will those analyses which have not included this additional cost. Analysts who include this factor would be expected to report less severe underinvestment than would the latter group. This is consistent with what has been reported in the literature. Those analysts who have incorporated interdependencies between costs and incidences of r and E have, in fact, found that underinvestment was less severe than those studies which failed to incorporate this interaction. However, it is clear from conditions (10a) and (11a) that analyses based upon (11d) have still not fully captured the nature of the underlying causal mechanism (Ruttan 1982; Lichtenberg and Zilberman 1986; Alston et al. 1988; Oehmke 1988). As a result, such analyses are likely to have arrived at faulty interpretations of the empirical evidence.

The hypothesis of this chapter implies that a correct interpretation of the available empirical evidence requires recognition of the implicit welfare weights and of the joint nature of the choice of r and E. As is evident from (10b), the government will only choose to provide subsidies to producers ($r > 0$) if producer welfare is weighted more heavily than consumer welfare ($w_2 > w_1$). This being the case, the fact that E is chosen according to (11a) has implications regarding whether or not analysts who implicitly assume that w_2 equals w_1 will discover what appears to be unaccounted for underinvestment.

It is useful to consider several cases in examining the implications of (11a).

4.4.1 Case 1

Assuming $w_2 > w_1$, if the marginal impact of E upon producer profits is negative at the chosen level of E and r; i.e., if $\pi_E = SP_E - C_E < 0$, then it must also be the case that:

$$-C_E > 1 + r(dS/dE).$$

This implies that the chosen and observed level of E will be below that which would be described as the appropriate level of investment by analyses performed under the assumption that $w_2 = w_1$ and which did not take into account the consequences for the chosen E of the joint determination of E and r. Simply stated, such analyses would describe the observed E as underinvestment.

4.4.2 Case 2

Again assuming that $w_2 > w_1$, if the marginal impact of E upon producer profits is positive at the chosen combination of E and r (i.e., if $\pi_E = SP_E - C_E > 0$), it follows that

$$-C_E < 1 + r(dS/dE).$$

The implication is that the chosen and observed level of E will be above that which would be described as the appropriate level at investment by analyses which were performed under the assumption that $w_2 = w_1$ and which did not treat E and r as jointly chosen. The evidence would cause such analyses to describe the observed E as overinvestment.

4.4.3 Case 3

Once again, working under the assumption that $w_2 > w_1$, if at the margin E has no effect upon producer profits at the chosen level of E and r (i.e., if $\pi_E = SP_E - C_E = 0$), then it must also be the case that

$$-C_E = 1 + r(dS/dE).$$

In this special case, analyses which have assumed that $w_2 = w_1$ would be led to believe, correctly so if for the wrong reason, that the appropriate level of E has been chosen. It would appear to such analyses that E has been chosen at a level at which it generates normal returns to invested resources. In this special case such analyses would be correct in their conclusion.

The three cases outlined above suggest that empirical analyses of the returns to public investment in agriculture which have correctly measured all costs and benefits but which have been guided by an underlying framework which implicitly assumed that $w_1 = w_2$ would interpret the evidence to imply the existence of unexplained underinvestment (overinvestment) whenever $w_2 > w_1$ and $\pi_E < (>)\ 0$.

Two factors, the implicit weights which reflect the way in which government values the relative welfare of producers and consumers and the direction in which producer profits respond at the margin to the results of research, combine to determine whether or not traditional studies will have come to a conclusion of underinvestment.

Producer profits are affected by *E* at the margin in accordance with

$$\pi = SP_E - mC_E \gtrless 0.$$

At the margin, producer profits respond positively to increases in research expenditures if revenues fall by less than total costs fall or if revenues actually rise (de Gorter and Zilberman 1990). In the analysis developed here, the results of research are never allowed to increase costs for a given level of output. Therefore, if revenues to producers actually rise as a consequence of research results, profits to producers must rise. Substitution from above reveals the following expression for the determinants of marginal response of producer profits to changes in research expenditures:

$$SP_E - C_E \gtreqless 0 \quad as \quad \frac{d(C_E)}{dQ}\frac{Q}{C_E}\left(\frac{\eta^s}{\eta^s - \eta^d}\right) \lesseqgtr 1. \tag{12}$$

Factors which work in the favor of farmers' profits at the margin include a large absolute value of demand elasticity, an inelastic supply schedule, and a small absolute value of the elasticity of the marginal research effect on cost with respect to the level of output. The elasticity of the marginal effect of research on cost with respect to output is constructed here. This approximation, together with the available information about the demand and supply elasticities, is used to assess the sign of the marginal response of producer profits to changes in research expenditures. The ratio of marginal to average cost (using the definition of cost adopted above) is used as an approximator for the elasticity of the marginal effect of research on cost with respect to output. The direction in which this estimator is biased is shown in Table 4.1. The ratio of marginal to average cost is constructed from the annual cost of production estimates reported by the Economic Research Service of the US Department of Agriculture (ERS). The average price received

Table 4.1. Direction of bias incurred in using C_QQ/C as an approximator for $C_{QE}Q/C_E$

	Owned inputs' share of revenue		
Returns to scale	Falls with E	Constant E	Rises with E
Decreasing	<	<	Ambiguous
Constant	<	$\frac{C_Q Q}{C} = \frac{C_{QC} Q}{C_E}$	>
Increasing	Ambiguous	>	>

Table 4.2. Ratio of marginal cost to average cost of the unowned inputs for important agricultural commodities

	Year																	
Commodity	1972	1973	1974	1975	1976	1977	1978	1979	1980	1981	1982	1983	1984	1985	1986	1987	1988	Average
Milk	1.50	1.42	1.32	1.32	1.43	1.46	1.56	1.55	1.48	1.49	1.48	1.40	1.38	1.38	1.26	1.32	1.16	1.41
Rice	[a]			1.39	1.37	1.83	1.49	1.75	1.77	1.51	1.45	1.48	1.71	1.77	1.97	1.89	1.62	1.64
Cotton				1.11	1.36	1.16	1.05	1.24	1.03	1.31	1.25	1.19	1.47	1.50	1.33	1.61	1.41	1.29
Wheat				1.96	1.42	1.26	1.82	1.82	1.41	1.34	1.48	1.69	1.65	1.65	1.61	1.77	1.41	1.59
Sorghum				1.66	1.45	1.75	2.07	1.64	1.32	1.34	1.34	1.21	1.46	1.93	2.09	2.29	1.73	1.66
Barley				1.88	1.78	1.38	1.30	1.33	1.32	1.23	1.32	1.26	1.33	1.21	1.38	1.37	0.78	1.35
Corn				1.90	1.62	1.49	1.81	1.91	1.59	1.34	1.57	1.33	1.63	1.78	1.95	1.97	1.36	1.66
Oats				2.39	2.45	2.25	1.80	1.52	1.60	1.63	1.25	1.26	1.22	1.02	1.10	1.01	1.11	1.54
Soybean				2.24	2.80	2.55	2.60	2.37	2.07	1.74	1.55	1.93	1.56	1.75	1.51	1.62	1.74	2.00
Beef	1.06	1.14	0.75	0.65	0.70	0.81	1.25	1.54	1.19	0.93	0.90	0.86	0.88	0.89	0.96	1.17	1.09	0.99
Sheep	1.24	1.19	1.11	1.18	1.23	1.60	1.78	1.68	1.38	1.19	1.14	1.15	1.38	1.57	1.62	1.75	1.51	1.39
Pork	1.25	1.15	0.93	1.32	1.20	1.16	1.34	1.08	0.95	0.97	1.30	1.00	1.06	1.13	1.32	1.54	1.05	1.16

[a] Blanks indicate data not available.
Calculated.

Table 4.3. Estimates of own-price elasticities for demand and supply of agricultural commodities reported in the literature

Reported by	Category	Time period	Estimate
Elasticities of demand, η^d			
Tweeten (1967)	Long-run aggregate		
	demand for:	1948–1965	
	food at farm level:		
	domestic		–0.1
	total		–1.11
	food and feed:		
	domestic		–0.076
	export		–0.841
Brandow (1961)	Farm level demand for:	1955–1957	
	corn		–0.03
	wheat		–0.02
	barley		–0.07
	beef		–0.68
	pork		–0.46
	chicken		–0.74
	turkey		–0.92
	milk and cream		–0.14
	sugar		–0.18
	lamb		–1.78
Hertel et al. (1989)	Long-run farm level		
	demand for:	1977	
	dairy		–0.513
	poultry		–0.205
	red meats		–0.407
	cotton		–0.997
	food grains		–0.956
	feed grains		–1.021
	sugar crops		–0.044
	oil seeds		–0.793
	aggregate		–0.53
Wohlgenant (1989)	Farm-level demand for:	1956–1983	
	beef and veal		–0.76
	pork		–0.51
	poultry		–0.42
	eggs		–0.15
	dairy		–0.61
	vegetable		–0.43
Elasticities of supply, η^s			
Tweeten (1979)	Long-run farm level supply of:	1926–1959	
	crops		1.56
	livestock		2.90
	total aggregate supply		1.79
Johnson (1973)	Long-run farm output	1952–1961	2.96
Hertel (1989)	Long-run farm output	1977	3.31
Ball (1985)	Long-run farm output	1948–1979	3.6

(inclusive of government program payments) is used to approximate marginal cost. Average cost is calculated as the sum of all costs other than those allocated as "return to owned inputs" in the cost of production estimates. Thus, using this methodology, the difference between marginal cost and average cost represents the per-unit return to owned inputs as calculated by the ERS. These owned inputs include land, capital, unpaid labor, and management. The constructed ratio of marginal cost to average cost is reported in Table 4.2 for several of the important commodities of US agriculture for each of the years from 1972 through 1988.

Empirically, the price elasticity of supply for agricultural products is generally assumed to be somewhat greater than the price elasticity of demand. Empirical demand estimates for agricultural products generally fall within the inelastic range. However, there is significant variance in the reported estimates of elasticities of supply and demand. Because the effects of research accrue over an extended period of time, long run elasticities are appropriate here. Table 4.3 presents several examples of relevant elasticities which have been reported in the literature. Table 4.4 presents a matrix of possible values for the left-hand side of the second expression in (12) above. It seems more than plausible, according to condition (12) and the information presented in Tables 4.2, 4.3, and 4.4,[1] that farmers' welfare is injured at the margin from technical improvements which are produced by *E*. This conclusion is fully consistent with the analysis developed herein.

Table 4.4. Matrix of sample calculated values [$(\eta^s - \eta^d)$/hs calculated at sample values]

	η^s						
η^d	0.5	1.0	1.5	2.0	2.5	3.0	3.5
–0.10	1.20	1.10	1.07	1.05	1.04	1.03	1.03
–0.25	1.50	1.25	1.17	1.13	1.10	1.08	1.07
–0.50	2.00	1.50	1.33	1.25	1.20	1.17	1.14
–0.75	2.50	1.75	1.50	1.38	1.30	1.25	1.21
–1.00	3.00	2.00	1.67	1.50	1.40	1.33	1.29

Note: η^s = own-price elasticity of supply; η^d = own-price elasticity of demand.

[1] Note that the information contained in Table 4.1 might be interpreted to indicate that the estimates in Table 4.2 are underestimates of the quantities which they are designed to approximate. Although there is much variation in reported estimates, it is not inconsistent with the bulk of the agricultural economics literature to consider agriculture to be a constant returns-to-scale industry. It is also commonly known that the share of returns to owned inputs in total revenues of farmers declines with development and, therefore, presumably with technical advancement in agricultural production (Johnson 1973). As can be seen from Table 4.1, together these observations imply that the constructed approximations in Table 4.2 underestimate the actual value of the right-hand side of the inequality in expression (12).

4.5 Are Subsidies and Research Expenditures Complements?

It is clear that, as has been argued in Ruttan (1982), Lichtenberg and Zilberman (1986), Alston et al. (1988), and Oehmke (1988), the existence of a commodity subsidy increases one component of the social costs associated with investments in agricultural research. This has made it tempting to suggest that the impact of the results of research upon the costs of subsidy programs might imply that lowering subsidies would lead to increases in research expenditures. According to this line of reasoning, this would happen because reducing subsidy levels would effectively increase the social return to such research expenditures by reducing one component of their associated costs.

Although it is clear how this conclusion might easily be drawn, the analysis here suggests that such a conclusion might well be empirically misleading in many cases. Further, in those cases in which the conclusion itself was found to be empirically accurate, when evaluated within the systematic structure of our model, the analytical path which was employed in arriing at such a conclusion can be seen to have been logically misleading. The potential mistake is a consequence of a failure to adequately capture the full nature of the allocation problem facing the government. The problem arises if the analytical framework does not reflect the fact that the factors which motivate and determine the extent to which governments engage in funding agricultural research also influence the incentives which determine the governments' choices with regard to instruments of commodity policy.

As has been argued throughout this chapter, this interaction necessitates that the two types of policy instruments be chosen jointly if the government is to rationally pursue its objectives. When this interaction is not explicitly recognized and when only the choice of the level of research expenditures is considered as endogenous, to the extent that the analytical framework developed in this chapter is appropriate, the government's choice of E will be necessarily a positive function of the level of the subsidy, r. This result is guaranteed by the concavity of the government's objective function. The result implies that, if one's analysis were to ignore the interactions which necessitate that the two policy instruments be chosen jointly, the correct implication for the nature of the relationship between the two policy instruments is that they would be chosen as if they were complements by a government which rationally pursues its objectives.

As is evident from the above discussion, this result is in direct contrast to the conclusion which has been thought to have been informally implied by previous studies (Lindner and Jarrett 1978; Alston et al. 1988; Lichtenberg and Zilberman 1986). The result reflects the fact that, when the level of the subsidy is exogenously increased, the government is able to use the increase in the level of the subsidy to compensate producers for the losses which will accompany increases in E. In effect, when operating according to these conditions, it utilizes the two instruments as if they were complements at the margin.

When the analysis is allowed to reflect the joint determination of the two instruments, r and E, the issue of whether or not r and E behave as complements to each other is not as easily resolved. The information contained in the first-order conditions (11a) and (11c) can be employed to examine how the choice of E is conditional upon the prevailing level of r, given that r has been chosen to satisfy (10a). The exact response of E to changes in r is dependent upon the values of the parameters and the functional forms which describe the demand and supply sides of the industry. Total differentiation of expression (11a), after utilizing (10a) to substitute for the level of r, allows the slope of the choice of E as a function of r to be determined in the neighborhood of the optimally chosen pair (r,E). This slope is described by expression (13):

$$\frac{dE}{dr} = \frac{\dfrac{dQ}{dE}\left\{(\gamma - 1)\left[\dfrac{Q\dfrac{d^2S}{dpdE}}{S_p S_E}\right] - \gamma\dfrac{dQ}{dr}\right\}}{\dfrac{d^2V}{dE^2}}, \tag{13}$$

where $\gamma = w_2/w_1$.

The sign of $dE(r)/dr$ at the chosen level of r and E is determined by the nature of the tradeoffs which exist at the margin between weighted consumer and producer welfare. If $\gamma > 1$ – i.e., if producer welfare is weighted more heavily than is consumer welfare – the government's reaction function which determines E as a function of r will be positively sloped unless the marginal costs to consumers of supporting these policies become large enough to dominate the marginal benefits to producers.

Consumer costs can dominate producer gains and force $dE(r)/dr$ to be negative when the expression within the brackets in (13) takes a positive value. It is possible for this expression to become negative when the supply response to marginal increases in the subsidy become large. This can happen in two ways. One condition which can contribute to this possibility is characterized by significant concavity of the supply schedule with respect to output price. A second characteristic which can contribute to costs to consumers more than offsetting producer benefits at the margin is associated with the particular form of the response of the supply schedule to research-induced technical change. A pivot-like shift in the supply schedule, which is consistent with a large impact of the technical improvement upon variable costs relative to the impact upon fixed costs of production, is consistent with consumer costs which are capable of offsetting producer benefits at the margin.

Both of these conditions contribute toward a supply function which is particularly responsive to the two policy instruments at the margin. Although the existence of both of these conditions is not sufficient to ensure functional substitutability between the two instruments at the margin, a strong combination of these two conditions can result in public research expenditures responding positively to decreases in subsidy levels. Otherwise, research expenditures will respond to changes in r as if they were complementary to production subsidies in the neighborhood of the selected combination of the two instruments.

It should be noted that several commonly adopted functional forms yield this latter outcome of observed complementarity between the two instruments. For example, a linear supply function, or a constant elasticity supply function with elasticity less than one, combined with technical change that shifts the supply curve in a parallel fashion ensure that r and E behave as complements. These examples are of particular interest because they represent the cases which have been most frequently analyzed in previous studies (Lindner and Jarrett 1978; Lichtenberg and Zilberman 1986; Alston et al. 1988).

As indicated, under the conditions in which the two policies behave as complements, the opportunity to choose r and E jointly allows a larger E to be chosen than otherwise would have been chosen. Under these circumstances, losses to producers which may be incurred as a result of research results can be offset through the use of the income redistributing subsidy instrument, r. This leads to the suggestion that commodity policies which subsidize producers may not be as costly to society as has generally been thought. In providing a way for the government to compensate producers for the losses which they suffer as a result of research policies, commodity policy allows the government to invest more heavily in agricultural research endeavors than it otherwise would. For this reason, it is possible that, when government (and/or society) values producer welfare more highly than consumer welfare, the use of subsidies in this way may actually be Pareto-improving.

This possibility is illustrated in Fig. 4.1, which is constructed to portray the effects of the policy instruments upon the welfare of both consumers and producers. At point 0, neither instrument is employed, i.e., both r and E equal zero. Movement from point 0 along the path, T_0, represents the welfare effects of increasing E in the absence of a subsidy. The level of expenditures upon research (E) increases with the distance from point 0 along path T_0. If the welfare weights assigned to consumers are equivalent to those assigned to producers, the level of E will be chosen which results in achieving point A on path T_0.

A producer subsidy achieved through altering price incentives has the effect of improving producer welfare and lowering consumer welfare at the social expense of incurring deadweight loss. Thus, the effect of such a subsidy in the absence of research expenditures could lead to an outcome such as that described by point X. Given equal welfare weights between producers and consumers, point X is not preferred to point 0. However, given welfare weights which are skewed in favor of producers, this may no longer be the case. Furthermore, when welfare weights favor producers, point A no longer corresponds to the chosen level of research expenditures along path T_0. If the subsidy instrument is not available, policymakers who favor producers over consumers will prefer a lower level of research expenditures than characterize point A. A point such as B will be chosen at which the ratio of the welfare weights is equal to the slope of the path T_0.

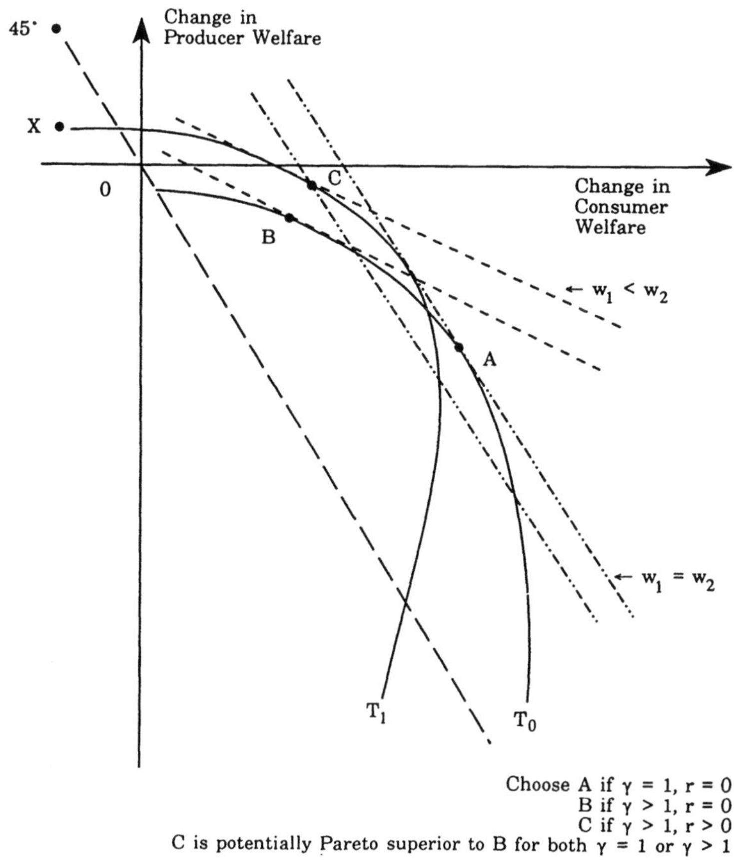

Fig. 4.1. Impact of policy instruments upon producer and consumer

If both policy instruments are available, a combination of the two instruments generally will be preferred to the use of only one instrument when producer welfare is favored over that of consumers. A combination of the two instruments will allow policymakers to achieve a point such as C along path T_1. Path T_1 has been constructed such that point C is Pareto preferred to point B. Given unequal welfare weights, point C is preferred to both points A and B. Due to deadweight losses associated with the subsidy scheme, point A continues to represent a potential Pareto improvement from the outcome at point C.

4.5.1 The Case of US Agriculture Policy

The extent to which governments intervene in commodity markets can be approximated by Producer Subsidy Equivalents (PSEs) calculated by the US Department of Agriculture. These data for several commodity sectors in the United States are reported in column (1) of Table 4.5. The commodities are separated into three groups according to their elasticity of demand. Sugar, milk, rice,

Table 4.5. PERT versus PEST subsidy equivalent in US agriculture, (1982–1986 average)

	(1) Total subsidy		(2) PERT subsidy		(3) PEST subsidy		(4) Ratio of PERT to PEST subsidies (%)
	Percent unit value	Dollars per ton	Dollars per ton	Percent unit value	Dollars per ton	Percent unit value	
Sugar	77.4	221.0	17.5	6.1	203.5	71.3	8.6
Milk	53.9	152.0	11.8	4.2	140.2	49.7	8.4
Rice	45.0	145.0	9.4	2.9	135.6	42.1	6.9
Wheat	36.5	57.0	7.7	4.9	49.3	31.6	15.7
Sorghum	31.5	36.4	5.3	4.6	31.1	26.9	17.0
Barley	28.8	32.0	6.7	6.0	25.3	22.8	26.5
Corn	27.1	31.5	5.6	4.8	25.9	22.3	21.6
Oats	7.6	7.3	4.5	4.7	2.8	2.9	160.7
Soybean	8.5	18.0	13.4	6.3	4.6	2.2	291.3
Beef	8.7	175.8	97.7	4.8	44.5	3.9	125.0
Poultry	8.3	82.0	53.3	5.4	28.7	2.9	285.7
Pork	5.8	85.4	67.6	4.8	17.8	1.0	380.0
Average	28.3	87.0	25.0	4.9	61.8	23.1	112.3

Source: US Department of Agriculture (1988).

and wheat are categorized as being very inelastic in demand, compared to the other commodities reported in Table 4.5. The feed grains, on the other hand, are regarded as modestly inelastic while the soybean and red meat sectors are viewed as least inelastic among agricultural commodities.

In terms of percent unit value, the inelastic demand sectors of sugar, milk, rice, and wheat have by far the highest level of support, averaging 53.2% which is well above the average of 28.3%. The feed grains have an intermediate level of support as shown in Table 4.5 while the sectors with less demand curves, such as soybean and beef, have the lowest level of support.

The breakdown of support between PERT and PEST policies are given in Table 4.5 by the corresponding "subsidy equivalents" in columns (2) and (3), respectively. The data show that the level of PERT and PEST subsidy equivalents are inversely related. Furthermore, the level of PERT policy interventions are higher than those of PEST policies for the crops least inelastic in demand but are much lower for the most inelastic crops in demand. Notice that the productive equivalents, as a percent unit value of production, are stable across all commodity sectors at around 4% to 6%. On the other hand, PEST equivalents are very high as a percent unit value of production for the most inelastic sectors and fall well below that of PSEs for the least inelastic sectors.

Column 4 depicts the ratio of PERT to PEST subsidy equivalents in terms of percent. The data indicate that, as a percent of PEST equivalents, PERT equivalents are lower in the inelastic demand sectors (e.g., 8.6% for sugar), while the elastic sectors have the highest ratio of PERT subsidy equivalents (e.g., 380% for pork).

As shown in the previous section, the relative producer gains to research increase as the elasticity of demand is increasing (in absolute value), the elasticity of supply is decreasing, and the output elasticity of the marginal impact of research on cost is decreasing. The data in Table 4.5 partially confirms the theoretical model predictions. Studies have reported supply elasticities greater than one for wheat and milk (Chavas and Klemme 1986; Burt and Worthington 1988). Hence, one can conclude that the demand elasticities are low and relatively higher supply elasticities exist in agriculture. It is confirmed in Table 4.5 that these same sectors have relatively (and in absolute terms) higher levels of PEST expenditure equivalents and lower levels of PERT equivalents. Three studies (Furtan 1983; Judd et al. 1983; Ruttann 1983) indicate that the allocation of research among commodity groupings are inconsistent with economic efficiency with field crops underfunded relative to livestock. This evidence, combined with that in Table 4.5, partially confirms the predictions of the model. Furthermore, our theory indicates would result in a more pronounced degree of underinvestment in PERT policies.

4.6 Concluding Remarks

In this chapter we have attempted to provide a framework for interpreting why so many studies have concluded that agricultural research is the victim of chronic underinvestment. We have also indicated that the underlying structure which helps to explain this phenomenon suggests that, in some (perhaps important) cases, commodity policy and research expenditures may be expected to be chosen as complementary instruments. This suggestion carries with it the further implication that production subsidies may not be as detrimental to social welfare as has commonly been argued. Rather, by providing a vehicle through which to compensate producers for losses incurred as a result of research expenditures, production subsidies may be necessary components of potentially Pareto-improving portfolios of policy instruments.

The analysis as it currently stands is presented using a stylized model of an agricultural market. It is our belief, however, that our formulation is fairly robust under a number of potential elaborations that might be designed to increase the model's realism. Nevertheless, it is useful to list here several of the notable simplifications which have been adopted in the current version of this model.

Our analysis has not attempted to explain how the political process works to determine the structure of the government's objective function. That the government behaves as though it were attempting to satisfy the demands of this objective function is simply taken as given. Further, the model as presented does not address the intertemporal nature of the relationship between E and the improvements in productivity which are eventually produced as a result of E. This feature of the model does not appear to alter the qualitative results substantially.

Nor has our analysis dealt with the issue of entry and exit of producers. It is likely that both policy instruments r and E have an impact upon entry and exit of producers through their impact upon producer profits and welfare. This observation necessarily highlights the rather casual treatment of the concept of profit and rents to factors of production as they have been utilized in this chapter. A complete treatment of these issues demands an explicit treatment of dynamic relationships and is left to further research.

Finally, only one type of commodity policy has been explicitly considered here. A more comprehensive treatment of the various policies which characterize intervention in agricultural markets would be instructive. Nevertheless, this analysis is presented in the belief that the fundamental patterns of interactions and tradeoffs which are highlighted in the presentation of the decision problem when a production subsidy is the instrument of commodity policy are likely to be illustrative of the patterns which would determine the outcomes of situations characterized by alternative commodity policy instruments.

A natural test of the robustness of the basic premises of this analysis could be constructed by examining the nature of agricultural policies and markets in developing countries. The literature reports that, as in the United States, underinvestment in agricultural research also characterizes the situation in these countries. However, in a number of important ways, agriculture in the developing countries operates in a different environment from that found in the United States and many industrial nations. In contrast to the situation faced by agriculture in industrial countries, stylized facts have emerged to suggest that welfare weights favor consumers over farmers in developing countries. This is evident in the consistent subsidization of the consumption of food as well as taxation of agriculture which occurs in the developing countries.

Our framework implies that, when political preferences are such that commodity policy is used to redistribute income away from agriculture toward consumers, underinvestment in agricultural research should be expected to prevail when farmers benefit at the margin from technical change. In terms of the model developed in this chapter it can be seen from expression (11b) that, when $w_2 < w_1$, the condition,

$$-mC_E > 1 + r\left(\frac{dS}{dE}\right),$$

i.e., underinvestment in agricultural research, will be observed to be the case when it is also the case that $\pi_E > 0$.

The model developed in this chapter provides a natural interpretation of such an outcome. Development efforts in developing countries often entail investing in research projects designed to improve the productivity of agriculture. Typically, such investments must be provided at the expense of consumers and taxpayers. However, the political strength of the urban sector relative to that of agricultural ensures that these projects will likely be undersupported from a cost-benefit point of view, precisely because they tend to be detrimental to consumer welfare. Further, commodity policy which redistributes income away from agriculture and toward consumers may be adopted in order to partially compensate consumers for shouldering the expense of providing agricultural research. Based upon our framework, these are the patterns that we would expect to emerge. Evidence confirming the prediction of the model that agriculture should be observed to tend to benefit from agricultural research in developing countries, while commodity policy redistributes wealth away from agriculture, provides strong support for the empirical validity of the ideas advanced in this chapter.

References

Alston J, Edwards G, Freebairn J (1988) Market distortions and benefits from research. Am J Agric Econ 70:281–288

Arndt TM, Dalrymple DG, Ruttan VW (eds) (1977) Resource allocation and productivity in national and international agricultural research. University of Minnesota Press, Minneapolis

Ball VE (1985) Output, input, and productivity measurement in US agriculture, 1948–79. Am J Agric Econ 67:475–486

Becker GS (1983) A theory of competition among pressure groups for political influence. Q J Econ 98:371–400

Brandow GE (1961) Interrelations among demands for farm products and implications for control of market supply. Bltn No 680, Pennsylvania Agricultural Experiment Station, University Park

Burt OR, Worthington VE (1988) Wheat acreage supply response in the United States. West J Agric Econ 13:100–111

Chavas JP, Klemme RM (1986) Aggregate milk supply response and investment behavior on US dairy farms. Am J Agric Econ 68:55–66

de Gorter H, Zilberman D (1990) On the political economy of public good inputs in agriculture. Am J Agric Econ 72:131–700

Fishel WD (1971) Resource Allocation in Agricultural Research. University of Minnesota Press, Minneapolis

Furtan WH (1983) Economics of crop and oilseed research in Canada. Paper presented to the Outlook Conference, Ottawa, Canada, December

Gardner BL (1989) Price supports and optimal spending on agricultural research. Working Paper, Department of Agricultural and Resource Economics. University of Maryland, College Park

Hertel TW (1989) Technology and factor mobility. Am J Agric Econ 71: 559–573

Hertel TW, Ball VE, Huang KS, Tsigas ME (1989) Farm level demand elasticities for agricultural commodities. Res Bull 1988, Agricultural Experiment Station, Purdue University, West Lafayette

Hertford R, Schmitz A (1977) Measuring economic returns to agricultural research. In: Arndt TM, Dalrymple DG, Ruttan VW (eds) Resource allocation and productivity in national and international agricultural research. University of Minnesota Press, Minneapolis, pp 148–167

Johnson DG (1973) World agriculture in disarray. Macmillan St. Martin's Press, London

Judd MA, Boyce JK, Evenson RE (1983) Investing in agricultural supply. Discussion Paper No 442, Economic Growth Center, Yale University, New Haven

Lichtenberg E, Zilberman D (1986) The welfare economics of price supports in US agriculture. Am Econ Rev 76:1135–4111

Lindner RK, Jarrett FG (1978) Supply shifts and the size of research benefits. Am J Agric Econ 60:48–58

Magee SP, Brock WA, Young L (1989) Black hole tariffs and endogenous policy theory. Cambridge University Press, Cambridge

Mueller DC (1989) Public choice II. Cambridge University Press, New York

Oehmke JF (1988) The calculation of returns to research in distorted markets. Agric Econ 2:291–302

Peltzman S (1976) Towards a more general theory of regulation? J Law Econ 19: 211–240

Rausser GC (1982) Political economic markets: PERTs and PESTs in food and agriculture. Am J Agric Econ 64:821–833

Rausser GC, de Gorter H (1990) Endogenizing policy in models of agricultural markets. In: Maunder A (ed) Agriculture and governments in an interdependent world. Gower, Aldershot

Rausser GC, Foster WE (1990) Political preference functions and public policy reform. Am J Agric Econ 72; 3:642–652

Rausser GC, Freebairn JW (1974) Estimation of policy preference functions: An application to US beef import policy. Rev Econ Stat 56:437–449

Ruttan VW (1982) Agricultural research policy. University of Minnesota Press, Minneapolis

Ruttan VW (1983) Agricultural research policy issues. Hortic Sci 108:809–818

Tweeten LG (1967) The demand for US farm output. Food Res Inst Studies 7:343–369

Tweeten LG (1979) Foundations of Farm Policy. University of Nebraska Press, Lincoln

Wicksell K (1967) A new principle of just taxation. In: Musgrave RA, Peacock AT (eds) Finanztheoretische untersuchungen, Jena, 1896. Reprinted in Classics in the theory of public finance. St. Martin's Press, New York

Wohlgenant MK (1989) Demand for farm output in a complete system of demand functions. Am J Agric Econ 1:241–252

US Department of Agriculture. (1988) Estimates of producer and consumer equivalents: Government intervention in agriculture. ATAD Staff Report No AGES 880127, Economic Research Service, Washington, DC

Chapter 5

Coalition Breaking and Policy Reform

William E. Foster and Gordon C. Rausser

5.1 Introduction

In the previous chapter, de Gorter, Nielson, and Rausser address the possible complementarity of public good investments and wealth transfers to agriculture. Their model assumes that producers are homogeneous and concentrates on the tradeoffs between general transfers and public good expenditures. In this chapter we address the more realistic case of heterogeneous producers and derive the implications for the choice of the wealth transfer mechanisms.

The motives behind the choice of distorting transfers over nondistorting transfers are not fully appreciated. Indeed, faced with the seeming omnipresence of distorting policies, many political-economic analyses of governmental intervention tend to dismiss altogether neutral policies as impractical and to ignore the fundamental question of why states opt for socially wasteful transfers. The analytical concentration on the selection of the level of transfers between groups, to the neglect of the choice of the level of distortion, is due to conventional rent-seeking approaches to explaining production subsidies. These frameworks focus on the relative organization power of groups receiving or granting wealth transfers (e.g., Stigler 1971; Peltzman 1976; Becker 1983). One important element of these frameworks is that groups struggle over a limited amount of potential total wealth, or surplus. The greatest level of wealth available to all parties together is defined by the ideal of freely operating markets, where no rent seeking takes place and where, of course, subsidies and taxes necessarily waste some of this potential.

In such models, wealth transfers do not serve the public interest; they are only the rewards of political maneuver, pork barrel, and the consumer's (and taxpayer's) unwillingness or inability to resist interventions. Two corollaries are notable. First, the degree to which groups gain directly from these transfers is a measure of their political clout. Second, transfer mechanisms would tend to be the most efficient, or least distorting, in the sense of minimizing deadweight losses, because all groups could share in an efficiency gain (e.g, Becker 1983; Gardner 1987). The willful choice of a more socially wasteful transfer method over a less wasteful method cannot be inferred from this model. Such selections can only be generated by reference to additional assumptions regarding the characteristics of transfer methods. For example, in a world of limited information, distorting policies may be less transparent, concealing the level of transfers, and thus serve to circumvent political opposition (Magee et al. 1989).

An alternative view of wealth transfers holds that policies that increase total social welfare may have to be accompanied by subsidies, or they will not be implemented because of obstruction by potentially losing groups (Rausser 1992). A potentially winning group taxes itself in order to mitigate the losses suffered by another group whose political strength lies in its ability to veto a move from the status quo. If threatened with sufficient harm, the members of the latter group would form a blocking coalition that obstructs the implementation of a new policy. Distorting wealth transfers, compared to neutral transfers, may actually serve the purpose of overcoming this veto more efficiently by targeting members of the losing coalition who suffer less. These members who suffer less are able to take advantage of the new policy to some degree. In effect, the taxed group is in control of the policy choices, including the method of wealth transfer, and the subsidized group merely sets constraints on the feasible choices. This model offers an alternative hypothesis to the traditional view of rent seeking: wealth transfers flow to the politically weaker group (weaker in the sense that it loses in the move from the status quo), and these transfers serve to secure increases in total social welfare.

As discussed in the previous chapter, unproductive wealth transfers do not exist in isolation from a larger set of government activities, some expanding total social welfare and others promoting waste. Economic policies may be usefully divided into two types: (1) those which are meant to correct market failures, or provide public goods, and are ostensibly neutral with respect to their effects on the distribution of society's economic surplus; and (2) those which are meant to redistribute wealth between groups and are ostensibly independent of the question of strictly economic efficiency (Rausser 1982; Mueller 1989). The distinction between the two types of policies is briefly summarized by the popular metaphor of the economy as a pie: The first type of policy expands the size of the pie, and the second type allocates the portions served to various groups.

The expansion of the social-welfare pie does not guarantee that each group's portion will also grow. If social groups must cooperate, at least to some extent, then wealth transfers and increases in total social welfare are politically inseparable. For example, a group that gains from the investment in public goods may promote transfers to groups that suffer from the investment, so that these groups receiving the transfer will acquiesce to the public good. The wealth transfer may appear as an inefficient, wasteful, rent-seeking policy conditioned on a given level of public good investment. In actuality, however, it may be a means of securing the welfare-increasing investment, and thus the transfer is a crucial and Pareto-improving component of general policy. An important point that follows from this model is that the true social costs of any policy cannot be measured in isolation. The benefit of a wasteful policy may lie in the public good which it allows to exist; and the benefit of investment in a public good may be less than those observed directly, carrying with it inefficient transfer schemes necessary to assure its political viability.

Wealth transfers to a large group, such as agricultural producers, typically are shared unequally by firms; and it is in consumers' and taxpayers' interests that this be so. In the context of a supply-enhancing public good (such as technological research and development), some producers are harmed less than the industry average because they can take greater advantage of the advance. Wealth transfers weighted in favor of these *innovators* would serve to break producer coalitions obstructing change with less expense to consumers and taxpayers. Those who expand production or cut costs to a greater degree simply need less transfer payments to be made indifferent to the investment in the public good. Nonneutral policies target payments according to either production levels or input use. Therefore, a transfer based on production which makes those innovators expand their output just as well of as without the advance would transfer less to those who take less advantage of the public good. A transfer based on the use of an input encouraged by use of the public good will have a similar effect (see Chap. 6). The popularity of nonneutral (i.e., coupled) payments in agriculture especially may be explained by this property of targeting transfers from consumers to innovators; that is, to those firms less harmed by investments in public goods and, thus, to those most cheaply divided from a coalition that might obstruct moves from the status quo.

In Section 5.2, we present the basic model of coalition breaking in order to gain acceptance of a public-interest policy, or a public good. We present the model as a conflict between two groups, producers and consumers/taxpayers, over the release of a price-decreasing technical change. Section 5.3 presents the choice by consumers/taxpayers of the means of wealth transfer. We consider the continuum of transfer mechanisms which are combinations of two polar cases that, on the surface, do not differentiate between firms with respect to ability to take advantage of the public good. The two polar schemes are a per-unit-output subsidy, which distorts producer and consumer prices, and a production-neutral payment, which the producer cannot affect by choice of output level. Here, the conditions under which consumers/taxpayers would prefer price distortion are demonstrated. In Section 5.4 we address the particular case of constant-elasticity supply and demand to illustrate in concrete form the more general analysis of the previous sections. Finally, some concluding remarks are offered.

5.2 Interest Group Structure

Suppose there are two interest groups in society – consumers/taxpayers and producers. Individual members of these groups behave competitively in the market place, but they may cooperate with other groups' members in political choice. Each group is composed of many members; and there is some rule for weighting the votes of individual members to decide each group's position on a policy, as well as whether or not the group will expend effort opposing a particular policy. For pedagogical purposes, we will take the particular public-interest policy to be the release of a technical innovation that will increase production but, by doing so, also harm enough producers by decreasing output price such that the release without compensation will be vetoed.[1]

Producers are endowed with different levels of ability to utilize the new technology, due to firms differing in location, vintage of capital, and endowments of human capital and entrepreneurial talent; therefore, some producers will suffer more than others with the innovation release. To formalize the concept of ability to take advantage of the innovation, let a be some index of producer attributes. Define $\pi_0(a)$ as the rent accruing to a-type firms prior to the release of the innovation and $\pi_1(p,a)$ as the rent accruing to those firms under output price, p, after the release. Also define $y_0(a)$ as the pre-release output level and $Y(p,a)$ as the post-release level at price p. Assume the profit functions are well-behaved in p and a. Without placing any sign on the derivatives of π_0 and π_1, with respect to a, we define ability to take advantage of the release as

$$\frac{\partial \pi_1(p,a)}{\partial a} > \frac{\partial \pi_0(a)}{\partial a} \quad \textit{for all } p. \tag{1}$$

Intuitively, condition (1) implies that firms of higher level ability gain relatively more, or lose relatively less, from the innovation release. If producers are homogeneous prior to the innovation release, then $\partial p_0/\partial a = 0$. Note that, with homogeneous firms prior to release, condition (1) implies that $\pi_1(p,a)$ is a strictly increasing function of ability. Similarly, the gain in output level, y, for a producer due to the innovation release is a positive function of a:

$$\frac{\partial y(p,a)}{\partial a} > \frac{\partial y_0(a)}{\partial a} \quad \textit{for all } p, \tag{2}$$

where $\partial \pi_0/\partial p = y_0$ and $\partial \pi_1/\partial p = y(p,a)$. In other words, a producer's ability to profit, or suffer less, from the innovation release is positively correlated with the change in output level.

Let $f(a)$ be the proportion of firms of a-level ability over the number of producers. Define V as the minimum number of producer votes needed to have the producer group support the innovation release. Define the index, c, such that

$$V = \int_c^\infty f(a)da. \tag{3}$$

If the c-type producer is just indifferent to voting against the release (i.e., if $\pi_1(p,c) = \pi_0(c) - k$, where k is a cost of lobbying to prevent release), then all firms with ability greater than c will benefit from the innovation's release. Therefore, any generic compensation scheme that affects the producer price or offers a per-producer, price-neutral payment need only make indifferent the c-type producer in order to gain the producer group's acquiescence to the policy.

[1] It should be emphasized that other public-interest policies, such as investment in transportation systems or public utilities, could replace "technical innovation." Moreover, the roles of consumers and producers could be reversed.

5.3 Targeting Payments Under Heterogeneous Adoption

Given that some form of wealth transfer is necessary to gain acquiescence of a supply-enhancing public-interest policy, the question becomes that of determining the least costly means of breaking the potential coalition of producers. We narrow our attention to a priori rules that effect the size of the political coalition. We may think of such rules as being announced at the same time as the promised consequences of the technical advance, but prior to the actual dissemination of the advance. This approximates the situation in the case of agriculture in the United States, where rules of wealth transfer are in place and production growth is anticipated to be supported by a structured and on-going system of R&D and dissemination. Aggregate production is expected to grow due to future innovations and discoveries, the particulars of which are unknown to all but, perhaps, a few.

Of a priori rules, we consider two schemes: a nondistorting payment promised to all producers, and perhaps based on initial output levels, and a distorting per-unit-output payment. The key features of these a priori rules is that they are generic in the sense that they do not distinguish directly between producers. Consumers/taxpayers do not target payments to specific producers, either because there exist high transaction costs to the identification of those with superior abilities or because there exist political constraints to transfers based on overtly personal criteria. Nevertheless, while per-unit-output payments do not directly target a group, they do, in effect, tend to concentrate transfers on those who make the greatest relative use of the supply-expanding public-interest policy. The cost to consumers/taxpayers of concentrating transfers on those with the greatest ability is the inefficient level of production brought about by a producer price higher than the market-clearing price.

Specifically, consider the following price-distorting and production-neutral payment schemes. Consumers/taxpayers seek to choose the levels of two generic payments: a production-neutral payment of b dollars per-unit-output on the initial (prerelease) level of a producer's output; and a nonneutral subsidy $(P_T - P_1)$ on the producer's change in output due to the release. The term, P_T, is the targeted producer price and P_1 is the equilibrium market-clearing price-paid by consumers. Representing supply as $S(P)$ and demand as $D(P)$, P_1 is such that $D(P_1) = S(P_T)$. In order to assure breaking of the producer coalition, P_T and b are chosen such that the c-level firms are indifferent to the innovation release.

Represent the a-type producer cost of output level y by $e(y,a)$. The instruments, P_T and b, are chosen such that

$$(P_T - P_1)[y(P_T,c) - y_0(c)] + P_1 y(P_T,c) + b y_0(c) - e(y,c) = \pi_0(c) - k. \tag{4}$$

Note that the firm makes production decisions based on the target price, P_T. Therefore, the point of indifference of the c-level firm with both the innovation release and the transfer payments may be written

$$\pi(P_T,c) + (b + P_1 - P_T) y_0(c) = \pi_0(c) - k \tag{5}$$

where $\pi(P_T,c) = P_T\, y(P_T,c) - e(y,c)$ – the familiar profit function satisfying Hotelling's lemma: $\partial\pi_c/\partial P_T = y_c(P_T)$.

Consumer/taxpayer welfare gains under the innovation release and the compensation schemes may be measured by the sum of the consumers' Marshallian surplus and the total taxpayer outlays, i.e.,

$$CS = \int_{P_1}^{P_0} D(P)dP - \{[P_T - P_1][S(P_T) - S_0] + bS_0\} \tag{6}$$

where $S(P_T) = \int_a y(P_T,a)f(a)da$ and $S_0 = \int_a y_0(a)f(a)da$. Payments may be either nondistorting, in the sense that the consumer and producer prices are equal (i.e., $P_T = P_1$), or distorting (i.e., $P_T > P_1$). The extent to which payments are distorting, or coupled to production decisions, depends on the degree to which consumers/taxpayers rely on the price subsidy to make indifferent the c-type producers to the innovation release, i.e., to satisfy condition (5). If $P_T = P_1 + b$, then the payments are entirely of the distorting kind; and, if $P_T = P_1$, then payments are entirely of the nondistorting kind.

The first result relates to the suboptimality of a completely nondistorting payment scheme.

If the output of the marginal coalition-breaking firm (the c-type producer), relative to its initial level $[y_c(P)/y_{c0}]$, is greater than the average relative output increase $[\bar{y}(P)/\bar{y}_0]$ of the industry, then a distorting payment scheme is preferred by consumers/taxpayers.

To prove the result, consider the nondistorting case where $P_T = P_1$ and all transfers are accomplished by the nondistorting per-unit-output payment, b, on the initial output. A marginal increase in P_T and a decrease in b satisfying $D(P_1) = D(P_T)$ and (5) will increase net consumer/taxpayer gains due to the release-with-compensation policy, if $\partial CS/\partial P_T > 0$. From (6),

$$\frac{\partial CS}{\partial P_T} = -D(P_1)\frac{\partial P_1}{\partial P_T} - \left[(P_T - P_1)\frac{\partial S(P_T)}{\partial P_T} + \left(1 - \frac{\partial P_1}{\partial P_T}\right)(S(P_T) - S_0) + \frac{\partial b}{\partial P_T}S_0\right]. \tag{7}$$

From (5),

$$b = \frac{\pi_0(c) - k - \pi(P_T,c)}{y_0(c)} - P_1 + P_T, \tag{8}$$

implying

$$\frac{\partial b}{\partial P_T} = \frac{y(P_T,c)}{Y_0(c)} - \frac{\partial P_1}{\partial P_T} + 1. \tag{9}$$

From the equilibrium condition, $\partial P_1/\partial P_T$ is such that

$$\frac{\partial D(P_1)}{\partial P_1}\cdot\frac{\partial P_1}{\partial P_T} = \frac{\partial S(P_T)}{\partial P_T} \tag{10}$$

or

$$\frac{\partial P_1}{\partial P_T} = \frac{P_1}{P_T} \cdot \frac{\eta}{\varepsilon},$$

where ε is the aggregate supply elasticity and h is the elasticity of demand.

Using (9) and (10) in (7),

$$\frac{\partial CS}{\partial P_T} = -D(P_1)\frac{P_1}{P_T}\frac{\eta}{\varepsilon}$$
$$-\left\{\left[\left(1-\frac{P_1}{P_T}\right)\varepsilon S(P_T) + \left(1 - \frac{P_1}{P_T}\frac{\eta}{\varepsilon}\right)[S(P_T)-S_0] + \frac{y(P_T, c)}{y_0(c)}S_0 - \frac{P_1}{p_T}\frac{\eta}{\varepsilon}S_0 + S_0\right]\right. \tag{11}$$

Using $D(P_1) = S(P_T)$ and $P_1 = P_T$,

$$\frac{\partial CS}{\partial P_T} = -S(P_T) + \frac{y(P_T,c)}{y_0(c)}S_0 = S(P_T)\left[\frac{y(P_T,c)}{y_0(c)}\frac{\bar{y}_0}{\bar{y}(P_T)} - 1\right]. \tag{12}$$

And thus, from an initial nondistorting policy where $P_1 = P_T$, $(\partial CS)/(\partial P_T) > 0$ if and only if

$$\frac{y(P_T,c)}{y_0(c)}\frac{\bar{y}_0}{\bar{y}(P_T)} = \rho > 1 \tag{13}$$

where $\bar{y}_0 = S_0/N$ and $\bar{y}(P_T) = S(P_T)/N$, and ρ is the measure of relative output increase due to the innovation release. This result demonstrates that there are simple and plausible conditions under which one would expect to observe distorting payment policies, even if consumers/taxpayers had complete control over the selection of those policies.

The above result relies on heterogeneous producers, where the marginal defector from the blocking coalition increases supply by a greater percentage than the industry average. The relative level of the marginal defector's output increase to the industry's average is a measure of the degree to which consumers/taxpayers can optimally target payments via nonneutral transfers. More generally, if there is a mix of price distorting and nondistorting payments, then the optimal level of price distortion is a function of these relative rates of output increases due to the innovation release and payment scheme.

If both price-distorting and nondistorting payment mechanisms are optimal, then the rate of price distortion, measured by $(1 - P_1/P_T)$, is proportional to the rate of increase in the c-type firm's output relative to the industry's average:

$$(1 - P_1/P_T) = \varepsilon^{-1}\left[\frac{y_c(P_T)}{y_{c0}} \cdot \frac{\bar{y}_0}{\bar{y}(P_T)} - 1\right] = \varepsilon^{-1}(\rho - 1), \tag{14}$$

where ε is the aggregate supply elasticity.

The result is simply the first-order condition for maximizing the consumer/taxpayer welfare given by (6). The optimal degree of price distortion is an increasing function of the rate of output increase of the marginal defector, and is a decreasing function of the aggregate supply elasticity. The measure of relative output increase due to the innovation release (r) is an indicator of how easily one can target payments to defecting producers via a per-unit-output payment. The supply elasticity indicates the degree to which resources will be misallocated due to the nonneutral payments. The optimal level of price distortion, therefore, is a function of both the heterogeneous ability to take advantage of the innovation release *and* the inefficiency caused by the coupling of payments. This implies that, in those industries where producers are fairly homogeneous in their adoption of new technologies, one would be less likely to observe nonneutral, "inefficient" payments.

One particularly noteworthy result is that the optimal degree of price distortion is not directly dependent on the degree of technical change: It is the heterogeneity of the technology's adoption that is important, not the size of the supply increase. The absolute degree of the potential supply shift will determine whether or not consumers/taxpayers seek to break the producer coalition – a small supply that may generate insufficient Marshallian surplus to justify the implementation costs of any transfer payment scheme. The absolute degree of the supply shift will also affect the total amount of compensation that must be offered to the marginal defector.

Of course, condition (14) presumes an interior solution, but a corner solution (i.e., $P_T = P_1 + b$) to the consumer/taxpayer problem is possible. The conditions under which the consumers/taxpayers would rely solely on distorting payments would depend on the specific functional forms of the producers' supplies and the demand function. Nevertheless, if producers are sufficiently heterogeneous, in the sense that the c-type firm's output increase is great relative to the industry average, then no nondistorting transfers are made. The broadest condition is that, if

$$(1 - P_1/P_T < \varepsilon^{-1}(\rho - 1) \text{ for all } P_T, \tag{15}$$

then a corner solution is optimal; P_T is chosen such that $\pi(P_T,c) = \pi_0(c) - k$, and no nondistorting payments are made. A sufficient condition is given by the following proposition:

> *If the percentage difference between the marginal defector's output increase and the industry's average is greater than the supply elasticity at all levels of P_T that satisfy (5), then consumers/taxpayers prefer the exclusive use of distorting payments to assure coalition breaking.*

The foregoing result is simply a stronger version of condition (4), where the left-hand side of the inequality has been replaced by unity.

5.4 A Particular Case: Constant Elasticity Supply and Demand

To draw out implications from the above theoretical results, we explore specific cases in this section. We do this in two parts. First, a graphic example is presented of the case of perfectly inelastic supply. This case reveals the important factors determining output-based payments. Second, we turn to constant elasticity supply and demand and two types of producers, innovators and noninnovators. For this case, the explicit algebra determining price distortions that achieve effective coalition breaking is presented.

5.4.1 The Case of Perfectly Inelastic Supply

Consider the example illustrated in Fig. 5.1. This is an extreme case of two types of producers, innovators and noninnovators, who must be unanimous in opposition in order to obstruct the dissemination of a technical advance. Initially, their individual supply curves are identical, perfectly inelastic, and given by the curves labeled I_0 and N. Aggregate supply is given by $S_0 = I_0 + N$; demand, by D; and initial equilibrium price, by P_0. Now, if the technical advance is implemented, innovators would have the new supply curve of I_1; the new aggregate supply would be $S_1 = I_1 + N$; and the new equilibrium price would be P_1. Both types of producers would lose rents given by area a + b + c due to the fall in equilibrium price. Innovators, however, would gain by area d due to expanded production. Here, area a is defined to equal area d. The potential net loss to innovators is, therefore, area b + c, which must be the least per-producer payment promised in order to prevent obstruction of the technical advance. The total consumer transfer to producers of 2(b + c) would leave the innovator just as well off as without the advance. The noninnovator would be a net loser of area a – the output-expanding benefit to the innovator. Consider now the use of a "target" price that, when announced prior to the implementation of the technical advance, would guarantee to innovators that they would remain as well off as without the advance. The level of this producer price, P_T, is such that area b equals area e. Area e can be interpreted as benefits accruing to the innovator who responds to the target price by adopting the technical advance. With the coupled policy, the innovator loses area a + b + c due to the price fall, gains area d due to the technical advance, and gains area c + e due to the support price. The noninnovator, however, gains only area c from the support price, implying a net loss of area a + b with the coupled policy. Under the coupled policy, consumers need only transfer the amount of 2c + b to producers in order to gain the benefits of technical advance. A coupled policy, in this case of perfectly inelastic supply curves, benefits consumers by the amount b relative to the decoupled policy.

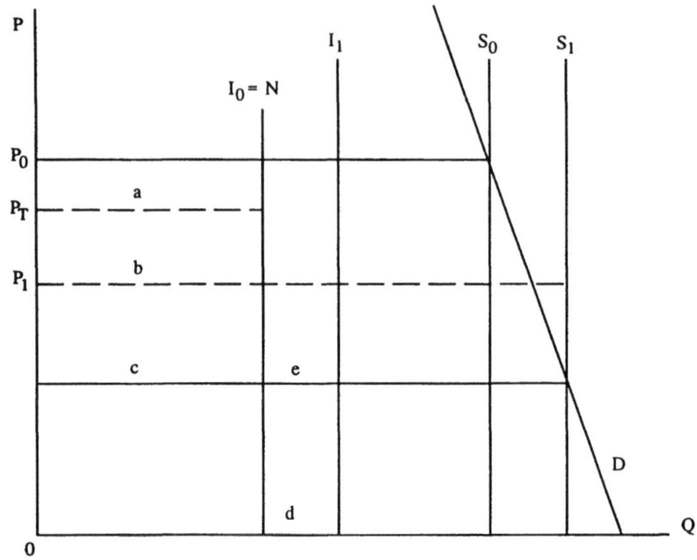

Note: $S_1 = I_1 + N$, $I_0 = N$.

Fig. 5.1. The optimality to consumers of coalition-breaking per-unit output payments under perfectly inelastic supply

This example clearly demonstrates that coupled transfer schemes distinguish those who would lose less under output-expanding changes in production. Coupled transfer schemes are better targeted at those who are the most easily divided from the obstructing coalition. The case of inelastic supply curves and the similarity of innovators and noninnovators are the special features of this example that make apparent the superiority of a per-unit-transfer policy. As the proportion of innovators enlarges, the relative consumer gain from using per-unit rather than per-firm payments declines. At the extreme, where all identical firms would adopt the technology, the total amount of transfers is the same under both types of policies; and under perfectly inelastic supply curves the consumer would be indifferent between either scheme.

When supply is perfectly inelastic, no distortionary consequences result from a support price policy being coupled to the level of production. In essence, the policy is decoupled from production decisions because production is divorced from all price considerations. And, in this sense, there is no cost of distinguishing innovators and noninnovators with a coupled policy. As the supply curves move from being perfectly inelastic to having some price responsiveness, the cost of the transfer grows. This example illustrates clearly that there are two elements crucial to determining for consumers the best means of transferring benefits to producers – the relative proportion of innovators to noninnovators and the responsiveness of supply to coupled policies.

5.4.2 The Case of Constant Elasticity Supply and Demand

Consider that there are only two types of producers – innovators who would make use of a future technical advance and noninnovators who would not. Let the proportion of innovators be given by σ, and the proportion of *noninnovators* be given by $(1-\sigma)$. Let the initial profit and supply functions of both types of producers be given by the Cobb-Douglas constant-elasticity production model

$$\pi(P) = \frac{1}{1+\alpha}\left(\frac{1}{C}\right)^{\alpha} P^{1+\alpha}, \tag{16}$$

where

$$\frac{\partial \pi(P)}{\partial P} = y(P) = \left(\frac{1}{C}P\right)^{\alpha}. \tag{17}$$

To be specific, consider a technological advance that shifts innovators' supplies at every price by the proportion, $1/\theta$. An innovator's supply curve, after the technical advance, is given by

$$y = \frac{1}{\theta}\left(\frac{1}{C}P\right)^{\alpha}. \tag{18}$$

Total supply after the innovation is adopted, S_1 is the sum of innovator and noninnovator supplies, viz.,

$$S_1(P) = \sigma\frac{1}{\theta}\left(\frac{1}{C}P\right)^{\alpha} + (1-\sigma)\left(\frac{1}{C}P\right)^{\alpha}. \tag{19}$$

Take the constant-elasticity curve given by $D(P) = bP^{-\beta}$. Therefore, equilibrium price, P_1, is given by

$$S_1 = \frac{\sigma}{\theta}\left(\frac{P_1}{c}\right)^{\alpha} + (1-\sigma)\left(\frac{P_1}{c}\right)^{\alpha} = bP_1^{-\beta}, \tag{20}$$

implying

$$P_1 = P_0\left(\frac{\theta}{Z}\right)^{1/(\alpha+\beta)}, \tag{21}$$

where P_0 is the equilibrium price without the dissemination of the advance and $Z = \sigma + (1-\sigma)\theta \le 1$. The term, Z, may be given an intuitive meaning by noting that the percentage gain in an innovator's yield over the average yield of all producers can be expressed as

$$\frac{1}{\theta}\left(\frac{P_1}{c}\right)^{\alpha}(S_1 - 1)^{-1} = \frac{1}{Z} - 1. \tag{22}$$

The term, Z, is a measure of how well one can distinguish innovators from all other producers through production levels. As Z falls, an innovator's production level grows relative to the average production in the industry. Once the advance is adopted, an innovator profit is given by

$$\pi_1 = \frac{1}{\theta}\left(\frac{\theta}{Z}\right)^{(1+\alpha)/(\alpha+\beta)} \pi_0. \tag{23}$$

Without loss of generality, take the cost of lobbying in order to obstruct the dissemination of the technical information to be proportional to initial profits, viz., $l\pi_0$. The consumers/taxpayers will find it necessary to compensate at least the innovator if profits fall below that which could be obtained by obstructing the change. Define this level of compensation as B_d, where $B_d = \pi_0 - l\pi_0 - \pi_1(P_1) \geq 0$ and thus

$$B_d = \pi_0\left[(1 - l) - \frac{1}{\theta}\left(\frac{\theta}{Z}\right)^{(1+\alpha)/(\alpha+\beta)}\right] \geq 0. \tag{24}$$

This implies that, in order for transfers to serve the public interest, the level of technical change and the number of innovators must be such that

$$Z(1 - l) \geq \left(\frac{\theta}{Z}\right)^{(1-\beta)/(\alpha+\beta)}. \tag{25}$$

The intuition behind expression (25) is that for a transfer scheme to be necessary the residual demand facing the innovators must be sufficiently inelastic such that the price effect of the technical change is greater than the cost savings. For a given level of the advance, θ, as the proportion of innovators increases, the more inelastic the residual demand facing that group, the more depressed price will become and the more necessary will compensation become to avoid obstruction.

Under a decoupled policy, announced prior to discovering innovators, based on lump-sum, per-producer payments, the minimum total amount of wealth transfer, B_d, is given by expression (25). Gross consumer benefits of the technical advance with the decoupled policy CS_d, is represented by the area under the demand curve between prices P_0 and P_1, i.e.,

$$CS_d = \frac{bP_0^{1-\beta}}{1-\beta}\left[1 - \left(\frac{\theta}{Z}\right)^{(1-\beta)/(\alpha+\beta)}\right]. \tag{26}$$

Net consumer gains from the advancement are represented by $R_d = CS_d - B_d$.

Now consider the other polar case, where price-distorting, coupled policy is a "targeted" price, P_T, guaranteed to all producers, that will make the innovator just indifferent to obstructing the change:

$$\pi_1(P_T) = \frac{1}{1+\alpha}\frac{1}{\theta}\frac{P_T^{1+\alpha}}{c^{\alpha}} = \pi_0(1 - l), \tag{27}$$

implying

$$P_T = P_0[\theta(1 - l)]^{1/(1+\alpha)}. \tag{28}$$

The coupled policy will induce a greater level of production from all firms, implying a new equilibrium price, P_t, given by

$$bP_t^{-B} = S_t = \left(\frac{P_T}{c}\right)^{\alpha}\frac{Z}{\theta}, \tag{29}$$

$$P_t = P_0\left[[\theta(1 - l)]^{\alpha/(1+\alpha)}\frac{Z}{\theta}\right]^{-1/\beta}. \tag{30}$$

The total transfers to producers under the per-unit payment scheme are given by

$$B_c = (P_T - P_t)S_t, \tag{31}$$

which, after some algebraic manipulation, may be represented as

$$B_c = \frac{P_0^{1+\alpha}}{c^{\alpha}}\left\{(1 - l)Z - \left[[\theta(1 - l)]^{\alpha/(1+\alpha)}\frac{Z}{\theta}\right]^{-(1-\beta)/\beta}\right\}. \tag{32}$$

Gross consumer benefits from the coupled policy, CS_c, is given by the area under the demand curve between P_0 and P_t, viz.,

$$CS_c = \frac{bP_0^{1-\beta}}{1 - \beta}\left\{1 - \left[[\theta(1 - l)]^{\alpha/(1+\alpha)}\frac{Z}{\theta}\right]^{-(1-\beta)/\beta}\right\}. \tag{33}$$

Net consumer/taxpayer benefits from the coupled policy are given by $R_C = CS_c - B_c$.

The increase in consumer/taxpayer expenditures for an increase in the support price is given by

$$\frac{\partial P_T S_t}{P_T} = (1 + \alpha)\left(\frac{P_T}{c}\right)^{\alpha}\frac{Z}{\theta}N. \tag{34}$$

This increase in expenditures can be directly transferred through a decoupled program and can increase each innovator's welfare by $\partial P_T S_t / \partial P_T = (1 + \alpha)(P_T / c)^{\alpha} Z\theta^{-1}$. The associated increase in each innovator's profit with the coupled policy is given by $\partial p_1 / \partial P_T = (P_T / c)^{a}\, \theta^{-1}$. From the immediately preceding paragraph, the decoupled policy is clearly inferior if

$$\frac{\partial P_T s_t}{\partial P_T}\cdot\frac{1}{N} < \frac{\partial \pi_1}{\partial P_T}, \tag{35}$$

or if $1/(1 + \alpha) > Z$, or $1/Z - 1 > \alpha$.

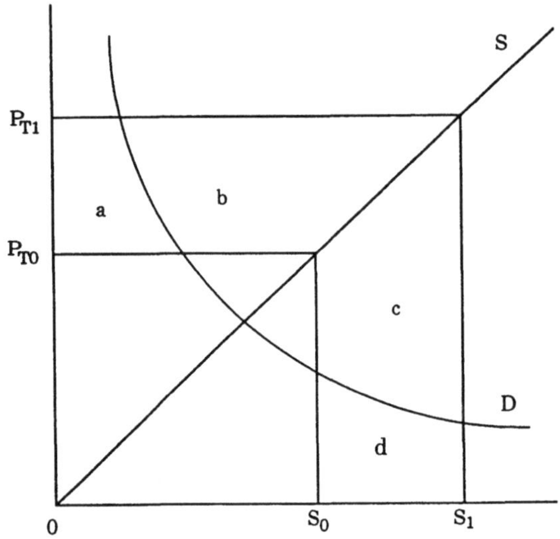

Fig. 5.2. Consumer gains from per-unit-output payment

In the case of a perfectly inelastic demand curve, the condition, $1/Z - 1 > \alpha$, is both sufficient and necessary for consumer preference of a coupled program transfer of an additional dollar to innovators. The intuition of the proposition can be illustrated in Fig. 5.2. Consider the additional transfer of $P_{T1}S_1 - P_{T0}S_0$ dollars to all producers, which is area a + b + c + d in Fig. 5.2. This could be done in two ways: through a $[P_{T1}S_1 - P_{T0}S_0]/N$ transfer to each innovator (as well as to each noninnovator) under a decoupled, per-firm payment scheme or through an increase in the target price from P_{T0} to P_{T1}. The target price increase would imply a total profit increase of area a + b for all producers taken together and a profit increase of $\pi_1(P_{T1}) - \pi_1(P_{T0})$ for each innovator. The condition, $1/Z - 1 > \alpha$, is simply that which assures a small increase from P_{T0} to P_{T1},

$$\frac{1}{N}\left[P_{T1}S_1 - P_{T0}S_0\right] < \pi_1(P_{T1}) - \pi_1(P_{T0}). \tag{36}$$

As it turns out, the condition, $1/Z - 1 > \alpha$ is sufficient for the superiority to consumers/taxpayers of coupled policies.

> *In the constant-elasticity supply-and-demand case, if a transfer policy is necessary to overcome obstruction (i.e., the condition in expression (25) holds) and the percentage gain in an innovator's level of production over the industry average is greater than the supply elasticity, then the coupled policy is optimal for consumers/taxpayers for all elasticities of supply and demand meeting these conditions.*

To show this, define $\rho = Z(1 - l) \cdot (\theta/Z)^{-(1-\beta)/(\alpha+\beta)}$ from expression (25) such that, if a transfer is necessary to accomplish the technical advance, then $\rho > 1$. The ratio of an innovator's production to average production is greater than the supply elasticity implies that $1/(1 + \alpha) > Z$. After some algebraic manipulations, the superiority to consumers of coupled over decoupled payments, $R_c > R_d$, implies

$$\rho > \frac{1/(1 + \alpha) - Z\phi}{1/(1 + \alpha) - Z}, \tag{37}$$

where

$$\phi = \frac{1 - \beta\rho^{\frac{-\alpha(1-\beta)}{\beta(1+\alpha)}}}{1 - \beta}. \tag{38}$$

If $\rho > 1$, then $\phi > 1$, and the right-hand side of expression (9) is certainly less than unity.

Expression (37) presents the necessary condition for the superiority of coupled relative to decoupled policies for breaking producer coalitions. We can re-express this condition as

$$\frac{1}{1 + \alpha} - Z \cdot \left[\frac{\rho - \phi}{\rho - 1}\right] \geq 0, \tag{39}$$

where $\phi \geq 1$ for all $0 \leq \beta \leq 1$ and $\rho \leq 1$. Note that, if $\beta = 0$, then $\phi = 1$ and the conditions in this result are both necessary and sufficient for $R_c > R_d$. As β grows positive, the term, $(\rho - \phi)\ (\rho - 1)$, decreases below unity, implying a tradeoff between $Z - \sigma + (1 - \sigma)\theta$ and the demand elasticity in assuring that the benefits are greater from a coupled transfer relative to a decoupled one. (One may also note that $\phi \leq \rho$ for all $\rho \geq 1$.)

The choice of public policies can be characterized by Fig. 5.3. The necessity of a transfer scheme is given by the inequality condition of expression (25). Let θ_1 be such that the equality in expression (25) strictly holds. For levels of θ above θ_1, producers are unwilling to form the coalition to obstruct the technical change, and a policy would include only a public good.[2] Where $\theta \leq \theta_1$, some transfer mechanism is necessary to break a producer coalition against the change. Note that as $l \to 0$, $\theta_1 \to 1$ and as $l \to 1$, $\theta \to 0$. Note also that, as demand becomes more elastic, that is, as β increases toward unity, θ_1 decreases (because the derivative of the right-hand side of expression (25) is positive).

The choice between policies is indicated by the inequality given by expression (39). Let θ_2 be such that $1/(1 + \alpha) = Z = \alpha + (1 - \alpha)\theta_2$. For $\theta < \theta_2$, and for some θ sufficiently near θ_2, a per-unit-output payment scheme is better for consumers/taxpayers. For θ sufficiently greater than θ_2, the per-firm scheme is superior.

As α (the elasticity of supply) decreases, the critical value, θ_2, increases toward unity; and, indeed, ϕ approaches unity also, implying that the "fuzzy" region above θ_2, where coupled policies are superior, vanishes. Note that for sufficiently

---

[2] This result presumes, of course, that the consumer benefits outweigh all costs associated with providing and distributing the public good. For further details, see Rausser and Foster (1990).

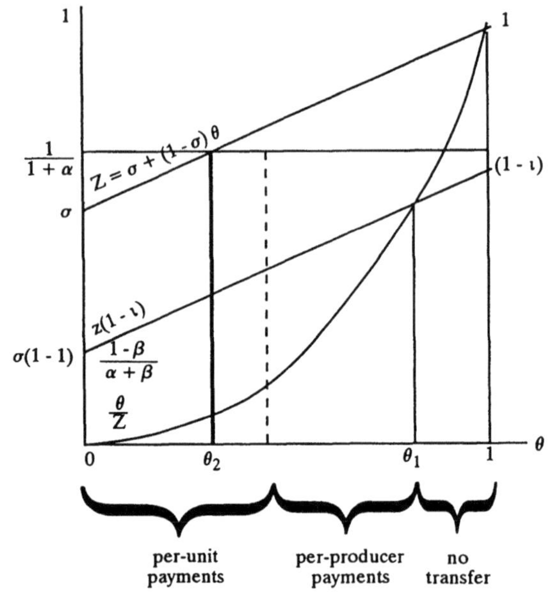

Fig 5.3. Optimal transfer scheme for levels of technical advice

high values of the proportion of innovators, σ, relative to the supply elasticity, $\sigma > 1/(1 + \alpha)$, no level of supply increase due to the technical change will result in a per-unit payment being superior to a per-firm transfer.

5.5 Conclusion

In this chapter it is demonstrated that a particular means of compensation may serve a purpose beyond that of simply transferring wealth. The motivating and underlying political-economic relationships that exist between social groups are not revealed by analyzing wealth transfers in isolation. Taxes and subsidies are a part of a larger portfolio of policies, all of which have some effect on the distribution of welfare. In the complete set of policies, wealth transfers may serve a remunerative function; in fact, recipients as a group may actually be losers when implementation of the larger portfolio is accounted for.

A less costly means of preempting coalitions that might otherwise obstruct the entire portfolio would be nonneutral, price-distorting payments. In the model presented, an output subsidy that distorts consumer and producer prices is useful to consumers/taxpayers because it effectively differentiates between decentralized producers; it counters the political opposition to, say, a supply-enhancing policy by dividing and conquering. This is in contrast to other models of political competition between groups that suggest that the transfer mechanism would tend to be the most efficient – in the sense of minimizing deadweight loss – because all groups could share in an efficiency gain (e.g., Becker 1983; Gardner 1987). Our analysis allows a governing group (consumers/taxpayers), through its choice of the compensation scheme, to overcome the problem of imperfect information (about the degree of ability to take advantage of the innovation).

Given the current debate over reform of agricultural policies, the framework presented here is particularly relevant. Many economists have approached this topic assuming wealth transfers to be the inefficient outcomes of chaotic rent seeking with their recommendations based on the belief that wasteful subsidies are the rewards of raw political power (or the consequence of consumer ignorance) and that a knowledgeable public would be concerned with gaining efficiency, if not with eliminating transfers altogether. Our framework, on the other hand, explains how a seemingly inefficient policy that appears to harm consumers could be, in fact, a rational component of a larger portfolio of policies that ultimately benefit consumers at the expense of producers.

References

Becker GS (1983) A theory of competition among pressure groups for political influence. Q J Econ (August) 98:371–400

Gardner BL (1987) Causes of US farm commodity programs. J Polit Econ 95:290–310

Magee SP, Brock WA, Young L (1989). Black hole tariffs and endogenous policy theory. Cambridge University Press, Cambridge

Mueller DC (1989). Public choice. Cambridge University Press, Cambridge, Chaps 12 and 13

Peltzman S (1976) Toward a more general theory of regulation. J Law Econ (August) 19:211–240

Rausser GC (1982) Political economic markets: PERTs and PESTs in food and agriculture. Am J Agric Econ (Dec) 64:821–833

Rausser GC (1992) Predatory versus productive governments: the case of US agricultural policy. J Econ Perspect 6, 3 (Summer):133–157

Rausser GC, Foster WE (1990) The evolution and coordination of US commodity and resource policies. In: Just RE, Bockstael N (eds) Commodity and resource policies in agricultural systems. Springer Berlin Heidelberg New York, pp 17–45

Stigler GJ (1971) The theory of economic regulation. Bell J Econ 2:3–21

Chapter 6

Public Goods and Welfare Transfer Tradeoffs

William E. Foster and Gordon C. Rausser

6.1 Introduction

The reliance on distorting interventions by governments is commonplace and largely impervious to reform, as argued in the previous chapter. In order to understand both governmental intervention in agricultural markets and the constraints to policy reforms, political-economic analysis must address two questions. The first regards the preference of distorting over nondistorting policies. The second regards the choice of the particular form of distorting policy; that is, the choice of the specific output or input markets that will carry the distorting taxes or subsidies. In this chapter we extend the analysis of the previous chapter examining the prevalence of nonneutral transfer schemes to both agricultural input and output markets. We demonstrate the conditions under which governments might select output subsidies, input subsidies, or lump-sum transfers. In our analysis, consumers and taxpayers have an incentive to target a certain level of wealth transfer to a specific and a priori unobserved set of producers within a broadly subsidized sector. Of two possible types of policies, per-firm versus per-unit output and input payments, the distorting policy may achieve the targeted transfer at least cost to consumers/taxpayers. The analytical subsidization framework is related to the optimal taxation and least-cost-subsidization literature (e.g., Mirrlees 1976; Chambers 1985).

We first present a conceptual model of wealth transfers from consumers and taxpayers to producers in the presence of other productive policies that increase potential total social welfare (Rausser 1992). Of the class of generic payments (per-capita or per-unit output), nonneutral payments are the most effective mechanism for overcoming a potential producer coalition which opposes moves from the status quo. We next present the optimal selection by consumers/taxpayers of the output and input subsidization levels as well as the level of per-capita (lump-sum) payments. The preference over subsidy mechanisms depends on the ability of nontargeted groups to take advantage of the mechanism and on the social costs of the associated market distortions. The analysis determines under what conditions input or output subsidies are likely to be observed. The concluding section discusses the implications of our results to the broader issue of policy reform.

6.2 A Portfolio of Policies: Supply, Expansion and Wealth Transfers

The political and economic motives behind the choice of distorting transfers over nondistorting transfers are not fully appreciated. Indeed, faced with the seeming omnipresence of distorting policies, many political economic analyses of governmental intervention tend to dismiss altogether neutral policies as impractical and to ignore the fundamental question of why states opt for socially wasteful transfers. As discussed in Chap. 5, these frameworks focus on the relative organization power of groups receiving or granting wealth transfers (e.g., Stigler 1971; Peltzman 1976; Becker 1983). These same frameworks are expanded here to examine a portfolio of policies.

Several examples of combinations of the two types of policies are available in both the economics literature and the popular press. Tirole (1990), for example, discusses privatization with safeguards for some social groups in formerly command economies. Newspaper articles relate the use of urban planning and the granting of zoning variances in exchange for the supply of local public goods (e.g., Lueck 1989). Labor economics literature examines special worker adjustment compensation for industries facing increased international competition (e.g., Richardson 1982; Kruse 1988; Corson and Nicholson 1989). Mitchell (1989, p. 495) notes that the motivation for the Trade Adjustment Assistance Act "has been viewed as a device to foster adoption of trade liberalizing bills by Congress – or to fend off protectionist legislation." Finally, in many countries there is the combination of farm income supports and publicly supported, price-decreasing agricultural projects, such as research and development, and water delivery systems. Technical innovations have been supported by state and federal agricultural research and dissemination evolving since the Hatch Act of 1887. These innovations have benefited or harmed commodity sectors to differing degrees and many have suspected them of leaving at least some farmers worse off – e.g., the treadmill hypothesis of Cochrane (1979).

Wealth transfers to a large group, such as agricultural producers, are typically shared unequally by firms; and it is in consumers' and taxpayers' interests that this be so. In the context of a supply-enhancing public good (such as technological research and development), some producers are harmed less than the industry average because they can take greater advantage of the advance. Wealth transfers weighted in favor of these innovators would serve to break producer coalitions obstructing change with less expense to consumers and taxpayers. Those who expand production or cut costs to a greater degree simply need less transfer payments to be made indifferent to the investment in the public good. Nonneutral policies target payments according to either production levels or input use. Therefore, a transfer based on production which makes those innovators who expand their output just as well off as without the advance, would transfer less to those who take less advantage of the public good. A transfer based on the use of an input encouraged by use of the public good will have a similar effect. The popularity of nonneutral (i.e., coupled) payments in agriculture especially may be explained by this property of targeting transfers from consumers to innovators, that is, to those firms less harmed by investments in public goods and, thus, to those most cheaply separated from a coalition that might obstruct moves from the status quo.[1]

6.3 A Model of Optimal Subsidies to Targeted Producers

The optimal subsidy rule is formalized here for consumers/taxpayers desiring to target a specific subset of firms, and thus prevent potential coalitions that may block change. Consider two interest groups in society – consumers/taxpayers and producers in an industry. Members of these groups behave competitively in the marketplace but may form coalitions with other group members in the political arena. Each group is composed of many members, and there is some rule for weighting the votes of individual members to decide each group's position on a particular policy. Producers are endowed with different levels of ability to take advantage of a public-interest policy. In this setting, ability may differ in location, vintage of capital, endowments of human capital, and so forth. We measure the welfare of each of N producers by the level of rent or profit to their ownership of a scarce resource, which is a function of output price, p, and input prices, w. We measure consumer welfare by the Marshallian surplus associated with consumption of aggregate output, and taxpayer welfare by the total outlays associated with all price subsidies and lump-sum payments. The particular public-interest policy may be, for example, the release of a technical innovation that will increase pro-

---

[1] Along with per-firm or per-unit output payments, other forms of identifying least-harmed producers in order to prevent an obstructing coalition are conceivable. We have chosen the most prevalent stylized subsidy/tax mechanisms. For example, the government might uncover the characteristics of each firm and provide firm-specific, neutral compensation to the smallest number of firms necessary to break the coalition. The administrative cost of such a scheme would tend to increase as the number of firms grows and as the technical abilities of the government decrease.
There may be forms of auction mechanisms that provide self–revelation of firms' minimal willingness-to-accept for defection from a coalition. It has been pointed out to us that an auction mechanism might be established where less-harmed losers would identify themselves by submitting bids on the level of compensation that would make them indifferent to the introduction of the producer-harming public good. Like the per-firm payment, this particular auction mechanism would entail identifying firms within an industry but would not involve assessing each firm's minimum compensation for indifference to implementing the public good. Furthermore, in the absence of firm coordination in bid submission, by only compensating the minimum number of bidding firms to refrain from entering a blocking coalition, the government pays no more than the per-unit output scheme. Unfortunately, given that this auction mechanism is in place and no firm coordination exists, the only stable outcome is a zero bid for each firm. This implies that a coalition of firms opposing this auction in the first place would arise. With cooperation between firms in bidding, the final cost to the government may be greater than the per-unit output scheme. In addition, auction mechanisms would generally require greater administrative costs and expertise than either the per-firm or per-unit output cases.

duction but, by doing so, will also harm enough producers by decreasing output price that a coalition will form to block the release.

To formalize the concept of ability to take advantage of the public good, let a_i be a vector of the ith firm's attributes. Define $\pi_0(a_i)$ as the rent accruing to the ith firm prior to the implementation of the public good policy; and $\pi_i = \pi(p, w, a_i)$ as the rent accruing with the release under output price, p, and input prices, w. We assume that producers own the fixed factors and that all other inputs are free to move across industries with perfectly elastic supplies facing the industry in question. The ranking of firms, $i = 1, 2, ..., N$, is chosen to define ability (to take advantage of the public good) in terms of rents lost or gained:

$$\pi_0(a_i) - \pi(p, w, a_i) \leq \pi_0(a_j) - \pi(p, w, a_j) \tag{1}$$

for all $i \leq j$, and for p and w in reasonable ranges. Intuitively, condition (1) implies that firms of higher level ability gain relatively more, or lose relatively less, from the public good. In this way we may indicate that, if the cth producer expects to be indifferent to the public good policy, then at least c number of producers are indifferent to, or desirous of, the new policy. And, if $\pi(p, w, a_c) > \pi(p, w, a_{c+1})$, then only the first c producers are at least indifferent, the remainder being harmed.

Turning to supply and input demands, the individual production levels and input demands are found by standard application of Hotelling's lemma:

$$y_i = \frac{\partial \pi_i}{\partial p}, \; and \; x_{ki} = \frac{\partial \pi_i}{\partial w_k}, \; k = 1, K. \tag{2}$$

Aggregate supply, S, and aggregate or total input utilization is simply the sum of the N individual firms' behavior:

$$S = \sum_i y_i, \; and \; X_k = \sum_i x_{ki}, \; i = 1, K. \tag{3}$$

In order to illustrate the effects on production of a public good and the wherewithal of a particular firm to take advantage of it, consider an agricultural water resource project. Suppose prior to the provision of the public good all firms are identical and face the same input and output prices. The project has the effect of altering input prices facing firms, some firms enjoying a greater decrease in input price than others. Prior to the project's provision, firms face a common price for input k of w_k, and the provision has the result of reducing the price on average by λ_k. With the project's provision, the ith firm faces a firm-specific input price of $\hat{w}_{ki} = w_k - \lambda_k a_{ki}$; where a_{ki} measures the ability of the ith firm to take advantage of the price decrease,

$$\left(\sum_{i=1}^{N} a_{ki} = 1 \right).$$

Condition (1) certainly holds if $a_{ki} > a_{kj}$ for all k and $i < j$.

To be even more specific, firms initially pump groundwater from their own wells and a common technology. The project delivers surface water at an effective per-unit cost that is a function of a firm's distance from the source, the firms closest enjoying the greatest price decrease. Moreover, there exist some firms altogether unable to take advantage of the project, because of their geographical location (i.e., firms with $a_{ki} = 0$). Water projects tend to raise agricultural output (and to increase the demand for other inputs, such as fertilizer), which in turn tends to reduce the output's price to all producers.[2] The project's effect on a firm's profit, therefore, depends on the equilibrium output price effect as well as on the firm's individual input-price decrease. To formalize, let $w_1 - \lambda a_i$ represent the per-unit water cost to the ith firm, and consider an incremental decrease in the average price of water from $\lambda_0 = 0$ with no other input price effects. The ith firm's profit is given by $\pi_i = \pi(p, w_1 - la_i)$, and the effect of the incremental decrease in average water price is given by $\partial\pi_i/\partial\lambda = y(p, w_1)\cdot(\partial p/\partial\lambda) + a_i x_1(p, w_1)$. With demand, $D(p)$, the effect on equilibrium output price depends on the water demand elasticity with respect to output price, η_{1p}, and the sum of the absolute values of the supply and demand elasticities $(\varepsilon^p + \gamma)$: that is, $\partial p/\partial\lambda = (x_1/y) \cdot (-\eta_{1p})/(\varepsilon_p + \gamma)$.[3] And therefore we may determine whether or not a firm is a net loser due to the project in terms of its "ability" a_i and parameters characterizing aggregate market behavior. Specifically, the ith firm is a net loser if $a_i < \eta_{1p}/(\varepsilon_p + \gamma)$.

---

[2] For an extensive overview of the economics and management of water and drainage in agriculture, see the volume edited by Dinar and Zilberman (1991). For an analysis of the distributional welfare implications of irrigation water subsidies, see Foster, et al. (1986).

[3] Equilibrium price, p, is found by equating aggregate supply and demand:

$$S(p,w;\lambda) = \sum_i y(p,w_1 - \lambda a_i) = D(p).$$

Totally differentiating with respect to λ yields

$$\sum \frac{\partial y}{\partial p}(p,w)\cdot\frac{dp}{d\lambda} - \sum \frac{\partial y}{\partial w_i}(p,w)a_i = \frac{\partial D}{\partial p}\frac{dp}{d\lambda}.$$

Using the equation, $-\partial y/\partial w_1 = \partial x_1/\partial p$, and making the appropriate translations of derivatives into elasticities [i.e., $\eta_{1p} = (p/x_1)(\partial x_1/\partial p)$, $\varepsilon_p = (p/y)(\partial y/\partial p)$, and $\gamma = -(p/D)(dD/\partial p)$], the effect on equilibrium output price can be written

$$\frac{\partial p}{\partial\lambda} = \frac{-\eta_{1p}}{\varepsilon_p + \gamma}\cdot\frac{x(p,w_1)}{y(p,w_1)}.$$

Note that a firm can be a net loser when it has greater than or less than the average ability to take advantage of the project. Furthermore, using this input-price-altering formulation of a project's effect one can succinctly describe the changes in individual supply relative to the industry average. The percentage increase in the ith firm's and the industry's average output levels are given by $d\ ln\ y_i /d\lambda = [\varepsilon^p(dp/d\lambda) = [\varepsilon^p(dp/d\lambda) = (x_i /y_i)\eta_{1p}a_i]/p$, and by $d\ ln\ \bar{y}/d\lambda = [\varepsilon^p(dp/d\lambda) - (\bar{x}/\bar{y})\eta_{1p}]/p$. (Note that, for an incremental decrease in average water price, the ith firm's output and input use equal the industries averages: x_{1i}, and $\bar{x}_1$.) Therefore, the increase in the ith firm's output due to the project relative to the increase in the industry average depends on the firm's relative ability to take advantage of the water price decrease. If $a_i > 1$, then the firm's output level increases relative to the industry average; such a firm may be termed an *innovator*. If $a_i < 1$, then the firm's output decreases relative to the average; such a firm is a *noninnovator*.

Whether a water project or any public good is provided depends on the degree of producer support. Define V as the minimum number of producers needed to support the provision of the public good project. Under a weighted-vote rule, $w(i)$, define the index c such that

$$V = \sum_{1}^{c} w(i), \tag{4}$$

where the weighting rule could be based on the initial level of output. Therefore, any generic compensation scheme that affects the producer output or input prices, or offers a per-producer, price-neutral payment need only make indifferent the cth producer in order to gain the producer group's acquiescence to the policy. In the following, we will be making use of the following notation. The percent deviation of the cth producer's production level, y_c, from the industry's average, $\bar{y}$, is represented by

$$\rho_0 = y_c/\bar{y} - 1, \tag{5a}$$

and the percent deviation of the cth producers input use, x_{kc}, from the industry's average, $\bar{x}_k$, is represented by

$$\rho_k = x_{kc}/\bar{x}_k - 1,\ i = 1, K. \tag{5b}$$

For the case of the water project altering firm input prices, we can characterize the effect on the cth firm's input use, relative to the industry norm, due to a bias toward the particular input. Consider two projects affecting the prices of inputs (substitutes) x_1 and x_2, but each having the same effect on equilibrium output price. For all firms, the first project reduces the first input price by λ_0 and the second input by θ_0. The second project alters incrementally the first input price to the ith firm by $\lambda_0 + \lambda a_{1i}$, and the second input price by $\theta_0 + \theta a_{2i}$. From an initial value of $\lambda = 0$, an incremental increase in λ is associated with a change in θ such that the equilibrium output price remains unchanged. For normal inputs, the second project's bias towards the first input will lead to an increase in the cth firm's relative use of that input (i.e., $\partial\rho_k/\partial\lambda > 0$), if the firm is unambiguously better able to take advantage of the project compared to the industry average (i.e., if $a_{ki} > 1$

for all k). The opposite result holds (i.e., $\partial\rho_k/\partial\lambda < 0$, if the cth firm is unambiguously less able to take advantage of the project.[4]

Suppose that a blocking coalition of producers would obstruct the implementation of the public good without some form of compensation, but that the coalition is avoided by making at least indifferent to the change a fixed percentage of all producers. Let c index the smallest percentage of producers that consumers/taxpayers, acting as the government, must make at least indifferent to the change in order to prevent its obstruction. Of course, for some public goods the cth producer may by better off than prior to the change: $\pi(p_0, w_0) \le \pi(p_1, w_1; a_c)$, implying that no wealth transfers are necessary to obtain the policy. But for other public goods some set of per-unit output and input price subsidies, s and d, and lump-sum payments, r, are required to break the coalition. The consumer/taxpayer

---

[4] To see this result, write the cth firm's demand for the first input as $x_{1c} = x_1(p, w_1 - \lambda_0 - \lambda a_{1c}, w_2 - \theta_0 - \theta a_{2c})$, and the change in the use of the input with respect to the measure of bias as

$$\frac{dx_{1c}}{d\lambda} = -\frac{\partial x_{1c}}{\partial w_1} a_{1c} - \frac{\partial x_{1c}}{\partial w_2} a_{2c} \frac{d\theta}{d\lambda}.$$

Let $\bar{w}_1 = w_1 - \lambda_0$ and $\bar{w}_2 = w_2 - \theta_0$. From $\lambda = 0$, increasing λ and decreasing θ such that output price remains unchanged leads to

$$\frac{d\theta}{d\lambda} = -\frac{\bar{w}_2}{\bar{w}_1}\frac{\varepsilon_1}{\varepsilon_2},$$

where $\varepsilon_k < 0$ is the supply elasticity with respect to the kth input price (assuming normal inputs). Therefore, combining the above two equations yields the percentage change in x_{1c}

$$\frac{dx_{1c}}{d\lambda}\frac{1}{x_{1c}} = \frac{\varepsilon_1}{\bar{w}_1}\cdot\left[\frac{\eta_{12}a_{2c}}{\varepsilon_2} - \frac{\eta_{11}a_{1c}}{\varepsilon_1}\right] > 0,$$

where η_{ij} is the demand elasticity of the ith input with respect to the jth input price. Similarly, the percentage change in the industry average input use is given by

$$\frac{d\bar{x}_1}{d\lambda}\frac{1}{\bar{x}_1} = \frac{\varepsilon_1}{\bar{w}_1}\cdot\left[\frac{\eta_{12}}{\varepsilon_2} - \frac{\eta_{11}}{\varepsilon_1}\right] > 0.$$

The cth firm's use of the first input, relative to the industry norm, will increase with a project bias toward the first input, if $\partial\rho_k/\partial\lambda > 0$ or if $d \ln x_{1c}/d\lambda > d \ln \bar{x}_1/d\lambda$. This latter result will hold
if

$$\frac{\varepsilon_1}{\bar{w}_1}\left[\frac{\eta_{12}}{\varepsilon_2}(a_{2c} - 1) = \frac{\eta_{11}}{\varepsilon_1}(a_{1c} - 1)\right] \rightarrow 0, \quad or\ if$$

$$\frac{\eta_{12}}{\varepsilon_2}(a_{2c} - 1) < \frac{\eta_{11}}{\varepsilon_1}(a_{1c} - 1).$$

A sufficient condition, in the case of noninferior substitutes, is simply that $a_{1c} > 1$ and $a_{2c} > 1$.

will choose a set (s, d, r) such that the cth producer is just indifferent to the change. Representing producer output prices by $\hat{p} = p + s$, and input prices by $\hat{w}_i = w_i - d_i$, the consumer/taxpayer is constrained to set

$$\pi_0 = \pi(\hat{p}, \hat{w}; a_c) + r, \tag{6}$$

where π_0 is the cth producer's pre-public-good level of welfare. We represent the subsidy rates on prices as $\sigma = s/(p + s)$ and $\delta_i = d_i/(w_i - d_i)$. The consumer/taxpayer's problem is to select the various subsidy levels in order to target the cth producer in the most efficient manner. We must recognize that prices may be altered by the subsidies, because all producers respond to changes in prices, expanding supply or changing input demand with increases in s and d. We take output price, p, to be endogenous and input supplies facing the industry to be perfectly elastic.[5] Market output price will equilibrate aggregate supply and demand:

$$S(p + s, w - d) = D(p). \tag{7}$$

For ease of exposition, we represent the response of output prices to the subsidy as

$$\frac{\partial p}{\partial s} = -\frac{\partial S}{\partial p} / \left(\frac{\partial S}{\partial p} - \frac{\partial D}{\partial p} \right) = -\frac{\varepsilon_p}{\left[\varepsilon_p + \gamma(1 - \sigma)^{-1} \right]} , \tag{8}$$

where ε_p is the elasticity of aggregate supply with respect to output price and γ is the absolute value of the demand elasticity. Similarly,

$$\begin{aligned} \frac{\partial p}{\partial d_i} &= \frac{\partial S}{\partial w_i} / \left(\frac{\partial S}{\partial p} - \frac{\partial D}{\partial p} \right) = -\frac{\partial X_1}{\partial p} / \left(\frac{\partial S}{\partial p} - \frac{\partial D}{\partial p} \right) \\ &= -(X_k/S)(-\eta_{kp})[\varepsilon + \gamma/(1 - \sigma)] \end{aligned} \tag{9}$$

where η_{kp} is the elasticity of total utilization of input k with respect to output price. Total government outlays, G, are given by

$$G = sS + \Sigma d_i X_i + N[\pi_0 - \pi_{,}(\hat{p}, \hat{w}; a_c)], \tag{10}$$

where $sS + \Sigma d_i X_i$ represents the level of distorting payments due to output and input subsidies, and $N[\pi_0 - \pi(...)]$ represents the level of per-capita payments. The consumer surplus, CS, is measured by the area under the demand curve for the output:

$$CS = \int_p^{\infty} D(z) dz. \tag{11}$$

---

[5] Making all prices endogenous would complicate the analysis, but not alter the basic results.

The use of both output and input subsidies is redundant, if all prices can be subsidized. This is because subsidizing via the output market is equivalent to equiproportional input subsidies (Chambers 1985). Typically, however, governments do not have access to all input markets in order to affect prices. Indeed, governments tend to concentrate their subsidization efforts in one or two markets, usually in the output market.[6] We first turn our attention to such subsidy schemes.

6.3.1 Output Subsidies

With only an output subsidy and lump-sum payments, the optimal levels of s and r are found by maximizing the consumer/taxpayer's criterion function given by

$$\max_s W = \int_p^{\infty} D(z)dz - sS(p + s, w) - N[\pi_0 - \pi(\hat{p}, w; a_c)]. \tag{12}$$

The optimal level of s is given by the first-order condition

$$\frac{\partial W}{\partial s} = -D(p)\frac{\partial p}{\partial s} - S - s\frac{\partial S}{\partial p}\left(\frac{\partial p}{\partial s} + 1\right) + Ny_c\left(\frac{\partial p}{\partial s} + 1\right) = 0. \tag{13}$$

Defining the industry's average production by $\bar{y} = D/N$, we can solve (13) for the optimal subsidy level as

$$\frac{s}{p + s}\frac{\partial S}{\partial p}\frac{p + s}{S} = \frac{y_c}{\bar{y}} - 1, \quad or, \tag{14a}$$

$$\sigma = \frac{\rho_0}{\varepsilon_p}. \tag{14b}$$

This result demonstrates:

> *The optimal rate of distortion in the output market is equal to the percent deviation of the targeted producer's output from the industry's average weighted by the reciprocal of the supply elasticity.*

Intuitively, the value, ρ_0, measures the output of the targeted producer relative to the industry's average, and thus it measures the distinguishability of the cth producer. *Ceteris paribus*, the greater the targeted producer's new output relative to the industry's average – that is, the easier it is to identify the targeted firm by its output – the greater will the government rely on nonneutral payments, and the greater the optimal rate of distortion. In contrast, the greater the price responsiveness of supply to increases in the subsidy rate, the less the government would rely on coupled payments, and the more on per-firm, neutral payments.

Expression (14b) illustrates that the use of distorting subsidies takes advantage of the heterogeneity of the public good project's effect on producers, if some pro-

---

[6] We are abstracting from administrative costs that may vary across types of output and input markets.

ducers are able to expand output more than others. In the water project example, if the targeted firm is an innovator (i.e., if $a_{kc} > 1$), because it is located closer to the water source, then the project's provision would increase the output of the firm relative to the industry norm. If the targeted firm is a noninnovator, then the project's provision would tend to decrease the firm's output level relative to the average, making lump-sum payments more attractive. Expression (14b) also shows the tradeoff between identifying the targeted producer via an output subsidy and the inefficiency in production created by distorting prices.

The gain to consumers, due to the consumer-price decrease as producers expand output in response to the subsidy, makes the coupled scheme relatively more attractive than the per-firm scheme. The taxpayer interest alone may prefer coupled transfers, although the subsidy rate will typically be less. To see this, suppose there is a differential weight on consumer welfare as opposed to tax outlays. That is, let the consumer/taxpayer welfare be given by $W = (1 - \lambda)CS + G$. The rule for optimal tax subsidy may now be expressed as

$$\frac{\partial W}{\partial s} = (1 - \lambda)D(p)\frac{\partial p}{\partial s} - S - s\frac{\partial S}{\partial p}\left(\frac{\partial p}{\partial s} + 1\right) + Ny_c\left(\frac{\partial p}{\partial s} + 1\right) = 0. \tag{15}$$

This yields

The simple rule for optimal rate of price distortion is given by

$$\sigma = \frac{\rho_0}{\varepsilon_p} - \frac{\lambda}{\gamma}, \tag{16}$$

where the value of λ represents the weight on expenditures relative to that on consumer surplus, and γ represents the absolute value of the demand elasticity.

As the weight on expenditures increases, the subsidy rate decreases and there is a greater reliance on lump-sum payments. Even if the government places all weight outlays and zero on consumer surplus, there is still an incentive to have a positive subsidy rate, a distorting policy, if $\rho_0 > \varepsilon_p/\gamma$. The intuition here is that there is a tradeoff between the increase in supply in response to the coupled policy and the ability of consumers to absorb the extra production. If the distinction between the targeted firm's output and the industry's average is sufficiently great (i.e., a high ρ_0) or, regardless of the distinction, if the price effect of the coupled payment is sufficiently small (i.e., a low ε_p/γ), then a coupled policy is less expensive to taxpayers.

Coupled payments are worthless at distinguishing between producers when the producer being targeted is simply the representative firm. When producers are homogeneous in their response to change, there is little economic incentive for coupled transfers. Coupled, distorting payments are preferred when producers are likely to respond at different rates to change. If government represents only consumer and taxpayer interests and all producers immediately adopt some technical advance, or could make use of some other public good, then remunerative policies will be nondistorting. When some producers adopt or adopt more quickly than others, and some perhaps not at all, then distorting payments are optimal.

6.3.2 Input Subsidies

With only a single input subsidy, say, on the first indexed input, X_1, the level of the subsidy is found by maximizing the consumer/taxpayer's criterion function given by

$$\max_{d_1} W = \int_p^\infty D(z)dz - d_1X_1 - N[\pi_0 - \pi(p,\hat{w};a_c)] \ . \tag{17}$$

The optimal level of the input subsidy is given by

$$\frac{\partial W}{\partial d_1} = -D(p)\frac{\partial p}{\partial d_1} - X_1 - d_1\frac{\partial X_1}{\partial p}\frac{\partial p}{\partial d_1} + d_1\frac{\partial X_1}{\partial w_1} + Ny_c\frac{\partial p}{\partial d_1} + Nx_{1c} = 0, \tag{18}$$

implying

$$X_1\rho_1 + S\frac{\partial p}{\partial d_1}\rho_0 - d_1\frac{\partial X_1}{\partial p}\frac{\partial p}{\partial d_1} + d_1\frac{\partial X_1}{\partial w_1} = 0. \tag{19}$$

Hence,

The optimal subsidy rate on a single input, d_1, depends on the relative use by the targeted firm of the input, as well as the relative output level:

$$\rho_1 - \frac{\eta_{1p}}{\varepsilon_p + \gamma}(\rho_0 = \delta_1\varepsilon_1) + \delta_1\eta_{11} = 0, \tag{20}$$

where $\varepsilon_1 = (\partial S/\partial w_1)\cdot(w_1 - d_1)/S$, the elasticity of supply with respect to a change in the price of the first input, and where $\eta_{ij} = (\partial X/\partial w_j)\cdot(w_j - d_j)/X_j$, the cross-elasticity of demand for the ith input with respect to a change in the jth input price.

Expression (20) can be rewritten in terms of the price distortion:

$$\delta_1 = \left[\frac{\rho_1}{\eta_{1p}} - \frac{\rho_0}{\varepsilon_p + \gamma}\right]\cdot\left[\frac{\varepsilon_1}{\varepsilon_p + \gamma} - \frac{\eta_{11}}{\eta_{1p}}\right]^{-1}. \tag{21}$$

Assuming that the second-order condition holds for maximization, the denominator of (21) is positive. Again, as in the output subsidy case, there is a tradeoff between distinguishing the innovator via coupled payments and the social costs of distorting subsidies. *Ceteris paribus*, as the innovator uses more of an input relative to the industry's average use of that input, it is in the consumer/taxpayer's interest to rely more heavily on the input price rather than per-firm payments in order to accomplish the targeted payments. Note that a positive input subsidy rate will not hold for all targeted producers, as it does in the case of an output subsidy. The input subsidy is positive if and only if $\rho_1/\eta_{1p} > \rho_0/(\varepsilon_p + \gamma)$. The intuition is that the input subsidy drives output price down, harming all producers; and that innovators, by producing a greater output, are harmed disproportionately by the output price fall. The greater the relative output of the innovator, the smaller the input subsidy. On the other hand, the more responsive is the output to a reduction in the input price, the less the subsidy.

Input subsidies are two edged: they tend to increase directly the welfare of the targeted group, but they also tend to depress output prices and thus harm the targeted group. If there is no price response to the input subsidy, because demand is perfectly elastic (i.e., $\gamma = \infty$), then

$$\delta_1 = \frac{\rho_1}{\eta_{11}}, \tag{22}$$

and a positive subsidy level on the input price would prevail. If the government is restricted to only positive subsidies (no input taxation), then input subsidies are less likely than output subsidies because of their depressing effect on output prices. Indeed, input taxes may serve to minimize the cost of targeting a given level of wealth transfer, because reducing input use will decrease output and lead to an increase in output prices and profits. Thus consumers may be made worse off in order to reduce net total outlays for the industry subsidy. Input taxes and restrictions are therefore not inconsistent with targeting payments to groups of producers. In the case of input taxes, the less input used by the targeted producer relative to the industry average, the greater the government will reduce that input to raise output price.

In the case of a coupled scheme for two inputs, X_1 and X_2, the rules for optimal price distortion are determined by the first-order conditions for maximizing the consumer/taxpayer's criterion function in the same manner as that for one input, and are given by

$$\frac{\eta_{1p}}{\varepsilon_0 + \gamma}(\rho_0 + \delta_1\varepsilon_1 + \delta_2\varepsilon_2) = \rho_1 + \delta_1\eta_{11} + \delta_2\eta_{12} \tag{23a}$$

and

$$\frac{\eta_{2p}}{\varepsilon_0 + \gamma}(\rho_0 + \delta_1\varepsilon_1 + \delta_2\varepsilon_2) = \rho_2 + \delta_1\eta_{21} + \delta_2\eta_{22}. \tag{23b}$$

Writing $\alpha_i = \eta_{ip}/(\varepsilon_p + \gamma)$, we solve expressions (23a) and (23b) for the optimal rates of distortion:

$$\delta_1 = \Delta^{-1} \cdot [(\eta_{22} - \alpha_2\varepsilon_2)(\alpha_1\rho_0 - \rho_1) - (\eta_{12} - \alpha_1\varepsilon_2)(\alpha_2\rho_0 - \rho_2)] \tag{24a}$$

and

$$\delta_2 = \Delta^{-1} \cdot [(\eta_{11} - \alpha_1\varepsilon_1)(\alpha_2\rho_0 - \rho_2) - (\eta_{21} - \alpha_2\varepsilon_1)(\alpha_1\rho_0 - \rho_1)] \tag{24b}$$

where $\Delta = [(\eta_{11} - \alpha_1\varepsilon_1)(\eta_{22} - \alpha_2\varepsilon_2) - (\eta_{21} - \alpha_1\varepsilon_2)(\eta_{12} - \alpha_2\varepsilon_1)] > 0$, assuming that the second-order conditions for maximization hold. Again, if the innovator uses relatively more of one input than another, then that input would tend to be subsidized to a greater degree. If, however, the total industry demand for the input is responsive to price changes, then that input will tend to receive less subsidy. Expressions (24a) and (24b) also demonstrate the importance of the distortions to the use of all inputs created by a subsidy on a single input price. Even if an input is relatively unresponsive to changes in its own price, but the other input demands are very responsive to that input's price, then the subsidy will tend to be less on that input. Note that the subsidy rates can be both positive and negative, and optimal targeting of benefits to innovators can involve simultaneously the subsidization of some inputs and the taxation of others. Sufficient conditions for which both subsidy and tax hold are

$$\frac{\rho_1}{\eta_{1p}} > \frac{\rho_0}{\varepsilon_p + \gamma} > \frac{\rho_2}{\eta_{2p}}, \tag{25}$$

and the inputs are complementary: $\eta_{21}, \eta_{21} > 0$. The intuition is that the subsidy will be attached to the input more heavily used by the targeted group relative to the industry's average and the tax will be attached to the input least used relative to the industry's average. One should note that the first input need not be used in greater quantity than the second, nor have a larger cost share, but simply the innovator's use relative to the noninnovator's is greater. For example, with the introduction of the public good, the innovator may use less of one good, say, land (X_2), in order to produce the same level of output, but use more of another input, say, irrigation water (X_1). Then, optimal targeting would entail a tax or restrictions on land and a subsidy on water. Note also that the use of land by the targeted firm may be greater than the industry average (ρ_2), but the use of land relative to the average is less than the firm's output relative to the industry average. Thus, with targeting, a seeming contradiction in agricultural policies is dispelled: input subsidies and taxes may exist simultaneously. In general, taxes or restrictions will exist on traditional inputs, those less relied on by innovators to produce the same level of output. Subsidies will exist for "advanced" inputs, new technologies, or those inputs (although traditional) used in relatively greater amounts due to the public good, such as fuel or water. If only two inputs exist, then

$$\eta_{ip} + \eta_{ii} + \eta_{ij} = 0, \text{ for } i, j = 1, 2. \tag{26}$$

The case of a two-input production function simplifies the analysis considerably. In particular, we can rank subsidy rates by reference to two parameters: the level of relative use of the input by the targeted group, and the responsiveness of the total industry demands for the inputs with respect to output price. Applying (26) in (25), we may obtain the following,

for two inputs, $\delta_1 > \delta_2$ *if and only if* $\rho_1/\eta_{1p} > \rho_2/\eta_{2p}$.

Intuitively, each η_{1p} is a measure of the social costs of distorting the use of input i, and each ρ_1 is a measure of the ability to target a group via a subsidy on input i. Therefore, it is in the consumer/taxpayer's best interest to distort input prices according to the heterogeneity of firm input use weighted by the inefficiencies in production caused by the subsidies.

6.3.3 Both Output and Input Subsidies

Typically, governments distort the prices in both output and input markets, but rely predominantly on a single carrier of a subsidy. We may gain some insight into the selection of the portfolio of subsidies by examining the case of a subsidy on output and a single input. Now the consumer/taxpayer acting as the government has a choice over the three distinct carriers of the subsidy: output price, input price, and a per-capita transfer. Suppose the government has already selected the candidate for input subsidy, in this case, X_1. This selection can be made for extraneous reasons, such as ease of administration, or for reasons of efficiency: the input is the least distorting. The criterion function for selection of the optimal levels of s, d, and r is formally represented as

$$\max_{s_1 d_1} W = \int_p^\infty D(z)dz - sS(p+s,\ w-d) - d_1 X_1 - N[\pi_0 - \pi(p+s,\ w-d)], \quad (27)$$

where d is a vector of zeros except for the first element, d_1. The first-order conditions for selection of the optimal subsidy rates σ and δ_1 are given by

$$\rho_0 - \sigma\varepsilon_p + \delta_1\varepsilon_1 = 0 \quad (28a)$$

and

$$\rho_1 - \sigma\eta_{1p} + \delta_1\eta_{11} = 0. \quad (28b)$$

From expressions (28a) and (28b) the subsidy rates are explicitly solved:

$$\sigma = \frac{-\eta_{11}\rho_0 + \varepsilon_1\rho_1}{-\eta_{11}\varepsilon_p + \varepsilon_1\eta_{1p}} \quad (29a)$$

$$\delta_1 = \frac{-\eta_{1p}\rho_0 + \varepsilon_p\rho_1}{-\eta_{11}\varepsilon_p + \varepsilon_1\eta_{1p}} \quad (29b)$$

where $(-\eta_{11}\varepsilon_p + \varepsilon_1\eta_{1p}) > 0$ for the second-order conditions for a maximum to hold.

> *With intervention in both an input and output market, a positive input subsidy will occur if and only if* $\rho_1/\eta_{1p} > \rho_0/\varepsilon_p$, *which is a condition similar to the case with two input subsidies without output subsidy. If we allow negative input subsidies, then the input will be taxed if* $\rho_1/\eta_{1p} < \rho_0/\varepsilon_p$. *A positive output subsidy will exist if and only if* $-\rho_0\eta_{11} > -\varepsilon_1\rho_1$ *– which will certainly hold if* $\rho_1/\eta_{1p} < \rho_0/\varepsilon_p$, *that is, if inputs are taxed.*

Again, as before in the case of two input subsidies, intuitively the input will be subsidized if the targeted producer is more easily distinguished by his input re-

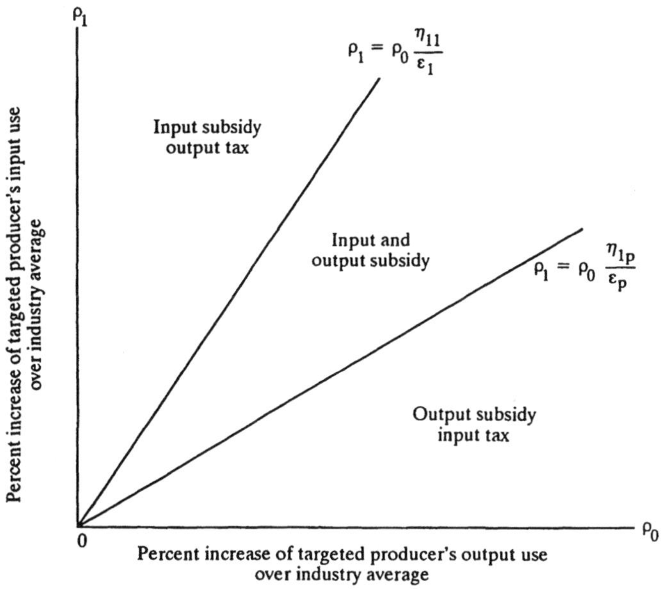

NOTE: η_{11} = Own-price demand elasticity for input 1.
η_{1p} = Demand elasticity for input 1 with respect to a change in output price.
ε_p = Supply elasticity with respect to output price.
ε_1 = Supply elasticity with respect to input price.

Fig. 6.1. Regions of optimal subsidy or tax on output and a single input as functions of the ability to target payments via coupled payments

sponse to the public good rather than his output response. If the input use by the targeted group relative to the industry norm is sufficiently greater than that for output, then the government will rely more on input subsidy, and less on output subsidy, perhaps even taxing output. Figure 6.1 illustrates the regions where input and output subsidies or taxes (if permitted) are optimally used as functions of the ability to target payments via coupled policies.

This discussion demonstrates that the use of seemingly contradictory policies of both distorting subsidies and taxes in agricultural sectors has a logic when government payments are made to target a certain set of producers. It has been noted that in many less-developed countries the government typically subsidizes input use and taxes output. The subsidy is more often than not attached to an input, such as fertilizer, used more heavily by innovators (International Bank for Reconstruction and Development 1986, pp. 94–103). If the output of the targeted group is small relative to the industry's average, then a tax is spread more evenly over producers, although the subsidy is concentrated. Generally, advances or policy changes that affect management or production practices of a targeted group, without significantly affecting the use of any specific input relative to the industry norm, are more likely to be accompanied by an output subsidy with zero input subsidies or an input tax. Countries relying more heavily on advances to management or on diffuse improvements in production will tend to have output subsidies. Countries where producers tend to be differentiated by management innovations would have more output subsidies relative to input subsidies. If, on the other hand, public research and development of new technologies or infrastructure are targeted to specific inputs, then subsidies will be tagged to those particular inputs, if it is likely that some producers will more readily innovate than others.

If a country's research and development (R&D) effort is broad, in the sense that any particular input is not expected to be used more by innovators than another input, then output subsidies are more likely to be observed. For example, in the United States a large R&D effort in agriculture may produce advances in a variety of input uses. It is difficult to predict where advances will occur, but that some advance will occur and output will increase are almost certain. Particular ρ_i/η_{ip} for inputs are likely to be small relative to the ρ_0/ε_p for the output, in an expectation sense, implying distorting output subsidies to assure the political support of the R&D effort. By contrast, in countries seeking to expand production by application of already developed technologies, where the particular input requirements of the advances are known, the use of input subsidies will be a more efficient means of targeting compensatory payments. If the use of the subsidized input by innovators relative to noninnovators is sufficiently great, then output would tend to be taxed, reducing net government outlays to the industry.

6.3.4 Complete Control Over Producer Prices

Suppose the consumer/taxpayer as the government can tax and subsidize producers at will, only constrained to keep the cth producer indifferent to some advance. The government's criterion function now becomes

$$\max_{sd} W = \int_p^\infty D(z)dz - sS - \sum d_i X_i - N[\pi_0 - \pi(p+s, w-d)]. \tag{30}$$

The first-order conditions for subsidies and taxes now take the form

$$S\rho_0 - s\frac{\partial S}{\partial p} - d_1\frac{\partial X_1}{\partial p} - d_2\frac{\partial X_2}{\partial p} - \ldots - d_K\frac{\partial X_K}{\partial p} = 0 \tag{31a}$$

$$X_i\rho_i + s\frac{\partial S}{\partial w_i} + d_1\frac{\partial X_1}{\partial w_i} + d_2\frac{\partial X_2}{\partial w_i} + \ldots + d_K\frac{\partial X_K}{\partial w_i} = 0, \ \textit{for all } i \tag{31b}$$

where again $\rho_0 = y_c/\bar{y} - 1$ and $\rho_i = x_{ci}/\bar{x}_i - 1$. Noting that $\hat{p}(\partial S/\partial p) + \sum \hat{w}_i(\partial S/\partial w_i) = 0$ and that $\hat{p}(\partial X_j/\partial p) + \sum \hat{w}_i(\partial X/\partial w_i) = 0$ for all j, we may sum the first-order conditions in (31) to find $\hat{p}S\rho_0 - \sum \rho_i \hat{w}_i X_i = 0$. This in turn implies

$$\pi(\hat{p},\hat{w}\backslash\alpha_c) = \bar{\pi}(\hat{p},\hat{w}). \tag{32}$$

That is, the government will set s and d such that the targeted firm earns the average profit. The lump-sum payment is then set to make up the difference between π_0 and $\bar{\pi}$, or to tax the producer: $r = \pi_0 - \bar{\pi}$. One should note that, with full control over all producer prices, at least one subsidy is redundant. In particular, the use of an output subsidy is like a proportional change in all input subsidies. Therefore, with full control over input prices, an output subsidy can be synthesized.

The case of *equiproportional* input subsidies can be extended to include the case of price effects of subsidies and the simultaneous choice between distorting and lump-sum payments. We may write the decision rules for input subsidies, without output subsidy, as

$$S\frac{\partial p}{\partial d_j}\left[\rho_0 - S^{-1}\sum d_i\frac{\partial X_i}{\partial p}\right] + X_j\rho_j + \sum d_i\frac{\partial X_j}{\partial w_i} = 0, \quad \textit{for all } i=1, K. \tag{33}$$

Note that with equiproportional input subsidies the following conditions hold.

$$\sum d_i\frac{\partial X_i}{\partial p} = -\sum\frac{d_i}{w_i - d_i}\frac{\partial S}{\partial w_i}(w_i - d_i)\delta\sum\frac{\partial S}{\partial w_i}(w_i - d_i) = \delta p\frac{\partial S}{\partial p} = \delta\,\varepsilon_p S. \tag{34}$$

And similarly, $\sum d_i(\partial X_j/\partial \omega_i) = -\delta\eta_{jp}X_j$. We may thus rewrite the first-order conditions in (33) as

$$\frac{S}{X_j}\frac{\partial p}{\partial d_j}(\rho_0 - \delta\varepsilon_0) + (\rho_j - \delta\eta_{jp}) = 0. \tag{35}$$

Recalling that $(S/X_j)\cdot(\partial p/\partial d_j) = -\eta_{jp}/(\varepsilon_p + \gamma)$ leads to the following proposition.

An equiproportional subsidy on inputs is consistent with maximizing consumer/taxpayer interest if and only if

$$\frac{\rho_j}{\eta_{jp}} = \delta\frac{\varepsilon_p}{\varepsilon_p + \gamma}\left(\frac{\rho_0}{\varepsilon_p} - \delta\right), \quad \textit{for all } j. \tag{36}$$

This proposition implies that the deviation of the targeted producer from average input use, relative to the elasticity of total utilization of the input with respect to output price, be a constant proportion across all inputs. The optimal rate of subsidy, δ, will be affected by the indirect influence of distortion on market price. The output price distortion, however, does not affect the conditions under which equiproportional subsidies are optimal. This is because the distorting affect on output price feeds back into the cost of the coupled scheme and producer profits, but it does not alter the marginal rate of technical substitution between inputs. Driving the market output price down, by increasing input subsidies proportionally decreases the profitability of any level of output, but the marginal rates of technical substitution remain unaffected by this feedback.

6.4 Concluding Remarks

There are several broad conclusions that may be drawn from the foregoing analysis. First, when encouraging policy reform of distorting policies, it is important to keep in mind that lump-sum, nondistorting transfer schemes are not necessarily in the best interests of those who *seem* to bear the brunt of costly redistribution policies, namely, consumers and taxpayers. Distorting payments allow a means of differentiating between firms that might otherwise be indistinguishable. Per-capita payments may be more efficient in the classic sense of maximizing social welfare when all other policies are held constant, but the actual objectives of governments may be much more complicated. In particular, when wealth transfers are used as remuneration in order to gain the acquiescence of a group of firms harmed by some other policy, then the optimal choice of subsidy involves some distortion.

Forcing nondistorting transfer schemes on a government may actually be counterproductive to the economic well-being of a country, because the use of nondistorting payments does not eliminate the desire for compensation in order to gain support for public goods that expand society's economic pie. Decoupling payments from production decisions may, in fact, raise the cost of assuring the political viability of largely beneficial advances that carry adverse distributional effects on some potent groups. Governments would then forego some of these beneficial policies that otherwise would be undertaken, and *total* social welfare would suffer.

Second, distorting coupled subsidies for the purpose of compensating subsets of firms are likely where producers are expected to be heterogeneous in their abilities to take advantage of a policy that harms the industry on average. Government sponsorship of technical advancement, whether in the form of R&D or adaptation of already developed practices, is a particularly notable example of a policy that promotes society's economic welfare but may harm particular groups of producers. Some firms are better able to make use of technical advances because of their geographic location, size, managerial abilities, and so forth. In order to judge the consumer/taxpayer's commitment to distorting subsidies – and thus the outsider's ability to encourage their elimination – one should examine the characteristics of the subsidized industry. If it is monolithic, or centralized, or highly concentrated, then the use of distorting payments is unlikely to make sense from the consumer/taxpayer's standpoint. If the industry is dispersed, heterogeneous, and decentralized, then distorting payments may actually have greater political support outside the industry itself.

Third, there are tradeoffs between distinguishing targeted producers via distorting payments and the social costs of distortion. There are three aspects to this point: (1) If the targeted producer uses one input relative to the industry average, more than another input, then that input will be more heavily subsidized relative to others. (2) If total industry demand is responsive to input price decreases, then that input will tend to receive less subsidy. (3) Even if an input demand is relatively less own-price responsive, but other input demands are responsive to a change in that price, then the input in question will tend to not have a subsidy.

Fourth, output subsidies will be preferred over input subsidies when input-specific advances are not anticipated. Input subsidies are likely when a government has complete information of the particulars of future advances. Developing countries that anticipate changes in their agricultural sectors due to the adaptation of technical advances developed elsewhere are likely to tailor compensating subsidies specific to those advances. This implies a greater reliance on input subsidies, perhaps with output taxes. As a country adopts the preexisting technical advances suitable to it, and (at least, in terms of agricultural production) catches up with the rest of the world, the government will probably begin anticipating unknown, as-yet-to-be-seen advances developed elsewhere and domestically. Therefore, as a country's agricultural sector advances, one is likely to see a change from input subsidization (perhaps with an output tax) to output subsidization (perhaps with input taxes or restrictions).

Fifth, and finally, reductions in output subsidies due to, say, foreign interventions, will increase input subsidies or lower taxes; the reduction in total market distortion will be less than the reduction in a particular market. Therefore, an outsider judging its success at encouraging policy reform in a country must not take the superficial reduction of output-price distortion as an accurate measure.

References

Becker GS (1983) A theory of competition among pressure groups for political influence. Q J Econ 98:371–400

Chambers RG (1985) Least-cost subsidization alternatives. Am J Agric Econ 67:251–256

Cochrane WW (1979) The development of american agriculture: a historical analysis. The University of Minnesota Press, Minneapolis

Corson W, Nicholson W (1981) Trade adjustment assistance for workers: results of a survey of recipients under the Trade Act of 1974. In: Research in labor economics, vol 4. JAI Press, Greenwich, CT, pp 417–469

Dinar A, Zilberman D (1991) The economics and management of water and drainage in agriculture. Kluwer, Boston

Foster WE, Calvin LS, Johns GM, Rottschaefer P (1986) Distributional welfare implications of an irrigation water subsidy. Am J Agric Econ 68:778–786

Gardner BL (1987) The causes of US farm commodity programs. J Politic Econ 95; No 2:290–310

International Bank for Reconstruction and Development (1986) World Bank development report, 1986. Oxford University Press for the World Bank

Lueck TJ (1989) The bulk–for–benefits deal in zoning. New York Times, July 23, sec. 10, p 1

Kruse DL (1988) International trade and the labor market experience of displaced workers. Industrial and Labor Relations Rev 41:402–417

Magee SP, Brock WA, Young L (1989) Black hole tariffs and endogenous policy theory. Cambridge University Press, Cambridge, MA

Mirrlees JA (1976) Optimal tax theory: a synthesis. J Pub Econ 6:327–358

Mitchell DJB (1989) Human resource management: an economic approach. PWS-Kent Publishing Co, Boston

Mueller DC (1989) Public choice II. Cambridge University Press, New York

Peltzman S (1976) Toward a more general theory of regulation. J Law Econ 19:211–240

Rausser GC (1982) Political economic markets: PERTs and PESTs in food and agriculture. Am J Agric Econ 64:821–833

Rausser GC (1992) Predatory versus productive governments: the case of US agricultural policy. J Econ Perspect, 6(3):133–157

Richardson JD (1982) Trade–adjustment assistance under the US Trade Act of 1974. In: Bhagwati JN (ed) Import competition and adjustment: theory and policy. University of Chicago Press, Chicago

Stigler GJ (1971) The theory of economic regulation. Bell J Econ 2:3–21

Tirole J (1990) Privatization and Eastern Europe: incentives and the economics of transition. Unpublished mimeo, Department of Economics, Massachusetts Institute of Technology, Cambridge

Chapter 7

Mobility, Diversification, and Sustainability of Trade Reform

William E. Foster, Richard Gray, and Gordon C. Rausser

7.1 Introduction

The discussion of the benefits of free trade is almost as old as the economics profession itself. Countries that have achieved free trade through policy reform almost inevitably succumb to protectionist policies after a period of free trade. Despite numerous successful initiatives for trade reform, interest groups seek and often eventually achieve protection. The trend toward increased protectionism continues until the economy-wide need for reform is sufficient to motivate a successful initiative for reform. The Uruguay round of negotiations of the General Agreement on Trade and Tariffs (GATT) represents a significant initiative in potential multilateral economic reform. The fact that it is the fourth round of negotiations for GATT is an indication of the difficulty in achieving reform which is sustainable.

The development of policies that lead to sustainable trade reform requires an examination of both the process of trade reform and the process of increased protectionism. In this chapter, we present three simple theses that relate trade policy to the diversification of ownership and the mobility of resources within an economy. We show that compensation requirements of trade reform are, at least in part, a function of the degree of the ownership diversification and the degree of resource mobility within the economy. Secondly, we show that the enhanced mobility of resources and the diversification of the ownership of resources can reduce the incentive for resource owners to seek protectionist policies. Finally, we argue that trade reform initiatives that compensate losers in such a way as to enhance the diversification of ownership and the mobility of resources within the economy are more likely to be sustainable.

In the current round of discussions for GATT, a framework has been developed that ranks policies. Policies which are regarded as distortionary and could be targeted for phase-out have been named *amber light policies*; those that are least distortionary and therefore acceptable from a international prospective, *green light policies*. This policy ranking scheme, at least in theory, is based on the degree of trade distortion each policy creates.

We suggest that, in addition to the degree of trade distortion, the political sustainability of the reform policy should influence the ranking. A reform policy that is likely to be politically sustainable should be ranked more desirable than an equally distortionary policy which is not sustainable. In this chapter we demonstrate policies which promote diversification of ownership. Moreover, mobility of the factors of production can reduce the incentive of agents to seek protectionist trade policies. Therefore, we argue that policies that promote the mobility of factors and the diversification of ownership of immobile factors should be ranked above those that do not.

Most economists would argue the role of government should be to correct market failures and to redistribute income in the most efficient way possible. Despite the espoused benefits of this limited role of government intervention, government policies often introduce significant distortions in the economy and redistribute wealth from the poor to the rich. This undesirable aspect of government policy has been described as government failure. Economists have developed models of rent seeking that have been used successfully to explain government failure (Tullock 1967; Bhagwati 1971; Krueger 1974; Becker 1983; Mayer 1984; among others). Within these models it is in the interest of groups to seek and promote government policies which increase their wealth at the expense of other's wealth and the size of the economy as a whole. Given that government failure decreases the size of the economy, if anyone is to gain from the policy, the rents from the policy must go to a limited number of individuals in the society. All rent-seeking models share this *limited access* characteristic. In most models, limited access to policy-generated rents is usually derived from the ownership of some fixed resource but could also come from a particular set of preferences that differentiates one group from another in society.

These rent-seeking models have been successful in explaining government policy; however, in general they lack any prescriptive implications. In these models, behavior reflects a political-economic equilibrium that, in turn, is a function of the constitution of the country and the resources within its borders (Rausser and Zusman 1992). The variety of distortionary policies across countries lends a great deal of support to this notion. Extending this argument to an extreme, the constitution itself could be viewed as an outcome of the political process. Here, we suggest that the constitution and the policies that are pursued are a function of the mobility of the resources within the economy and, in particular, the distribution of ownership of those resources. Policies that affect the mobility of the resources and the distribution of ownership of those resources can permanently alter the political-economic equilibrium path of development in the economy.

For very similar reasons, resource ownership and mobility also play a major role in the process of trade reform. For a small country, trade reform represents a potential Pareto improvement. Protectionist price-distorting policies are widely criticized from a standpoint of economic efficiency; and nondistorting policies are widely supported. Despite the espoused benefits of trade reform, trade protection exists in most, if not all, countries of the world. Not all citizens of a country moving toward free trade will benefit from the move. The extent to which individuals within an economy undergoing trade reform would have to be compensated for the policy change to be Pareto improving is a function of preferences, resource ownership of each individual, and the mobility of those resources. We term this level of individual compensation the Pareto-fulfilling compensation requirement.

A measure of the Pareto-fulfilling compensation requirement is relevant for several reasons. From a normative point of view, the Pareto-fulfilling levels would indicate whether or not a proposed change unambiguously increases the welfare of the individual citizen; and, if the change creates losers, the levels would indicate whether or not the distributional consequences of the change are practically important. From a positive political-economic perspective, an understanding of Pareto-fulfilling compensation would form the basis for assessing the likelihood of reform taking place (Rausser and Foster 1990). Given that losers from a policy change can delay, restrict, or even block the change, the Pareto-fulfilling compensation represents a real barrier for the trade reform process. There are several examples where governments have made payments to individuals or groups in compensation for moves toward free trade. For instance, the Canadian government made payments to Canadian wine producers in compensation for the removal of trade barriers brought about by the 1989 Canada-US Free Trade Agreement (FTA). This compensation scheme promised wine producers the full value of their assets employed in wine production in the event the FTA was ratified. Other groups in Canada were less successful in receiving direct compensation for the FTA but did receive concessions in terms of lengthy phase-in periods for trade reform. Another example of compensation to gain acquiescence to trade reform is the case of Australian milk protection quotas, which artificially inflated domestic prices through production controls. Rather than simply do away with the protection, the government purchased the production quotas from producers and then eliminated the program. A measurement of the magnitude of the Pareto-fulfilling compensation may provide the upper bound of the budgetary expenditure required to have a contemplated economic reform widely accepted within an economy.

The remainder of this chapter is organized as follows: In Section 7.2 we examine the effects of diversification and resource mobility on the Pareto-fulfilling compensation requirement of trade reform. A simple general-equilibrium framework is presented for examining the interaction of diversification and resource mobility on the Pareto-fulfilling compensation associated with trade liberalization. Using this framework, it is possible to identify a priori the conditions under which the potential compensation requirements of trade reform are likely to be large. In Section 7.3, we introduce a simple rent-seeking model to explain the creation of protectionist policies within a similar framework. In particular, we show that the degree of resource mobility and ownership diversification can affect the incentives for interest groups to seek protectionist policies. In Section 7.4, we address the issue of sustainable reform through a corollary to the arguments presented in Sections 7.2 and 7.3. We show that trade reform initiatives which are combined with polices that promote ownership diversification and factor mobility within the economy are less subject to rent seeking and are, therefore, more sustainable. Section 7.5 summarizes the chapter.

7.2 The Welfare Effects of Trade Reform

The study of the welfare effects of trade reform has long been a part of international trade theory. Some studies focus on the effects at the industry level (Meade 1955; Corden 1969; Bhagwati 1971), while others concentrate on the implications of reform on the distribution of return to resources employed in production (Stopler and Samuelson 1941; Heckscher 1949; Jones 1970). The effects of resource mobility have been studied extensively and have become central to neoclassical trade theory. In more recent literature, the effects of resource mobility have been studied in a many-goods economy (Jones 1975).

Another important factor from a policy perspective is the diversification of resource ownership in the economy. While resource mobility may be determined by technical relationships, both diversification and mobility of factors in the economy may be a function of the regulation and subject to political influence. Government promotion of diversification of resource ownership, and the existence of markets for contingent claims may be vehicles for increasing the propensity for trade reform. In most economies, the degree of mobility is also restricted through institutional and regulatory barriers. For example, state or provincial laws may reduce the movement of factors within a country. One of the most significant economic effects in Canada of the Canada-US Free Trade Agreement will be the dismantling of interprovincial trade barriers.[1]

Generally, the motive for diversification is to reduce risk or the variability of returns. By owning a portfolio of risky assets an individual can reduce and effectively manage risk. Yet we observe many resources or sector-specific assets which are not widely held in the economy. This lack of diversification may be explained by two economic forces. First, the underlying risk may have been virtually eliminated with the existence of insurance, financial instruments, and government policy which stabilize income. Thus, the existence of government compensation schemes may impede the *natural* level of diversification. Second, high transaction costs associated with diverse ownership may make diversification prohibitively expensive. These high transaction costs may be the result of a market failure and also are often the result of some underlying principal-agent problems which make efficient contracts prohibitively expensive to enforce. The classic example of such a case is the ownership and operation of farms. The vast majority of the world's grain production continues to be grown by largely undiversified owner-operator enterprises. This industry and similar industries with similarly high transaction costs may resist attempts to promote the diversification of ownership in these industries.

---

[1] For example, state or provincial laws may reduce the movement of factors within a country. One of the most significant economic effects in Canada of the Canada-United States Free Trade Agreement will be the dismantling of inter-provincial trade barriers.

7.2.1 A Small-Country Trade Model

In order to examine the Pareto-fulfilling compensation levels under trade reform, we consider the case of a small country. For a small country, free trade would be optimal based on the criterion of maximizing total social surplus but individual groups gain and lose disproportionately with reform. The instrument of trade distortion examined is a quota on imports which controls the domestic price level relative to the given world price level. The rents generated by the quota are allocated to each factor in the import industry in proportion to the factor shares. In effect, the quota is equivalent to a tariff on imports and exports where the tariff revenue is returned to the effected sectors. This specification of income is consistent with the existence of an import monopoly in which the profits are distributed to the owners of resources employed in the production of the import good.

The import monopoly could be a state organization, a state-sanctioned organization, or a private monopoly. In the case of a state organization, the quotas are set with the government objectives in mind. In the case of a private monopoly, the quota level will be set to maximize the sum of the return from production and the import quota rents.[2] With the former, trade reform would entail the removal of the import quota; in the latter, it may require a change in the regulation of the monopoly. In either case, the move to free trade represents a potential Pareto improvement in the economy and the losers of the reform process may have to be compensated for the reform. In a two-sector, open economy with a balance-of-payments constraint it is irrelevant whether the trade policy is imposed on exports or imports.

As a further simplification, we have presumed that the economy is competitive with no other distortions. Without this assumption, the effect of removing one distortion may not be even potentially Pareto improving. This difficulty in determining the effects of piecemeal policy reform was pointed out in trade theory by Meade (1955) and Viner (1950).[3]

Conditions under which trade reform without compensation is Pareto improving and conditions under which compensation to some individuals is required to meet the Pareto criterion are presented in this section. These conditions are derived using a simple two-sector trade model that emphasizes diversification of resource ownership and thus provides a different focus than previous models (Mussa 1982; Mayer 1984; Eaton and Grossman 1985; Staiger and Tabellini 1987).

The economy of N individuals is represented in a two-sector (goods A and B) open-economy model. The country produces A_x and B_x, consumes A_c and B_c, imports $A_c - A_x$, and exports $B_x - B_c$. Imports and exports are carried on with rest of the world at fixed world prices. Prices are determined in competitive markets, implying that the domestic and world prices of good B will be equal in equilibrium,

---

[2] We are ignoring the additional welfare losses due to rent-seeking activities as described by Tullock (1967); Krueger (1974); and Hillman and Riley (1989).

[3] These arguments were formalized in a General Theory of Second Best by Lipsey and Lancaster (1957).

regardless of quota (or tariff) on good A. The wedge between the domestic price of A and its world price will reflect the restrictions on trade. There exists a quota, Q, on the imports of good A; that is, $Q \geq A_c - A_x$. Prices are normalized such that the world and domestic price of good B is equal to one; and P_w and P represent the world and domestic price of good A.

The production of A_x and B_x takes place in competitive sectors of the economy with identical and homogeneous degree-one production functions. Production takes place in both sectors using positive levels of two inputs: a mobile resource, L, the returns to which equilibrate across sectors; and an immobile resource, K. For example, the two types of resources may be thought of simply as labor and capital, or as capital (perhaps human capital) not specific to an industry and capital specific to each industry. The inputs devoted to the import industry, L_A and K_A, and those devoted to the export industry, L_B and K_B, are constrained by the total resource available to the economy: $L_A + L_B \leq L$ and $K_A + K_B \leq K$. To be explicit, the production relationships are given by

$$A_x = f(L_A, K_A) \text{ and } B_x = f(L_B, K_B). \tag{1}$$

In equilibrium, decentralized income is a function of the general level of prices (which are functions of the level of quota employed) and the share of each resource in each sector. The gross income in sectors A and B, respectively, are:

$$Y_A = PA_c - P_w(A_c - A_x) \tag{2a}$$

or

$$Y_A = PA_x + (P - P_w)(A_c - A_x) \tag{2b}$$

and

$$Y_B = B_x. \tag{2c}$$

Note that, in this formulation, the quota value is attributed to income in each sector. Given a binding quota, national income is defined by

$$Y = PA_x + (P - P_w)Q + B. \tag{3}$$

The per-unit wages earned by the mobile resource, w, is common across industries; and firm managers use the input until its marginal product equals the wage: $P\partial A_x/\partial L_A = w = \partial B/\partial L_B$.

Using the Euler equations, factor shares of L_A, K_A, L_B, and K_B, are:

$$\alpha_L = \frac{wL_A}{PA_x}, \quad \alpha_K = \frac{r_A K_A}{PA_x}, \tag{4a}$$

$$\beta_L = \frac{wL_B}{Y_B}, \quad \beta_K = \frac{r_B K_B}{Y_B}. \tag{4b}$$

The quota value is attributed to income in the import sector and divided among the inputs according to factor shares. This is equivalent to an import tariff or an export tariff where the tariff is returned to the resource owners in each sector in proportion to their income.

In this economy, the consumption and the utility of individuals is determined by their preferences, the prices they face, and their incomes. In a general-equilibrium setting, their incomes will be determined by the resources each individual owns and the aggregate output and trade of the general economy. The output of the economy will, in turn, be a function of prices, which is a function of the level of import quota. Using these linkages, it is possible to determine the effect trade policy has on individual utility. The primary tool of analysis will be the indirect utility function of individuals. Let V_j represent the indirect utility function of individual, j:

$$V_j = V_j\,[P\,(Q),\, Y_j(Q)]. \tag{5}$$

The effect of an incremental change in the quota level on V_j is found by using the chain rule:

$$\frac{dV_j}{dQ} = \left[-A_{cj}\frac{\partial P}{\partial Q} + \frac{\partial Y_j}{\partial Q}\right]\cdot\frac{\partial V_j}{\partial Y_j} \tag{6}$$

where the individual demand for the import good, $A_{cj} = -\,(\partial V_j/\partial P)/(\partial V_j/\partial Y_j)$.

The standard gains-to-trade argument for a competitive economy holds – that total income is maximized where $P = P_w$, that is, when the quota is nonbinding. If a quota is in place and removal of the quota constraint is Pareto improving (i.e., all individuals are at least as well off), then reform will certainly take place. The Pareto-fulfilling compensation requirement can be derived in the following manner. Suppose there is a change in the quota level from Q_0 to Q_1 resulting in a change in the price-income pair from (P_0, Y_{j0}) to the pair (P_1, Y_{j1}). The level of compensation necessary to make the jth individual just a well off as without the policy change is the familiar compensating variation measure, CV_j:

$$V_j(P_1, Y_{j1} + CV_j) = V_j\,(P_1, Y_0). \tag{7}$$

For incremental changes in Q, the change in the level of CV_j – represented by C_j – is given by:

$$C_j = -A_{cj}\frac{\partial P}{\partial Q} - \frac{\partial Y_j}{\partial Q}. \tag{8}$$

7.2.2 Preferences, Resource Ownership and Mobility, and Pareto-Fulfilling Compensation

The above framework will be used to derive propositions which describe the relationship between preferences, resource ownership, and resource mobility on the Pareto-fulfilling compensation associated with reform. First, consider the usual case that demonstrates the potential gains from free trade. In a two-sector economy, individuals with identical preferences and equal resource ownership are sufficient to show that moves to free trade are Pareto improving. As a result, the potential compensation requirement is zero for all individuals.

The condition of equal resource ownership, however, is not necessary. Take a representative individual with consumption levels of A and B equal to some fraction of total consumption: $A_{cj} = \gamma_{Aj}A_c$ and $B_{cj} = \gamma_{Bj}B_c$. The individual consumer's assets are diversified in the sense that the consumer has a claim to a share of national income that is invariant to changes in the quota or prices. That is, the individual's income is some fraction of national income:

$$Y_j = \omega_j\,[PA_x + (P - P_w)\,Q + B_x]. \tag{9}$$

For homothetic preferences, identical across consumers, $\omega_j = \gamma_{Aj} = \gamma_{Bj}$. The effect of relaxing the quota on an individual's welfare can be examined using the effect on the individual j's indirect utility, V_j. From expression (9), the effect of a change in quota on national income is given by:

$$\frac{dY}{dQ} = A_x\frac{\partial P}{\partial Q} + P\frac{\partial A_x}{\partial Q} + Q\frac{\partial P}{\partial Q} + \frac{\partial B_x}{\partial Q} + (P - P_w). \tag{10}$$

Given that a change in the quota only affects use of the mobile resource in each sector, and that the total amount of the resource is unaffected by the change, we note that

$$P\frac{\partial A_x}{\partial Q} + \frac{\partial B_x}{\partial Q} = P\frac{\partial A_x}{\partial L_A}\frac{\partial L_A}{\partial Q} + \frac{\partial B_x}{\partial L_B}\frac{\partial L_B}{\partial Q} = w\frac{\partial L_A}{\partial Q} - w\frac{\partial L_A}{\partial Q} = 0. \tag{11}$$

Therefore, noting that $A_c = Q + A_X$, the change in the jth consumer's income is given by

$$\frac{dY_j}{dQ} = \omega_j\left[A_c\frac{\partial P}{\partial Q} + (P - P_w)\right]. \tag{12}$$

Using (6) and (12), the effect on the jth consumer's welfare is given by:

$$\frac{dV_j}{dQ} = \left[(1 - \gamma_{Aj}/\omega_j)A_c\frac{\partial P}{\partial Q} + (P - P_w)\right]\cdot\omega_j\frac{\partial V_j}{\partial Y_j}. \tag{13}$$

With identical homothetic preferences, the first item in the square brackets vanishes and $dV_j/dQ > 0$ when $P > P_w$; or utility increases with increases in quota until $dV_j/dQ = 0$ when $P = P_w$, that is, when the quota is nonbinding. Stated more formally,

Under identical, homothetic preferences, complete diversification in ownership of all resources assures the Pareto optimality of trade liberalization.

Note that complete diversification is implied by the invariance of w_j with respect to quota and prices; it does not imply that all individuals are identically endowed.

Identical endowments are those such that $\omega_j = 1/N$ for all resource owners. Diversification is not sufficient for relaxed quotas to be Pareto improving when preferences are not homothetic. When a consumer has a lower share of consumption of the imported good than of the national income, then the Pareto-fulfilling compensation level may be positive. In fact, as can be seen in Eq. (13), $(P - P_w)$ approaches zero as the import quota, Q, approaches the free trade level, and the compensation level becomes positive for anyone who spends less than an average share of his income on the import goods [i.e., $(1 - \gamma_A/\omega_j) > 0$]. Although national income will increase, the real income of these citizens will decrease.

In order to illustrate the combined effects of resource mobility and diversification, further structure must be added to the model. We begin with mobility assumptions employed by Jones (1970). The input, L, is perfectly mobile between sectors while the inputs, K_A and K_B, are not. The assumption of perfect mobility of one input is not unduly restrictive. If one wished to introduce imperfect mobility of a factor, such as has been often done with labor (Mussa 1982), the input, L, could represent fully transferrable skills and the immobile component could include human capital as well as other immobile assets. Diversification is introduced into the model by assuming that each individual in society holds a portfolio made of the mobile input and the sector specific inputs in each of the two sectors. The income of each individual is a function of the assets he holds and the returns accruing to each of those assets.

We assume the factor share of labor in the two sectors are identical, i.e., $\alpha_L = \beta_L = a$.[4] Without loss of generality, we make several notational simplifications and normalizations to clarify the relationship between diversification and resource mobility: The units of measure are chosen such that the total endowment to society is one unit of the mobile resource ($L = 1$), one unit of the immobile resource in the import-competing industry ($K_A = 1$), and one unit of the immobile resource in the export-oriented industry ($K_B = 1$).

Define λ_j as the jth individual's share of the labor resource: $\lambda_j = L_j/L$. From the Euler Eqs. (1) and (2), the jth individual's income from the mobile resource is given by $\lambda_j \alpha Y$. The income from the immobile resource is given by

$$(1 - \alpha) \cdot \left(\frac{K_{Aj}}{K_A} Y_A + \frac{K_{Bj}}{H_B} Y_B \right) = \theta_j 2(1 - \alpha)[\mu_j Y_A + (1 - \mu_j) Y_B], \tag{14}$$

---

[4] This effectively removes the Stopler-Samuelson effects regarding the differences in relative intensity of factor use between sectors. Accounting for these effects would tend to increase the levels of compensation for owners of resources used relatively more intensively in the import-competing sector. Any propositions regarding degrees of diversification would have to weight a person's portfolio of immobile resource by differences in the degree of relative factor intensities in the two sectors. Specifically, if the immobile factor is used relatively more intensely in the import-competing sector, then a perfectly balanced portfolio of immobile resources is not sufficient to guarantee a person's benefit from trade liberalization. Instead, the person's portfolio of immobile resources would have to be skewed toward ownership in the export-oriented sector, the degree of skewness depending upon the extent of relative factor intensity.

where $\mu_j = K_{Aj}/(K_{Aj} + K_{Bj})$ is the share of the jth individual's immobile resources in the A industry, and $\theta_j = (K_{Aj} + K_{Bj})/(K_A + K_B)$ is the jth individual's share of the total amount of the immobile resource. The jth individual's total income is given by

$$Y_j = \lambda_j \alpha Y + \theta_j 2(1-\alpha)[\mu_j Y_A + (1-\mu_j)\, Y_B]. \tag{15}$$

Complete diversification of the individual's ownership of immobile resources across industries A and B occurs where $\mu_j = 1/2$.

With complete diversification in the ownership of immobile resources, the jth income as a proportion of the total national income is invariant at the level of the quota.

To demonstrate this result, rewrite expression (15) in terms of the total incomes of the two industries, Y_A and Y_B,

$$Y_j = [\lambda_j \alpha + \theta_j 2(1-\alpha)] \cdot Y_A + [\lambda_j \alpha + \theta_j 2(1-\alpha)]\,(1-\mu_j) \cdot Y_B. \tag{16}$$

Setting $\mu_j = 1/2$ with complete diversification of ownership of the immobile resource between import-competing and export-oriented industries, the individual's income is proportional to the national income; and the proportion will not vary with the quota level,

$$Y_j = [\lambda_j \alpha + \theta_j 2(1-\alpha)] \cdot Y. \tag{17}$$

More generally, where $\mu_j \neq 1/2$, the change in income with respect to a change in the quota is equal to

$$\begin{aligned} \frac{dY_j}{dQ} &= [\lambda_j \alpha + \theta_j 2(1-\alpha)\mu_j] \cdot \left[A_c \frac{\partial P}{\partial Q} + P \frac{\partial A_x}{\partial Q} + (P - P_w)\right] \\ &\quad + [\lambda_j \alpha + \theta_j 2(1-\alpha)(1-\mu_j)] \cdot \frac{\partial B_x}{\partial Q}. \end{aligned} \tag{18}$$

Define the jth individual's shares of total income from the two industries, $\omega^A{}_j = [\lambda_j \alpha + \theta_j 2(1-\alpha)\mu_j]$ and $\omega^B{}_j = [\lambda_j \alpha + \theta_j 2(1-\alpha)(1-\mu_j)]$. The effect on the individual's utility due to an incremental increase in the quota without compensation is given by

$$\frac{dV_j}{dQ} = \left[(P - P_w) + \left(1 - \frac{Y_{Aj}}{\omega_j^A}\right) A_c \frac{\partial P}{\partial Q} + P \frac{\partial A_x}{\partial Q} + \frac{\omega_j^B}{\omega_j^A} \frac{\partial B_x}{\partial Q}\right] \frac{\partial V_j}{\partial Y_j} \cdot \omega_j^A. \tag{19}$$

The three terms in Eq. (19) represent three channels by which individual welfare is altered by a change in the quota level. First, there is an increase in revenue due to importing a unit for less than its domestic sale price. This is represented by the term $(P - P_w)$ and this gain would eventually vanish as the quota becomes non-binding. Second, there is a change in real purchasing power due to change in the import price that benefits the individual as consumer differently than it harms the individual as resource owner. This purchasing-power effect, represented by the first term in the square brackets, may be either positive or negative:

$$\left(1 - \frac{\gamma_{Aj}}{\omega_j^A}\right) A_c \frac{\partial P}{\partial Q} \gtreqless 0 \quad \text{as} \quad \gamma_{Aj} \gtreqless \omega_j^A . \tag{20}$$

This purchasing-power effect is positive as the individual consumes a greater share of the import goods and earns a lesser share of the revenues produced by the import industry. Although the individual has fewer dollars from the import industry, the purchasing power of those dollars has increased.

The third way in which individual welfare is affected by the change in quota is through the effect on income produced from changes in the domestic production of the two goods, apart from changes in relative prices. Income from the domestic production of the good A decreases as the quota increases and allows more imports to substitute for domestically produced goods. On the other hand, income from the export-oriented production increases. If the individual is completely diversified, $\omega^A{}_j = \omega^B{}_j$ and the loss of income from domestic production of the import-competing goods balance with the gain of income from production in the export-oriented industry. As the individual's assets are more concentrated in the import-competing industry, the less his gain from trade liberalization, or the greater his loss:

$$P \frac{\partial A_x}{\partial Q} + \frac{\omega_j^B}{\omega_j^A} \frac{\partial B_x}{\partial Q} \gtreqless 0 \quad \text{as} \quad \omega_j^B \gtreqless \omega_j^A . \tag{21}$$

To illustrate the relationship between the diversification of an individual's portfolio and the degree of compensation necessary to make that individual indifferent to a relaxation of the quota, consider the compensation required to balance an incremental change in Q, as defined in Eq. (8). To focus on the effects of resource ownership, as opposed to questions of the individual's consumption bundle, assume that one's share of total consumption of the imported goods is equal to one's share of the total revenue generated by the import-competing industry, $\gamma_{Aj} = \omega^A{}_j$. In general, the degree of compensation to the jth individual for an incremental increase in quota is given by utilizing Eqs. (8) and (19):

$$C_j = \left[-(P - P_w) - \left(P \frac{\partial A_x}{\partial Q} + \frac{\omega_j^B}{\omega_j^A} \frac{\partial B_x}{\partial Q} \right) \right] \cdot \omega_j^A . \tag{22}$$

Therefore, with complete diversification, a move toward free trade benefits the jth individual:

> *For $\omega^A{}_j = \omega^B{}_j$ or $\mu_j = 1/2$, the value of the compensation requirement, C_j, is negative, $C_j = -\omega^A{}_j (P - P_w)$ for any combination of mobile and immobile resources.*

As the individual's portfolio of immobile resources relies more heavily on the import-competing industry, the degree of diversification between mobile and immobile resources becomes important. Measuring the concentration of an individual's assets in the mobile resource by $\tilde{\lambda}_j = \lambda_j/(\lambda_j + \theta_j)$ allows the representation of the ratio, ω^B_j/ω^A_j, in terms of mobility, $\tilde{\lambda}_j$, and diversification, μ_j, namely,

$$\frac{\omega_j^B}{\omega_j^A} = \frac{\tilde{\lambda}_j\alpha + \left[(1 - \tilde{\lambda}_j)2(1 - \alpha)\right](1 - \mu_j)}{\tilde{\lambda}_j\alpha + \left[(1 - \tilde{\lambda}_j)2(1 - \alpha)\right]\mu_j}. \tag{23}$$

For an incremental increase from a given level of quota, Q, one can therefore write the level of Pareto-fulfilling compensation as a function of the concentration of assets in the mobile resource, $\tilde{\lambda}_j$, and the concentration of the immobile resources in the import-competing industry, μ_j:

The greater the individual's concentration of immobile resource ownership in the import-competing industry, the greater the Pareto-fulfilling compensation requirements for a relaxation of quota; and furthermore, for any given mix of immobile resource ownership the more mobile assets an individual owns, the smaller the Pareto-fulfilling compensation requirements.

To demonstrate these properties, note that (22) is affected by $\tilde{\lambda}_j$ and μ_j only through the ratio, ω^B_j/ω^A_j. Finally, note that the signs of the derivatives of C_j with respect to $\tilde{\lambda}_j$ and μ_j are opposite to the signs of the derivatives of (23). The relationship between compensation and diversification is graphically illustrated in Fig.7.1. Levels of compensation over values of μ_j in the domain [0.5, 1] are shown with given level of the mobility measure, $\tilde{\lambda}_j$.

Note that the individual may be unevenly diversified between industries (i.e., $\mu_j > 1/2$) and still require no positive level of compensation to remain at least indifferent to freer trade. Indeed, there are levels of concentration of one's portfolio

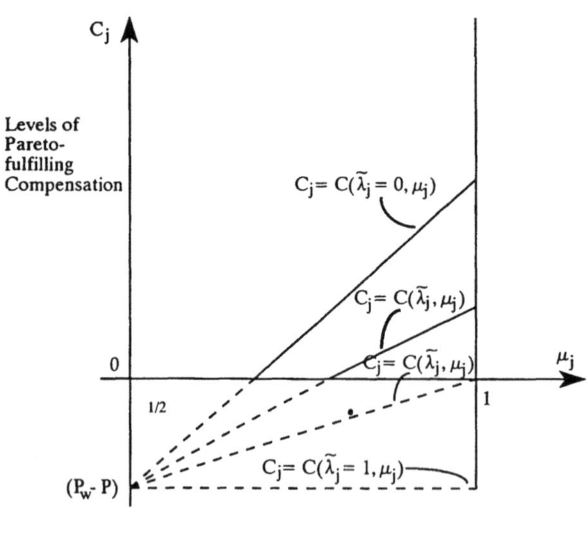

Fig. 7.1. Levels of Pareto-fulfilling compensation as a function of the proportion of ownership of fixed resources in the import competing sector, μ_j, conditioned on the degree of mobility, $\tilde{\lambda}_j$, and an initial quota, Q

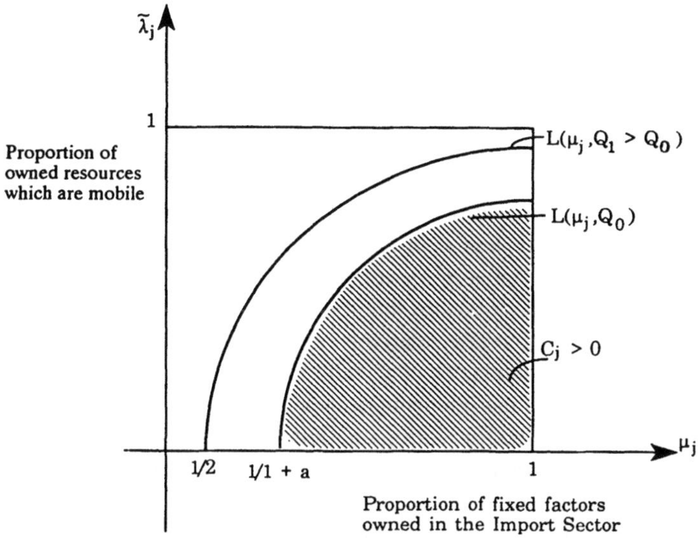

Fig. 7.2. The tradeoff between resource mobility, $\tilde{\lambda}_j$, and the proportion of fixed factors owned in the import sector, μ_j, in determining the region where Parety-fulfilling compensation is positive ($C_j > 0$) for an incremental increase in import quota, Q. Note: $a = 1 - \{[P(Q) - P_w] / (\partial B_x / \partial Q)\}$

in mobile assets such that the degree of diversification of immobile assets is irrelevant for positively compensating an individual for an incremental increase in quota. However, note as Q approaches the level where it is nonbinding (i.e. $P \to P_w$) the region vanishes where no compensation is required to make those individuals with an unbalanced mix of immobile resourcess indifferent. Stated more concisely,

> *As the effective quota approaches the nonbinding level, any resource ownership skewed toward the import-competing industry relative to the export-oriented industry will have a positive Pareto-fulfilling compensation.*

This rather strong result comes from the fact that the gains from trade are second-order small and the income effects or distributional effects are first-order.

Those individuals whose portfolio of assets is characterized by $\tilde{\lambda}_j$ and μ_j such that $0 < \tilde{\lambda}_j \leq 1$ and $1/2 < \mu_j < 1$ are those for whom there exists a potential for required compensation for an incremental increase in the level of some level of Q. Of course, for an effective quota (i.e., for a finite difference between domestic and world prices of good A), some individuals with certain pairs of $\tilde{\lambda}_j$ and μ_j will be indifferent to, or gain from, the relaxation of the quota. The shaded area in Fig. 7.2 represents those combinations of mobility and diversification that have a Pareto-fulfilling compensation for an incremental change in the quota. Thus, this shaded area represents a measure of the likelihood that a randomly selected individual will oppose freer trade without compensation.[5]

Suppose a country moves in incremental steps toward free trade, and compensates those who suffer losses at each stage. For each successive step, the number of individuals requiring compensation grows. This may at first appear paradoxical. As a country moves closer to free trade, the extent of individuals harmed increases. The intuitive explanation of this result may be expressed as follows. As the quota becomes proportionately less restrictive, gains to national income become less for further relaxations. Gains in income from all sources decrease to an individual, but at the same time the losses to income from domestic production in the import-competing industry becomes relatively more important. Income per unit from sales of the imported goods, $P - P_{w,}$ falls as the quota increases, and falls at a faster rate than the loss in income from domestic production.

This last point has implications for voting models of tariff formation (e.g., Mayer 1984) and Mayer and Riezman (1989). In the Mayer model, an individual's relative wealth in terms of capital ownership, along with the relative capital intensity of the import-competing industry, determines an individual's gains or losses from freer trade. There is an endogenously determined equilibrium quota that meets a majority-rule criterion (if each individual's relative vote weight is not equal to his relative gains or losses from quota imposition). The distribution of capital ownership thus determines the tariff (in Mayer's case) chosen by majority rule. In the present model, the distribution of individuals over the space defined by mobility, $\widetilde{\lambda}_j$, and diversification, μ_j, would determine the quota in a majority-rule voting model. If the government embarks towards freer trade, potential voter resistance would tend to grow as a continually greater number of individuals would be harmed without compensation. At initial stages of relaxing trade barriers, those harmed may suffer a great deal, but their numbers may be too small to compel compensation, as liberalization precedes, the political pressure for compensation would increase. If compensation carries with it additional costs (e.g., due to technical and administrative barriers), then exogenous pressures to relax trade restrictions may only be partially successful. Without a mechanism for compensation, a government would relax the quota just up to – but not beyond – the point where a successful voting block would prove obstructionist. Further trade liberalization would require some level of compensation.

7.3 Mobility, Ownership Diversification, and Rent-Seeking Incentives

In this section we examine the effect of the diversification and the mobility of resource ownership on the incentives for rent seeking using the framework developed above. We first consider two hypothetical economies – one in which all resources are mobile, and the other in which all nonmobile resources are held in

[5] One would need to specify a distribution function for $\widetilde{\lambda}_j$ and μ_j over persons in society.

equal proportions by all individuals. In either of these extreme configurations the incentive for rent seeking is eliminated. To consider the effect of the mobility on the incentive to rent seek, consider the extreme case where all output in an economy is produced by a single, perfectly mobile resource. As a further simplification, also assume all individuals have identical homothetic preferences but not identical resource endowments. Given the resource mobility, rental rates are equated across all sectors in the economy. The rental rate of the resource will have to equal the total value of output in the economy divided by the total number of units of the resource in the economy. Given this relationship, the rental rate is maximized when the total value of output in the economy is maximized.

If the economy is at a competitive equilibrium, any attempt to expand one sector of the economy at the expense of another in order to increase the returns in that sector will, in fact, lower returns. The increase in the returns to the resources employed in that sector will be more than offset by a movement of resources from shrinking sectors of the economy. This movement will occur until the rental rate within the expanded sector reflects the new lower rental rate in the distorted economy. Thus, even if rent seeking were a costless activity, the incentive to rent seek would be eliminated if resources were perfectly mobile. Therefore, the only type of policies which would be pursued within such a framework are those which will expand the size of the economy. Thus, perfect resource mobility would be sufficient to eliminate rent seeking.

Complete mobility, although a sufficient condition, is not a necessary condition for the nonexistence of incentives to seek rents. Diversification of ownership can also eliminate the incentives to rent seek. Consider the case where each individual owns a portfolio of resources in exact proportion to the economy as a whole. In this case, the income of each individual is directly proportional to national real income. Any activity which reduces national income will, therefore, also reduce any individual's income. Once again, with identical, homothetic preferences, the incentive to seek trade distortions is eliminated: Any rent-seeking activity which reduces national real income will also reduce the income of all individuals.

In these two rather extreme hypothetical situations, it is in the self-interest of all individuals to oppose policies which distort the economy. Nevertheless, these hypothetical constructs are polar cases which illustrate an important point. Simply stated, mobility of resources and diversification of ownership of immobile resources must, at least in the limit, reduce the incentive for agents in the economy to seek rents.

There are, of course, important qualifications to the above analysis. The first is that preferences are unlikely to be identical or homothetic. As trade restrictions are relaxed, the change in the relative prices of consumption goods may harm those whose consumption basket is most heavily weighted toward export-oriented goods. And likewise, trade liberalization would benefit those who consume relatively greater amounts of the imported good than the representative, "average" individual. In sum, generally personal rates of inflation are not the same as the economy's representative rate. The second qualification is that resources, although mobile, are not likely to be used in the same intensity in both sectors, nor are individual portfolios miniature versions of the economy. There will be gains or losses to individual portfolios.

7.3.1 Rent-Seeking Models

In order to analyze the marginal effect of ownership diversification and the mobility of resources on rent seeking and protectionism within an economy, we use a standard two-sector trade model advanced in Section 7.2, integrated with a very simple rent-seeking component. A single distortion (a level of import quota) is lobbied for by the owner of the factors in the import-competing sector. The owner of the resources in the export sector lobbies to have the quota relaxed. All import quota rents are returned to the factor owners in the import-competing sector in proportion to the factor ownership. A level of rent-seeking expenditure, E_A, has the sole purpose of lobbying for restrictive import quotas, while E_B has the sole purpose of lobbying for the relaxation of restrictive import quotas. It is assumed that the funds for the lobbying expenditure are raised through taxes in proportion to the income generated in the import-competing sector and the export-competing sector, respectively. This eliminates the free-rider problem that would occur if the rent-seeking expenditure were raised through voluntary contributions.

The introduction of lobbying expenditures, undertaken by both sectors, requires that (2a) and (2b) be revised to

$$Y_A = P\,A_x + (P - P_w)\,(A_c - A_x) - E_A \tag{24a}$$

and

$$Y_B = B_x - E_B. \tag{24b}$$

In this formulation, the import quota rents are a part of the income in the import-competing sector, A. Accordingly, the rent-seeking expenditure is deducted from the income in this sector. Note that once again it is assumed that these expenditures are deadweight losses and have no social value. Given a binding quota, national income is defined by

$$Y = P\,A_x + (P - P_w)\,Q + B_x - E_A - E_B. \tag{25}$$

In this economy, the consumption and the utility of individuals are determined by their preferences, the prices they face, and their incomes. In a general-equilibrium setting, their incomes will be determined by the resources each individual owns and the aggregate output and trade of the general economy and the expenditure on lobby effort, $E_A + E_B$. The output of the economy will, in turn, be a function of prices, which is a function of the level of import quota. Using these linkages, it is possible to determine the effect that trade policy has on individual utility. The effect of an incremental change in the expenditure on generic rent seeking, E, on V_j is found by applying the chain rule to (5), i.e.,

$$\frac{dV_j}{dE} = \left[-A_{cj}\frac{\partial P}{\partial Q}\frac{\partial Q}{\partial E} + \frac{\partial Y_j}{\partial Q}\frac{\partial Q}{\partial E} + \frac{\partial Y_j}{\partial E} \right] \cdot \frac{\partial V_j}{\partial Y_j} \tag{26}$$

where the individual demand for the import good, A_{cj}, is found by applying Roy's identity; viz., $-A_{cj} = (\partial V_j/\partial P)(\partial V_j/\partial Y_j)^{-1}$.

Define λ_j as the jth person's share of the mobile resource in the economy and θ_j as the share of the immobile resources in the economy. Further, let μ_j equal the proportion of immobile resources which the jth individual owns, K_A. The proportion of an individual's income originating from each sector can be defined as $\omega^A{}_j = \alpha\lambda_j + \theta_j \cdot (1-\alpha) \cdot \mu_j$ and $\omega^B{}_j = \alpha\lambda_j + \theta_j \cdot (1-\alpha) \cdot (1-\mu_j)$. Using this notation, expression (26) can be examined in parts. First note from expression (25);

$$\frac{\partial Y_j}{\partial Q} = \left\{\omega_j^A\left[\frac{\partial P}{\partial Q}A_c + P\frac{\partial A_x}{\partial Q} + (P - P_w)\right] + \omega_j^B \cdot \frac{\partial B_x}{\partial Q}\right\}. \tag{27}$$

Furthermore, from the assumption that the quota-reducing, rent-seeking expenditure E_A is raised as a tax on income in protected sector, A; and E_B, the quota increasing rent-seeking expenditure is raised in proportion to the income earned in the sector,

$$\frac{\partial Y_j}{\partial E_A} = -1 \cdot \omega_j^A, \quad \frac{\partial Y_j}{\partial E_B} = -1 \cdot \omega_j^B. \tag{28}$$

Using this information, and recognizing both forms of rent-seeking expenditure, we can rewrite expression (26) as,

$$\frac{dV_j}{dE_A} = \left[-A_{cj}\frac{\partial P}{\partial Q}\frac{\partial Q}{\partial E_A}\left\{\omega_j^A\left[\frac{\partial P}{\partial Q}A_c + P\frac{\partial A_x}{\partial Q} + (P - P_w)\right] + \omega_j^B \cdot \frac{\partial B_x}{\partial Q}\right\}\frac{\partial Q}{\partial E_A} - 1 \cdot \omega_j^A\right] \cdot \frac{\partial V_j}{\partial Y_j} \tag{29}$$

$$\frac{dV_j}{dE_B} = \left[-A_{cj}\frac{\partial P}{\partial Q}\frac{\partial Q}{\partial E_B} + \left\{\omega_j^A\left[\frac{\partial P}{\partial Q}A_c + P\frac{\partial A_x}{\partial Q} + (P - P_w)\right] + \omega_j^B \cdot \frac{\partial B_x}{\partial Q}\right\}\frac{\partial Q}{\partial E_B} - 1 \cdot \omega_j^B\right] \cdot \frac{\partial V_j}{\partial Y_j}. \tag{30}$$

By defining the individual share of national consumption, A_{cj}/A_c, as γ_j, the change in utility of any individual with respect to a change in the level of rent-seeking expenditure can be written as:

$$\frac{dV_j}{dE_A} = \left[\left[\left\{(P - P_w) + \left(1 - \frac{\gamma_j}{\omega_j^A}\right)\frac{\partial P}{\partial Q}A_c + \left(P\frac{\partial A_x}{\partial Q} + \frac{\omega_j^B}{\omega_j^A} \cdot \frac{\partial B_x}{\partial Q}\right)\right\}\frac{\partial Q}{\partial E_A} - 1\right] \cdot \frac{\partial V_j}{\partial Y_j} \cdot \omega_j^A \tag{31}$$

and

$$\frac{dV_j}{dE_B} = \left[\left\{(P - P_w) + \left(1 - \frac{\gamma_j}{\omega_j^A}\right)\frac{\partial P}{\partial Q}A_c + \left(P\frac{\partial A_x}{\partial Q} + \frac{\omega_j^B}{\omega_j^A} \cdot \frac{\partial B_x}{\partial Q}\right)\right\}\frac{\partial Q}{\partial E_B} - \frac{\omega_j^B}{\omega_j^A}\right] \cdot \frac{\partial V_j}{\partial Y_j} \cdot \omega_j^A . \tag{32}$$

These equations provide the necessary relationships to describe the desirable level of rent-seeking effort, E_A and E_B, for any individual in society as a function of resource mobility and ownership. Recognizing that E_A and E_B are constrained to be nonnegative, the following Kuhn-Tucker conditions describe the desired rent-seeking expenditure for a particular individual:

$$\frac{dV_j}{dE_B} \leq 0, \quad E_B \geq 0, \quad \frac{dV_j}{dE_B} \cdot E_B = 0 \tag{33}$$

$$\frac{dV_j}{dE_A} \leq 0, \quad E_A \geq 0, \quad \frac{dV_j}{dE_A} \cdot E_A = 0. \tag{34}$$

The four terms in the square brackets in Eqs. (31) and (32) represent the four channels by which individual welfare is altered by a change in the rent-seeking expenditure. Consider the case of a small increase in E_A, and a corresponding decrease in the level of the quota. First, there is a decrease in revenue due to importing a unit for less than its domestic sale price. This is represented by the term, $P - P_W$. This loss will vanish as the quota becomes nonbinding. Second, there is a change in real purchasing power due to change in the import price that benefits the individual as a consumer differently than it harms the individual as a resource owner. This purchasing-power effect is positive if an individual consumes a greater share of the import good than the share of the revenues earned in the import industry. If the import quota is relaxed, the individual has fewer dollars from the import industry, the purchasing power of those dollars has increased. The third way in which individual welfare is affected by the change in quota is through the effect on income produced from changes in the domestic production of the two goods, apart from changes in relative prices. Income from the domestic production of good A increases as the quota decreases and allows more imports to substitute for domestically produced goods. On the other hand, income from the export-oriented production of good B decreases. If the individual is completely diversified, i.e., $\omega^A{}_j = \omega^B{}_j$, the income effect is zero. The increase in income from domestic production of the import-competing good balances with the reduction in income from production in the export-oriented industry. If the individual's assets are more concentrated in the import-competing industry, i.e., $\omega^A{}_j > \omega^B{}_j$, the gain from increased protection will more than offset the loss in income from the export sector. The fourth terms (-1, $\omega^A{}_j/\omega^B{}_j$) are negative, reflecting the cost of the increased expenditure on rent seeking.

Note that the first three effects are a product of the change in quota with respect to a change in rent-seeking expenditure, E_A or E_B, and the quota effects are opposite in direction. Given the positive cost of rent seeking in either of these cases, a particular individual could favor a positive level of E_A or a positive level of E_B or a zero level for each but would never favor a positive level for both.

At the point where $dV_j/dE_A = 0$, the level of expenditure on the quota-reducing, rent-seeking activity is at an optimum for this individual. Consider the case where preferences are such that $\gamma_j/\omega^A_j = 1$, which removes the real price effect on the consumption. In this case, as shown in Fig. 7.2, $dV_j/dE_A = 0$ when

$$(P - P_w) + P\frac{\partial A_x}{\partial Q} + \frac{\omega_j^B}{\omega_j^A} \cdot \frac{\partial B_x}{\partial Q} = \frac{\partial Q^{-1}}{\partial E_A}. \tag{35}$$

The effects of changes in proportion of mobile goods in the portfolio, $\tilde{\lambda}_j = \lambda_j / \lambda_j + \theta_j$; and the effects of diversification in the fixed inputs, μ_j, can be derived by substituting ω^A_j and ω^B_j into expression (31) and dividing each by $1/\lambda_j + \theta_j$;

$$(P - P_w) + P\frac{\partial A_x}{\partial Q} + \frac{\alpha\tilde{\lambda}_j + (1 - \tilde{\lambda}_j) \cdot (1 - \alpha) \cdot (1 - \mu_j)}{\alpha\tilde{\lambda}_j + (1 - \tilde{\lambda}_j) \cdot (1 - \alpha) \cdot \mu_j} \cdot \frac{\partial B_x}{\partial Q} = \frac{\partial Q^{-1}}{\partial E_A}, \tag{36a}$$

and

$$\frac{\alpha\tilde{\lambda}_j + (1 - \tilde{\lambda}_j) \cdot (1 - \alpha) \cdot (1 - \mu_j)}{\alpha\tilde{\lambda}_j + (1 - \tilde{\lambda}_j) \cdot (1 - \alpha) \cdot \mu_j}\left\{(P - P_w) + P\frac{\partial A_x}{\partial Q}\right\} + \frac{\partial B_x}{\partial Q} = \frac{\partial Q^{-1}}{\partial E_A}. \tag{36b}$$

With either the ownership of only mobile resources or the complete diversification of the ownership of immobile resources, an individual will desire a zero level of expenditure on quota-reducing, rent seeking, E_A.

To prove this result, note that either complete mobility, $\tilde{\lambda}_j = 1$, or complete diversity, $\mu_j = 1/2$, implies $\omega^B_j/\omega^A_j = 1$. If ω^B_j/ω^A_j is equal to unity, either because of diversity or mobility, the left-hand side of the expression is equal to $(P - P_w)$, which is positive as long as the quota is binding and is equal to zero only when the quota is nonbinding. This implies a maximum of the indirect utility rent-seeking expenditure, E_A, equals zero, or where Q becomes nonbinding.

For any individual who has ownership of immobile resources concentrated in the import-competing sector ($\mu_j > 1/2$), and desires protection, an increase in the ownership in the mobile resources will decrease that individual's desired level of rent seeking, E_A, and level of protection in the sector.

This result is easily proven from Eq. (36a). For any $\mu_j > 1/2$, an increase in the diversification of the immobile resources, i.e., a larger $\tilde{\lambda}_j$, implies an increase in the ratio, ω^B_j/ω^A_j. Given that $\partial B_x/\partial Q$ is positive, this implies an increase in the left-hand side, which implies that the right-hand side must be less negative. Given that Q is decreasing in E_A and concave in E_A, this implies that the desired level of E_A must fall with an increase in $\tilde{\lambda}_j$. This reduction in the desired level of rent-seeking expenditure also implies a decrease in the desired level of protection or an increase in the desired level of import quota for this individual.

For any individual who has ownership of immobile resources concentrated in the export sector, an increase in the ownership in the mobile resources will decrease that individual's desired level of rent seeking, E_B.

This proposition is easily proven from Eq. (36b). For any $\mu_j < 1/2$, an increase in the diversification of the immobile resources, i.e., a larger $\tilde{\lambda}_j$, implies a decrease in the ratio, $w^B{}_j/w^A{}_j$. The term, $\{(P - P_w) + P(\partial A_w/\partial Q)\}$, which represents the change income from the import sector with respect to a change in import quota, is negative. This implies an increase in the left-hand side, which implies that the right-hand side must be less negative. Given that Q is increasing in E_B, and concave in E_B, this implies that the desired level of E_B must fall with an increase in $\tilde{\lambda}_j$.

The individuals whose ownership is concentrated in the import-competing sector favor a restrictive import quota. The point of tangency shown in Fig. 7.3 represents the point at which the marginal cost of reducing quota is just equal to its associated marginal benefit and, therefore, the utility maximizing level of quota and rent-seeking expenditure.

For any individual who has a positive desired level of rent-seeking expenditure, an increase in the diversification of the portfolio of the immobile resources will decrease the desired level of rent seeking.

To show this result, note that if $\tilde{\lambda}_j < 1$ and $\mu_j > 1/2$, an increase in the diversification of the immobile resources implies a decrease in the ownership of the immobile resource in the import competing sector, i.e., $\mu_j \rightarrow 1/2$. This implies an increase in the ratio, $\omega^A{}_j/\omega^B{}_j$. Given that $\partial B_x/\partial Q$ is positive, this increases the left-

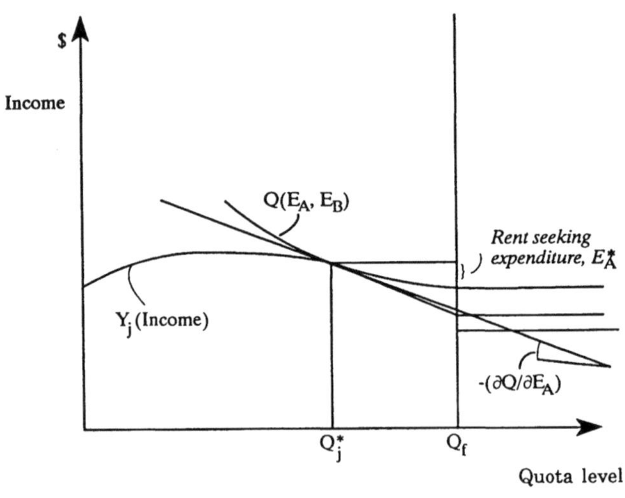

Q^* = desired import quota level for jth individual ω^A_j / ω^B_j, for a given E_B
Q_j = free trade level of import quota

Fig. 7.3. Desired rent-seeking expenditure

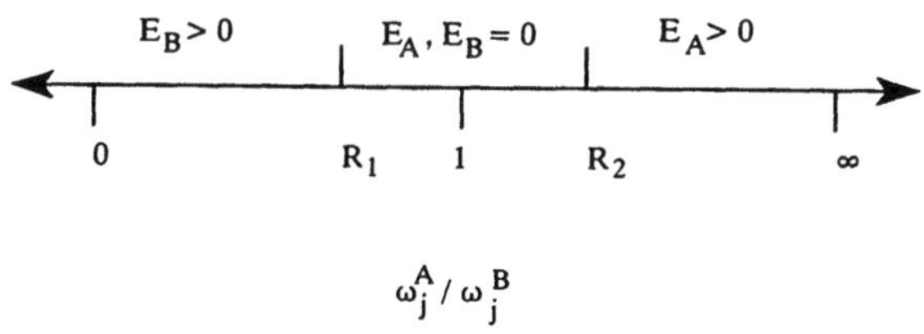

Fig. 7.4. Desired rent-seeking expenditure and the resource ownership ratio

hand side of Eq. (32), which then implies the right-hand side must be less negative. Given that Q is decreasing in E_A and concave in E_A, this implies that the desired level of E_A must fall.

Individuals who have ownership concentrated in the import sector, $\omega^A{}_j/\omega^B{}_j > 1$, will benefit from restrictive import quotas. If these individuals have enough to gain – that is, $\omega^A{}_j/\omega^B{}_j$ is sufficiently large ($> R_1$) – they will favor rent-seeking expenditure to reduce import quotas. Individuals who are diversified, $\omega^A{}_j/\omega^B{}_j = 1$, or have ownership concentration in the export sector, $\omega^A{}_j/\omega^B{}_j < 1$, achieve the highest level of utility at the free-trade level of import quota. These individuals, therefore, do not have an incentive to reduce quotas and may, in fact, oppose quotas and will desire a positive quota increasing rent-seeking expenditure, E_B. If quotas are binding and the marginal government response to this quota increasing rent-seeking expenditure is large enough, then these individuals will desire quota-reducing, rent-seeking expenditure, E_B. As shown in Fig. 7.4, if rent seeking is costly, there will be some levels of diversification, $\omega^A{}_j/\omega^B{}_j$, where the desired rent-seeking level for an individual is zero.

7.3.2 The Rent-Seeking Equilibrium

Individuals within the economy would, by the nature of their portfolio, desire different levels of rent-seeking expenditure. The equilibrium in the economy will be for a chosen aggregate level of expenditure, E_A and E_B. The aggregate levels of rent-seeking activity will, at least in part, be a function of individual preferences. The ability of those who desire rent-seeking expenditure to obtain the expenditure will be a function of the constitution, the industry structure, and other political economic variables. Certain configurations of ownership structure are more likely to achieve rent-seeking expenditure than others. For instance, if all individuals who owned resources in the import-competing sector had a portfolio balance where rent seeking was desirable, then their motivation follows immediately. A more relaxed condition for support may be the existence of positive, desired expenditure by the majority of owners or by the owner of the majority of shares in an industry. For the sake of simplicity we will assume that society is made up of two distinct groups, each with common resource-ownership ratios. Thus, all members in a particular group will have the same desired level of rent seeking. This is equivalent to assuming that the society is made up of two representative individuals. The desired level of rent seeking for these groups will be as described by the results of Section 7.3.1.

Given that the equilibrium level of the quota is a function of the level of rent-seeking expenditure, E_A and E_B, the level of rent seeking will not only be a function of the desired levels of each group but also the rent-seeking expenditure of the other. There are several possible strategies that could be used by each of the agents. A cooperative strategy, where the gains are shared, would be the welfare-maximizing, zero-expenditure equilibrium. If two groups were to engage in competitive strategy, it is possible that either group could act as a Stackelberg leader and incorporate the other group's reaction function into the choice of the level of rent seeking. In this model it will be assumed that each group acts in a Nash-Cournot way. That is, in making their decision, they assume that the level of rent seeking chosen by the other group is fixed and is not a function of their level of rent seeking, i.e., $\partial E_A/\partial E_B = 0$, and $\partial E_B/\partial E_B = 0$. The equilibrium in the economy will be the intersection of the reaction functions. With this configuration several outcomes are possible. The outcome will be a function of the economy, the political reaction function and, in particular, the degree of mobility and diversification of the two groups involved.

If all individuals in society hold either only mobile resources or are completely diversified in the ownership of immobile resources, there will be a zero rent-seeking expenditure, Pareto-efficient equilibrium in the economy.

From the first result of Section 7.3.1, with either the ownership of only mobile resources or the complete diversification of the ownership of immobile resources, an individual will desire a zero level of expenditure on quota-reducing rent seeking, E_A, for any level of expenditure, E_B. Thus, E_A will equal zero. Given the condition, $Q^* = q(0, E_B)$, the quota will be a free trade level. The reaction of those who favor free trade will also be a zero level of rent seeking. Given that the economy has no DUP activities and is a free-trade equilibrium, this is a Pareto-efficient outcome.

Consider the more interesting case where the groups do not hold the same portfolio. Group A has its portfolio biased toward the ownership of inputs in the import-competing sector, $\omega^A_j/\omega^B_j > 1$; and, by addition, the other group, B, has a portfolio biased toward the ownership of the resources in the export sector. Group A can benefit from some quota below the free trade level. Given the structure of the economy and the political support function, $q(E_A, E_B)$, the desired level of rent-seeking expenditure E_A will be a function of the portfolio of ownership and the level of rent-seeking expenditure, E_B, i.e., $E_A^* = g(\omega^A_j/\omega^B_j, E_B)$. The desired level of expenditure will be a nondecreasing function in the proportion of ownership in the import competing sector, ω^A_j/ω^B_j. The largest intercept would be obtained where this group only owned the fixed assets in this sector, $\omega^A_j/\omega^B_j = \infty$. The rent-seeking expenditure, E_A^*, will be an increasing function of the level of expenditure, E_B. An increase in E_B will increase the import quota, thus increasing the shadow value of its reduction to this group. The group, therefore, will always be willing to expend more at the margin for a quota-increasing rent-seeking expenditure. These two effects allow the construction of a reaction function for group A, R_A, as shown in Fig. 7.5.

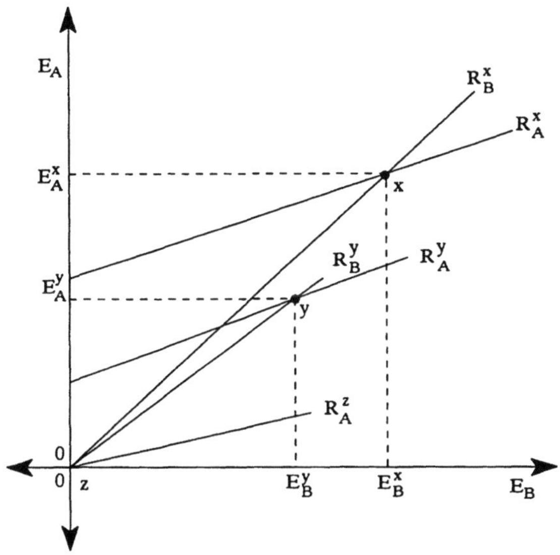

Fig. 7.5. Rent-seeking equilibriums in the economy

Group B, who by definition have some concentration in the ownership in the export sector, would desire free trade. They would be willing to pay to have the economy move in that direction. Given the structure of the economy and the political support function, $q(E_A, E_B)$, the desired level of rent-seeking expenditure, E_B*, will be a function of the portfolio of ownership, and the level of quota-reducing, rent-seeking expenditure, E_A, i.e., $E_B* = h(\omega^A{}_j/\omega^B{}_j, E_A)$. The desired level of expenditure of this group will be an increasing function of the concentration of ownership in the export sector or a decreasing function of $\omega^A{}_j/\omega^B{}_j$. E_B* will be an increasing function of E_A. The larger the E_A, the more restrictive the quota and the greater the marginal gain from an increase. This increase, combined with the concavity[6] of the political support function, implies that desired expenditure will increase. Note also that $E_B* = 0$ for any $\omega^A{}_j/\omega^B{}_j$ when $E_A = 0$. This comes from the condition, $Q* = f(0, E_B)$, or that free trade is present when $E_A = 0$; therefore any lobby to increase quota would have no value. These two effects allow the construction of a reaction function for E_B* as shown in Fig. 7.5. The quota level in this economy will be a direct result of the expenditures, $Q** = q(E_A**, E_B**)$. Possible equilibriums are shown in Fig. 7.5. The equilibrium, x, represents the extreme configuration where $\omega^A{}_j/\omega^B{}_j = \infty$ for group A. The equilibrium, y, represents the intersect of the reaction functions when $\omega^A{}_j/\omega^B{}_j$ is less extreme. In this configuration, both groups desire a lower level of rent-seeking expenditure. A movement in the equilibrium could be brought about by either increased diversification or by a greater proportion of mobile good within the economy.

---

[6] This framework would be fully consistent to the case where the other sector did engage in rent-seeking activity but not a function of the rent-seeking expenditure in the import-competing sector. For an example of a model where both groups simultaneously determined rent-seeking expenditure see Rausser and Foster (1990).

The Nash equilibrium in this economy will be where the reaction function for E_A and E_B intersect or $E_A{}^{**} = g(\omega^A{}_j/\omega^B{}_j, E_B{}^*)$, and $E_B{}^{**} = h(\omega^A{}_j/\omega^B{}_j, E_A{}^*)$, where $E_A{}^{**}$ and $E_B{}^{**}$ represent the equilibrium levels of rent-seeking expenditure. More specifically, for an internal equilibrium, when each group is at the desired level of rent seeking represented by the two first-order conditions as derived from Eqs. (31) and (32),

$$\frac{dV_j}{dE_A} = \left[\left\{(P - P_w) + \left(1 - \frac{\gamma_j}{\omega_j^A}\right)\frac{\partial P}{\partial Q}A_c + \left(P\frac{\partial A_x}{\partial Q} + \frac{\omega_j^B}{\omega_j^A}\cdot\frac{\partial B_x}{\partial Q}\right)\right\}\frac{\partial Q}{\partial E_A} - 1\right] = 0, \tag{37}$$

$$\frac{dV_j}{dE_B} = \left[\left\{(P - P_w) + \left(1 - \frac{\gamma_j}{\omega_j^A}\right)\frac{\partial P}{\partial Q}A_c + \left(P\frac{\partial A_x}{\partial Q} + \frac{\omega_j^B}{\omega_j^A}\cdot\frac{\partial B_x}{\partial Q}\right)\right\}\frac{\partial Q}{\partial E_B} - \frac{\omega_j^B}{\omega_j^A}\right] = 0. \tag{38}$$

If we assume preferences are identical and homothetic, and define $\omega^A{}_j/\omega^B{}_j$ for groups A and B as R_A and R_B, respectively, they can be rewritten as:

$$\frac{dV_j}{dE_A} = \left[\left\{(P - P_w) + \left(P\frac{\partial A_x}{\partial Q} + R_A\cdot\frac{\partial B_x}{\partial Q}\right)\right\}\frac{\partial Q}{\partial E_A} - 1\right] = 0, \tag{39}$$

$$\frac{dV_j}{dE_B} = \left[\left\{(P - P_w) + \left(P\frac{\partial A_x}{\partial Q} + R_B\cdot\frac{\partial B_x}{\partial Q}\right)\right\}\frac{\partial Q}{\partial E_B} - R_B\right] = 0. \tag{40}$$

Condensing the notation somewhat further by defining

$$\left\{(P - P_w) + \left(P\frac{\partial A_x}{\partial Q} + R_A\cdot\frac{\partial B_x}{\partial Q}\right)\right\} = \frac{\partial\psi_A}{\partial Q}\left\{(P - P_w) + \left(P\frac{\partial A_x}{\partial Q} + R_B\cdot\frac{\partial B_x}{\partial Q}\right)\right\} = \frac{\partial\psi_B}{\partial Q},$$

it becomes a straightforward task to do comparative statics with respect to some exogenous variable, X, that affects R_A and R_B on the rent-seeking Nash equilibrium:

$$\frac{\partial\psi_A}{\partial Q}\frac{\partial^2 Q}{\partial E_A^2}dE_A + \frac{\partial\psi_A}{\partial Q}\frac{\partial^2 Q}{\partial E_B\partial E_A}dE_B + \frac{\partial B_x}{\partial Q}\frac{\partial Q}{\partial E_A}\frac{dR_A}{dX}dX = 0, \tag{41}$$

$$\frac{\partial\psi_B}{\partial Q}\frac{\partial^2 Q}{\partial E_A\partial E_B}dE_A + \frac{\partial\psi_B}{\partial Q}\frac{\partial^2 Q}{\partial E_B^2}dE_B + \left\{\frac{\partial B_x}{\partial Q}\frac{\partial Q}{\partial E_B} - 1\right\}\frac{dR_B}{dX}dX = 0. \tag{42}$$

Rewriting Eqs. (41) and (42) in matrix notation, we have:

$$\begin{bmatrix} \frac{\partial \psi_A}{\partial Q}\frac{\partial^2 Q}{\partial E_A^2} & \frac{\partial \psi_A}{\partial Q}\frac{\partial^2 Q}{\partial E_B \partial E_A} \\ \frac{\partial \psi_B}{\partial Q}\frac{\partial^2 Q}{\partial E_A \partial E_B} & \frac{\partial \psi_B}{\partial Q}\frac{\partial^2 Q}{\partial E_B^2} \end{bmatrix} \begin{bmatrix} dE_A \\ dE_B \end{bmatrix} = \begin{bmatrix} -\left(\frac{\partial B_x}{\partial Q}\frac{\partial Q}{\partial E_A}\frac{dR_A}{dX}\right) \\ \left\{1 - \frac{\partial B_x}{\partial Q}\frac{\partial Q}{\partial E_B}\right\}\frac{dR_B}{dX} \end{bmatrix} [dX]. \tag{43}$$

Using the Cramer rule, the following comparative static results must hold:

$$\frac{dE_A}{dX} = \frac{\begin{vmatrix} -\left(\frac{\partial B_x}{\partial Q}\frac{\partial Q}{\partial E_A}\frac{dR_A}{dX}\right) & \frac{\partial \psi_A}{\partial Q}\frac{\partial^2 Q}{\partial E_B \partial E_A} \\ \left\{1 - \frac{\partial B_x}{\partial Q}\frac{\partial Q}{\partial E_B}\right\}\frac{dR_B}{dX} & \frac{\partial \psi_B}{\partial Q}\frac{\partial^2 Q}{\partial E_B^2} \end{vmatrix}}{\begin{vmatrix} \frac{\partial \psi_A}{\partial Q}\frac{\partial^2 Q}{\partial E_A^2} & \frac{\partial \psi_A}{\partial Q}\frac{\partial^2 Q}{\partial E_B \partial E_A} \\ \frac{\partial \psi_B}{\partial Q}\frac{\partial^2 Q}{\partial E_A \partial E_B} & \frac{\partial \psi_B}{\partial Q}\frac{\partial^2 Q}{\partial E_B^2} \end{vmatrix}} \tag{44}$$

$$= \frac{-\left(\frac{\partial B_x}{\partial Q}\frac{\partial Q}{\partial E_A}\frac{dR_A}{dX}\frac{\partial \psi_B}{\partial Q}\frac{\partial^2 Q}{\partial E_B^2}\right) - \left\{1 - \frac{\partial B_x}{\partial Q}\frac{\partial Q}{\partial E_B}\right\}\frac{dR_B}{dX}\frac{\partial \psi_A}{\partial Q}\frac{\partial^2 Q}{\partial E_B \partial E_A}}{\left[\frac{\partial \psi_A}{\partial Q}\frac{\partial \psi_B}{\partial Q}\right]\left[\frac{\partial^2 Q}{\partial E_A^2}\frac{\partial^2 Q}{\partial E_B^2} - \left(\frac{\partial^2 Q}{\partial E_B \partial E_A}\right)^2\right]}$$

$$\frac{dE_B}{dX} = \frac{\begin{vmatrix} \frac{\partial\psi_A}{\partial Q}\frac{\partial^2 Q}{\partial E_A^2} & -\left(\frac{\partial B_x}{\partial Q}\frac{\partial Q}{\partial E_A}\frac{dR_A}{dX}\right) \\ \frac{\partial\psi_B}{\partial Q}\frac{\partial^2 Q}{\partial E_A \partial E_B} & \left\{1 - \frac{\partial B_x}{\partial Q}\frac{\partial Q}{\partial E_B}\right\}\frac{dR_B}{dX} \end{vmatrix}}{\begin{vmatrix} \frac{\partial\psi_A}{\partial Q}\frac{\partial^2 Q}{\partial E_A^2} & \frac{\partial\psi_A}{\partial Q}\frac{\partial^2 Q}{\partial E_B \partial E_A} \\ \frac{\partial\psi_B}{\partial Q}\frac{\partial^2 Q}{\partial E_A \partial E_B} & \frac{\partial\psi_B}{\partial Q}\frac{\partial^2 Q}{\partial E_B^2} \end{vmatrix}} \tag{45}$$

$$= \frac{\frac{\partial\psi_A}{\partial Q}\frac{\partial^2 Q}{\partial E_A^2}\left\{1 - \frac{\partial B_x}{\partial Q}\frac{\partial Q}{\partial E_B}\right\}\frac{dR_B}{dX} + \left(\frac{\partial\psi_B}{\partial Q}\frac{\partial^2 Q}{\partial E_A \partial E_B}\frac{\partial B_x}{\partial Q}\frac{\partial Q}{\partial E_A}\frac{dR_A}{dX}\right)}{\left[\frac{\partial\psi_A}{\partial Q}\frac{\partial\psi_B}{\partial Q}\right]\left[\frac{\partial^2 Q}{\partial E_A^2}\frac{\partial^2 Q}{\partial E_B^2} - \left(\frac{\partial^2 Q}{\partial E_B \partial E_A}\right)^2\right]}.$$

We can use these comparative static results to prove the following result:

> *An increase in the factor share of the mobile resource in the economy with no change in the pattern of ownership will decrease the equilibrium level of rent-seeking expenditures E_A and E_B in the economy.*

To demonstrate this result, an increase in the factor share of the mobile resource in the economy with no change in the pattern of ownership will decrease the concentration of income derived from a particular sector, given that $\omega^B{}_j/\omega^A{}_j = \alpha\tilde{\lambda}_j + (1 - \tilde{\lambda}_j)(1 - \alpha)(1 - \mu_j)/\alpha\tilde{\lambda}_j + (1 - \tilde{\lambda}_j)(1 - \alpha)\mu_j$. Thus, an increase in the factor share, X, will reduce group A's concentration of income from the import sector, increasing R_A toward 1, $(dR_A/dX > 0)$, and simultaneously decrease group B's concentration of income from the export sector, decreasing R_B toward 1, $(dR_B/dX < 0)$. Given the direct income effects, $\partial\Psi_A/\partial Q < 0$, $\partial\Psi_B/\partial Q > 0$, and the properties of the rent-seeking function, $\partial Q/\partial E_A < 0$, $\partial Q/\partial E_B > 0$, $\partial^2 Q/\partial E^2{}_A > 0$, $\partial^2 Q/\partial E^2{}_B < 0$, $\partial^2 Q/\partial E_A \partial E_B > 0$, the denominators of Eqs. (44) and (45) are unambiguously positive. Given that $\{1 - (\partial B_x/\partial Q)(\partial Q/\partial E_B)\} > 0$, the gain in the income in the export sector must be sufficient to offset not only the expenditure on rent seeking but also the loss in income generated in the import-competing sector. Given the effect of the increased factor share, X, $dR_A/dX > 0$, $dR_B/dX < 0$, $dE_A/dX < 0$, $dE_B/dX < 0$. Thus, an increase in the factor share for a given pattern of ownership will reduce the desired level of expenditure.

It is also possible to demonstrate the following result:

An increase in the diversification of the portfolio of the immobile resources for groups within the economy will decrease the equilibrium level of rent-seeking expenditure, E_A and E_B, in the economy.

This result follows by recognizing that for group A, $\tilde{\lambda}_j < 1$ and $\mu_j > 1/2$, an increase in the diversification of the immobile resources implies a decrease in the ownership of the immobile resource in the import-competing sector, i.e., $\mu_j \to 1/2$. This implies an increase in the ratio, $\omega^B{}_j/\omega^A{}_j$. Thus, an increase in the diversification in the immobile resource, X, will reduce group A's concentration of income from the import sector, increasing R_A toward 1. Simultaneously, for group B, $\tilde{\lambda}_j < 1$ and $\mu_j < 1/2$, *an increase in the diversification in the immobile resource, X,* must decrease group B's concentration of income from the export sector, decreasing R_B toward 1. As shown earlier, if $dR_A/dX > 0$, and $dR_B/dX < 0$, then, from the comparative static results, $dE_A/dX < 0$, $dE_B/dX < 0$.

It is interesting to note that the last two results cannot be extended to show that the reduction in rent-seeking expenditure, E_A and E_B, due to an increase in the factor share of the mobile resources or an increase in the diversification in fixed factor, will not unambiguously lead to a reduced level of protection in the economy. The increase in import quota due to reduction in rent-seeking expenditure, could be more than offset by the reduction in import quota due to a reduction in the quota increasing expenditure, E_B. As shown earlier, this does hold in the limit as E_A falls to zero.

7.4 Conclusions

Resource mobility and portfolio diversification have a role in determining the distributional implications of trade reform. The four primary implications of the analysis presented in Section 7.2 are:

1. The degree of compensation required for an individual to remain indifferent to trade liberalization increases as an individual's share of consumption of the imported good decreases.
2. The Pareto-fulfilling compensation level increases as the individual's immobile assets are concentrated in the import-competing industry. In other words, the greater the deviation in the individual's portfolio from one of complete diversification of immobile assets toward concentration in the import-competing industry, the greater the compensation.
3. For an unbalanced portfolio of immobile assets which concentrates ownership in the import-competing industry, as the proportion of the individual's total wealth in immobile assets relative to mobile asset increases, the greater the Pareto-fulfilling compensation grows in association with freer trade.
4. Levels of an effective quota exist sufficiently close to the nonbinding level that, for a given level of mobility, if an individual is even somewhat skewed toward ownership of resources in the import-competing industry, then compensation will be required to make the individual indifferent to further quota

relaxation. Thus, as the trade restriction approaches the free trade level, a greater number of individuals will require such compensation.

These conclusions have some important policy implications. If interest groups have the power to block changes which make them worse off, then compensation may be a necessary part of the trade reform. Resource ownership diversification and resource mobility reduce the potential compensation requirements of trade reform. A government should, therefore, promote ownership diversification and mobility in order to facilitate trade reform without costly compensation. For many resources this may only require the correction of a market failure; while, in other sectors, where the transaction costs of diversified ownership are very large, such as agriculture, this may be a more formidable task.

In Section 7.3 we analyze the effect of diversification of ownership and resource mobility on rent-seeking incentives. We demonstrate that the diversification of ownership and enhanced resource mobility can reduce those incentives. This may provide some insight into the variation of rent seeking observed in different economies and between sectors within the same economy.

Governments seeking a reduction in wasteful rent-seeking activities should, therefore, pursue policies which tend to foster resource mobility and ownership diversification. A natural corollary is that governments should tend to discourage policies that tend to reduce mobility and increase the concentration of the ownership of resources. In the context of trade reform, these observations may be particularly relevant. In a fully open economy, price movements in the international economy cause both winners and losers within the domestic economy. The losers within the open market seek and obtain protection from the vagaries of the world market. If at any point in time the government is considering a trade-reform policy, then the policy should be chosen in such a way as to mitigate future rent seeking which leads to deadweight losses. Thus, policies which are sustainable should be preferred over those which are not. *A natural extension of the above argument is that, if trade reform is to be sustained, compensation for current reform should be paid in such a way as to promote mobility and diversification of ownership of immobile resources.*

In the current discussion of government policy, particularly trade-reform policy, mobility and/or diversification of ownership are seldom considered as relevant dimensions of government policy. In this chapter we present a very compelling argument for doing so. Simply put, resource mobility and diversification of ownership reduce the rewards to rent seeking in society. Reduced rent seeking reduces unproductive activities and economic distortions in the economy. Therefore, reform policies which foster diversification of ownership and mobility of resources should, *ceteris paribus*, be promoted over those which do not.

References

Becker G (1983) A theory of competition among pressure groups for political influence. Q J Econ 98:371–400

Bhagwati JN (1971) The generalized theory of trade and welfare. In: Bhagwati JN, Jones RW, Mundell RA, Vanek J (eds) Trade, balance of payments and growth. North Holland, Amsterdam

Corden WM (1969) Effective protection rates in the general equilibrium Model. Oxford Econ Pap 21:135–141

Eaton J, Grossman GE (1985) Tariffs as insurance: optimal commercial policy when domestic markets are incomplete. Can J Econ 18:258–272

Heckscher E (1949) The effect of foreign trade on the distribution of income. In: Ellis HS, Metzler LA (eds) Readings in the theory of international trade. Blackiston, Philadelphia

Hillman AL, Riley JG (1989). Politically contestable rents and transfers. Econ Polit 1:17–39

Jones RW (1970) Effective protection and substitution. J Int Econ 1:59–82

Jones RW (1975) Income distribution and effective protection in a multicommodity trade model. J Econ Theory, 11:1–15

Krueger AO (1974) The political economy of rent seeking. Am Econ Rev 64:291–303

Lipsey RG, Lancaster K (1957) The general theory of "second best." Rev Econ Stud 24:11–32

Mayer W (1984) Endogenous tariff formation. Am Econ Rev 74:970–985

Mayer W, Riezman R (1989) Tariff formation in a multidimensional voting model. Econ Polit 1:61–79

Meade JE (1955) Trade and welfare. Oxford University Press, London

Mussa M (1982) Imperfect factor mobility and the distribution of income. J Int Econ 12:125–141

Rausser GC, Foster WE (1990) Political preference functions and public policy reform. Am J Agric Econ 72:642–652

Rausser GC, Zusman P (1992) Public policy: explanation and constitutional prescription. Am J Agric Econ 74; 2 (May):247–257

Staiger RW, Tabellini G (1987) Discretionary trade policy and excessive protection. Am Econ Rev 77:823–837

Stopler, WF Samuelson PA (1941) Protection and real wages. Rev Econ Stud 9:58–73

Tullock G (1967) The welfare cost of tariffs, monopolies, and theft. West Econ J 5:224–232

Viner J (1950) The customs unions issue. Carnegie Endowment for International Peace, New York

Chapter 8

Modeling Policy Reform in the US Wheat and Feed Grain Sectors

Richard E. Just, Gordon C. Rausser, and David Zilberman

8.1 Introduction

A major objective of the GATT negotiation is an international agreement on permitted domestic agricultural policies referred to as *green light* policies (e.g., environmental and decoupled support policies) to replace nonpermitted or *red light* policies (e.g., trade distortion subsidies). To be implementable, these policies must improve the efficiency of resource allocation while being politically sustainable. Economists are challenged to develop quantitative estimates of the impacts of alternative policy proposals that serve both of these goals. Such impacts must be quantified not only for efficiency and distributional effects but also for short- and long-term effects over time. A conceptual and empirical modeling framework that meets this challenge is presented in this chapter. Specifically, the model is applied to assess the impacts of three alternative policy reform scenarios for the US feed grain, soybean, livestock, and wheat sectors.

Governments that sign a GATT agreement must be assured that reforms in agricultural subsidization will not dramatically reduce their political support (Rausser 1992). Analytical models are needed to evaluate reform paths which are both politically sustainable and welfare improving. A welfare-improving policy is politically feasible and sustainable if it can appease the interests of those who support existing policies. If a strong interest group is affected adversely by a new policy, the political sustainablity depends upon implementation of compensation schemes, which are implemented either through transfer schemes financed from general tax revenues (which have their own distortionary effects) or in the form of additional policies.

It is expected that the GATT negotiations will reduce the set of institutionally feasible policies. It is important to know whether allowable policies are politically feasible and whether welfare (economic efficiency) is improved. Unfortunately, available studies focus little attention on the costs (economic and political) of transition to free trade. One reason for the lack of attention to these issues has to do with the Lucas critique of econometric models of policy effects. That is, when a model is estimated or calibrated under a particular policy regime, the effects of the policy become embedded in the estimated parameters so that the model is not an accurate device for investigating the effects of altenative policies.[1] The model

[1] Note that this criticism also applies to using econometric models estimated under current policies for simulating the effects of free trade.

developed here suggests that if the role of policies is incorporated in the theoretical structure of econometric equations, then the effects of current policies can be sufficiently separated (Rausser and Just 1982).

Another difficulty with evaluating programs for phasing out current policies is that econometric models have not performed well in long-term dynamic forecasting, such as the 10-year horizon, that may be necessary for a politically feasible transition to free trade. However, developed in this chapter is a simulation approach in which policies directly affect endogenous variables, assure robustness of predictions, and improve long-term forecasting.

Here, we develop an operational econometric model of the US feed grain, soybean, livestock, and wheat sectors. We examine how a small selected set of permitted policies will affect the welfare of specific broad interest groups through a 10-year transition period. The focus of the analysis is on substituting green light policies (decoupled policies) for red-light policies (coupled distortionary policies) and/or partial phaseouts of red-light policies. The specific policy reforms considered include decoupled revenue-support programs, programs designed to protect the environment, and programs designed to retire land and facilitate soil conservation. The interest groups include grain producers, livestock producers, grain consumers, meat consumers, and environmental interests. The model demonstrates effects of policy reform on asset values as well as on profits and traditional short-term welfare measures. Some of the results are illustrated with regional detail.

8.2 Theoretical and Institutional Considerations in Policy Reform

For our purposes, the standard tools of welfare economics must be extended to incorporate political and institutional considerations. Traditionally, the analysis of economic welfare has focused on measurement of producers' and consumers' surplus. In traditional welfare-economic practice, if a policy makes the sum of the two increase, it is a potentially Pareto-preferred policy assuming that lump-sum redistribution is possible. This approach thus focuses primarily on economic efficiency and relatively little on equity.

In reality, costless lump-sum transfers are rarely, if ever, possible. An individual's behavior is affected by his/her wealth which, in turn, is affected by transfers. As a result, economic efficiency is altered by a transfer scheme so that the sum of producers' and consumers' surplus changes. Thus, the analysis of efficiency cannot be separated from the analysis of equity (Just 1987). This means that the policy problems of maximizing the size of the pie and distributing the pie must be considered simultaneously.

Once equity and distribution are introduced into policy analysis, problems of political feasibility must be integrated as a basic component. Many economic polices are not feasible or cannot be sustained without the simultaneous institution of compensation schemes that meet the distributional concerns of various special interest groups. Any operationally useful model must examine whether the available mix of instruments is sufficient to compensate all special interest groups sufficiently to make policy reform sustainable.

Sections 3.4.1 and 3.4.2. of Chap. 3 introduce a framework that explicitly recognizes political and institutional dimensions in economic-welfare analysis. This is the theoretical framework which guides the empirical analysis here.

To illustrate the conceptual approach in the context of the reform issues considered here, consumer demand is represented implicitly by $p = D(q)$, which implies that consumer surplus is $W_c = {}^{q}{}_{0}\int[D(q) - p]dq \equiv W_c(q)$. Let agricultural producer profits from regulated activities follow $\pi(p,r,w)$, where r is the rental rate for land and w is the cost of chemicals with the usual regularity assumptions applying to the profit function. Thus, agricultural supply follows $q = \partial p/\partial p \equiv S(p,r,w)$; acreage used in agricultural production follows $a = \partial\pi/\partial r \equiv a(p,r,w)$; and chemical use for agricultural production is $x = \partial p/\partial w \equiv x(p,r,w)$.

Suppose further that landowners possess a quantity of land, $\bar{a}$, that can be either in regulated agricultural production activities or in an alternative use with a private return of r_0 per acre. Assume that the alternative use can be pursued for land that enters a retirement program. If the government payment for land retirement is r_g per acre, the rental rate for land in agricultural production will be driven up to $r = r_0 + r_g$ and the total income for landowners will be $r\bar{a}$. The extent of the chemical-use restriction is represented by the shadow price of chemicals in agricultural production. Where chemicals are available in perfectly elastic supply at price w_0 and the shadow price of chemical use is $w = w_0 + w_g$, the aggregate welfare of agricultural producers and landowners is given by $W_p = \pi + r\bar{a} + w_g x$.

The latter term is an additional amount recovered by producers relative to the case where chemical price is w because, under the use restriction, they only pay price w_0 (assuming $w_g > 0$). (Producer plus landowner welfare is combined and called producer welfare for reasons noted later.)

8.2.1 The Pre-GATT Case

In this framework, consider initially a pre-GATT (current) policy scenario where policy instruments include a target price, p_t, received by producers along with a nominal payment for land retirement, r_g, which may reflect either payments for long-term land retirement or payments for short-term voluntary diversion under acreage reduction programs. Assume initially that no chemical use restrictions are in effect, i.e., $w_g = 0$. Production follows $q = S(p_t, r_0 + r_g, w_0)$; the market price is $p = D(q)$; consumer welfare is $W_c = W_c(q)$; and producer welfare is $W_p = \pi(P_t, r_0 + r_g, w_0) + (r_0 + r_g)\bar{a}$. The taxpayer cost necessary to pay deficiency payments and land retirement payments is $W_t = W_t\{-q[p_t - D(q)] - r_g(\bar{a} - a)\}$, where $W_t\{0\} = 0$ and $W_t \geq 1$ represents the inefficiency of taxation.

8.2.2 Decoupled Payments

One possibility for appeasing the producer interest group with elimination of target prices is to offer decoupled payments as a transitionary measure for political feasibility. With a return to the free market, the equilibrium condition is $p = D[S(p,r_0,w_0)]$; consumer welfare is $W_c[S(p,r_0,w_0)]$; producer welfare is $W_p = \pi(p,r_0,w_0) + r_0\bar{a} + K$ where K is the decoupled payment to producers, and the taxpayer welfare cost of decoupled payments compared to the free market case is $W_t = W_t(-K)$.

8.2.3 Land Retirement

Another possibility for appeasing producer interests is to replace target prices by higher land retirement payments which compensate producers directly with land retirement payments. If demand is inelastic, producers receive a further output price benefit because their production levels are lower. The equilibrium condition is $p = D[S(p,r_0 + r_g,w_0)]$; consumer welfare is $W_c[S(p,r_0 + r_g,w_0)]$; producer welfare is $W_p = \pi(p,r_0 + r_g,w_0) + (r_0 + r_g\bar{a})$; and the taxpayer welfare effect relative to the free market case is $W_t = W_t[-r_g(\bar{a} - a)]$. In this case, a nonmarket environmental benefit is also obtained as a function of the amount of land retired.

8.2.4 Environmental Preservation

Finally, consider the possibility of appeasing producer interests by imposing tight restrictions on chemical use. Studies of pesticide benefits generally find that producer welfare is increased by such restrictions because agricultural demand is inelastic (Lichtenberg and Zilberman 1988). Here the equilibrium condition is $p = D[S(p,r_0,w_0 + w_g)]$; consumer welfare is $W_c[S(p,r_0,w_0 + w_g)]$; producer welfare is $W_p = \pi(p,r_0,w_0 + w_g) + r_0\bar{a} + w_g x$; and the taxpayer welfare effect relative to the free-market case is $W_t = W_t(0)$. In addition, a nonmarket environmental benefit inversely related to chemical use is also obtained.

8.2.5 Comparison of Effects

The effects of these policies on market behavior are represented in Fig. 8.1. The effect of land retirement payments and/or pesticide/chemical use controls is to reduce supply from $S(0)$ to $S(g)$ where g is a general representation of the payment for land retirement or the magnitude of the chemical use restriction. The pre-GATT case is represented in Fig. 8.1a, where the target price raises the effective price received by farmers from the free-market equilibrium price, p_c to p_t. Where farmers would respond along supply $S(0)$ in absence of a land retirement program or chemical restrictions, the quantity supplied under the combination of programs is q_m as implied by supply $S(g)$. The resulting market price is p_m, determined by consumer demand, D.

An additional effect of a land retirement payment is illustrated in Fig. 8.1b. Available land is $\bar{a}$, and the demand for land for agricultural production is $a(p)$, where $p = p_t$ in the pre-GATT case. With no land retirement program, acreage used in agricultural production is a_0; but with land retirement, payment r_g declines to a_m. The total rent received by landowners increases by area j + k + l with the amount of land retirement payment burden to taxpayers represented by area k + l. Note that payments may be made for retirement of some land that would not

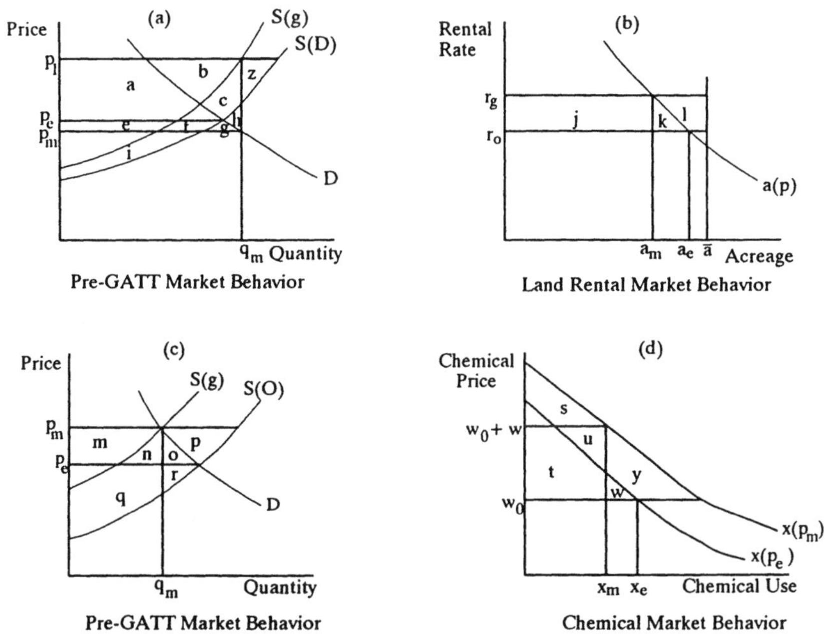

Fig. 8.1. Welfare effects of alternative policies

otherwise be used in production, $\bar{a} - a_0$. Relative to the free-market case, producer welfare (including landowners) is thus higher by area a + b − f − i + j + k + l; consumer welfare is higher by area e + f + g; and taxpayers lose area a + b + c + d + e + f + g + h + k + l. (For the remainder of this section, taxpayer effects are evaluated by budget impacts without consideration for the inefficiency of taxation.)

The post-GATT case with elimination of target prices is represented in Fig. 8.1c. A higher land-retirement payment or tighter chemical-use restriction shifts *S(g)* farther left than in Fig. 8.1a. The new market price to which both producers and consumers respond rises to p_m which may be above p_t if the leftward shift in supply is sufficient. In the case of a land-retirement policy, the welfare effect for producers cannot be assessed entirely in one market. Compared to the free-market case, profits from production in Fig. 8.1c are higher by area m − r − q, while, in Fig. 8.1b, the benefits to landowners are higher by area j + k + l. [Results in Just et al. (1982) show that area j + k in Fig. 8.1b can be accurately captured by area n + o + p + q + r in panel Fig. 8.1c, where the acreage demand in Fig. 8.1b is $a(p_m)$; thus, area j + k + l is somewhat larger.] Compared to the free-market case, producer welfare (including landowners) is higher by area m − r − q + j + k + 1, consumer surplus is lower by area m + n + o, and taxpayers lose area k + l. Note that with inelastic demand this producer gain can be made arbitrarily large by offering larger land retirement subsidies.

For the case of chemical-use restrictions, the producer welfare effect compared to the free-market case can be captured entirely in one market but only in the chemical market represented in Fig. 8.1d. In free-market equilibrium, the demand for chemicals is $x(p_e)$ with use x_e at price w_0. With a use restriction imposed at x_m, the shadow price of chemicals rises to $w_0 + w_g$ as the price in the output market increases to p_m with declining supply which, in turn, increases demand for chemicals to $x(p_m)$. Producer welfare relative to the free-market case increases by area s + u – w which, following Just et al., is equivalent to area m – q – r + t + u (profit falls by area t + w = area q + r with a rise in chemical price from w_0 to $w_0 + w_g$, rises by area s = area m with a rise in output price from p_0 to p_m, and finally improves by area t + u relative to the case of raising chemical price because the restriction is imposed in the form of a limit on use rather than a higher chemical price). If production is sufficiently sensitive to chemical use, then producers can be fully restored. In this case, consumer surplus is lower than the free-market case by area m + n + o, but there is no taxpayer effect compared to the free-market case.

From the graphic analysis in Fig. 8.1, some qualitative results are clear while others are ambiguous. If demand for agricultural output is inelastic and chemicals are sufficiently important in production, then each of the three post-GATT policy alternatives can be administered at levels that restore producers to pre-GATT welfare levels. Also, consumers clearly lose in moving to each of the three post-GATT alternatives but taxpayers gain. Because both consumer and taxpayer effects tend to be highly dispersed and, to some extent, fall on the same individuals, a crucial question is whether the sum of the consumer and taxpayer effects is positive or negative. The answer to this question cannot be determined from the general conceptual analysis. One can show that net consumer/taxpayer effects depend on elasticities. Presumably, however, elasticities would be such that consumers/taxpayers lose under the various post-GATT alternative policies if producers are fully restored because otherwise such policies would probably have been adopted without a GATT agreement.

The point of this conceptual analysis is that policy reform must take into account the institutionally feasible mechanisms for compensation and the implications of political feasibility and sustainability for policy choice. A useful analysis must depict the related effects on each of the major political interest groups that are distinctly and substantially affected by policy reform. The conceptual analysis in this section points out that if political feasibility/sustainability requires making producers whole, then restricting institutional feasibility to green-light policies may not improve overall social welfare as measured by the standard concept of economic efficiency. The problem of political sustainability of decoupled payments underscores the importance of more detailed empirical work to examine these issues in a dynamic context. The empirical model discussed in the next section attempts to depict the dynamic distribution of effects of the policy instrument reforms on each of the important special interest groups in US agriculture over a meaningful time horizon (10 years). The intent is to examine whether the available mix of instruments can compensate all special interest groups sufficiently to make policy reform sustainable while attaining an improvement in economic efficiency.

8.3 Necessary Econometric Principles for Modeling Policy Reform

Understanding both the distributional and temporal effects of agricultural policy is crucial to analyzing policy reform. This problem is complicated by the need to analyze frequent revisions, not only in the levels of policy instruments but also by changes in the active set of instruments. Rausser and Just (1982) and Just (1990) discuss several important principles necessary to model the role of alternative sets of policy instruments in determining agricultural commodity price and policy effects.

- First, a substantial a priori structure must be imposed in modeling the global properties and qualitative characteristics of functions governing policy response. Simply approximating the local properties of functions according to traditional econometric practice can be highly misleading when the estimated relationships are used to extrapolate beyond observed data. For example, as a target price is lowered to the point of ineffectiveness, a relationship estimated during a sample period where target prices are always effective can perform poorly and generate inappropriate implications.
- Second, as major changes in policies take place, some of the variables that are traditionally considered exogenous may require endogenous modeling. That is, as large changes in agricultural commodity prices are induced, prices in related markets may adjust substantially. Thus, prices of inputs, for example, may require endogenous consideration as opposed to the usual exogenous consideration they receive in many crop supply response models. If these adjustments in related markets are not considered, the estimates of effects in the markets in question may be misleading. Considerations of related markets may also lead to substantial dynamic implications as well, for example, as when livestock models generating feed demand are incorporated into grain supply-and-demand models.
- Third, heavily structured relationships need to be imposed in an econometric estimation. Kinked relationships associated with price supports and acreage controls are not likely to be reflected well by ad hoc functional forms without using excessive degrees of freedom. On the other hand, sufficient imposition of such functional structure can lead to a better global representation of policy instrument effects which is necessary in examining policy instrument phaseout.
- Fourth, an appropriate standard of comparison for development of heavily structured econometric models is the degree of fit obtained in ad hoc modeling efforts. Developing heavily structured models for empirical analysis is much like constructing theoretical models. In principle, an infinite set of alternatives

are possible. Without a standard of comparison, however, there is no reason to expect this process to converge on any representation or understanding of a market. Yet, when a heavily structured model performs comparably to a standard ad hoc or flexible model in the literature (supposedly the best economists have to offer), then an acceptable degree of empirical validity is achieved.

This approach was used in developing the model reported here. Its attraction and justification lie in the fact that flexible and unstructured econometric relationships cannot satisfy the demands of policy reform analysis, while noneconometric approaches such as programming models do not sufficiently fit a substantial history of observed data without artificial calibration.

8.4 Overview of the Model

This section describes an estimated market model of wheat and feed grains (which also necessitates modeling livestock markets) that depicts the global role of agricultural policy instruments. A number of innovations were introduced into conventional econometric models with the purpose of achieving robustness over a 10-year horizon. These innovations include logit equations to represent program participation decisions; theoretically imposed structural forms; slippage effects of government acreage reduction programs; specific types of government demand for stocks; composite convex pricing equations to enhance robustness; asymmetric equations for farmer-owned reserve and governmental stock accumulations versus sales; and intertemporally structured land-asset equations. The estimated model used sample data through 1986-87. The 10-year period over which the effects of policy reforms are evaluated is 1985 through 1994. Effects are evaluated by comparison to the case of pre-GATT policies as represented by 1986 policy instruments. Typical validation procedures reveal that the quantified model generates plausible ex ante simultaneous forecasts over periods for which data are now available.

The grain demand component disaggregates demands by consumption, market inventory, and exports following the specifications of Chambers and Just (1981). Demand for government stock and the farmer-owned reserve follows the work of Rausser (1985) and Love (1987) with considerably more structure incorporated to reflect the qualitative nature of policy instruments. The livestock component follows along lines used by Just (1981) with revisions to incorporate some refinements developed by Rausser et al. (1986). The grain supply model uses logit equations to represent program participation decisions following the spirit of the work by Chambers and Foster (1983) and later empiricized by Rausser et al. (1986). The acreage equations depart significantly from previous econometric practice and incorporate more structure among important program and market variables in the spirit of the intuitive and conceptual framework developed by Gardner (1988) and Lins (1988). They examine the gains and losses associated with the wheat and corn programs by means of a quantitative graphical analysis of the various policy instruments through which wheat and feed grain commodity policies are administered.

8.4.1 The Crop Supply Structure

The basic form of the acreage equations is as follows: First, acreage in a market free of government programs is assumed to follow

$$A_f = A_f(\pi_n, \pi_a), \tag{1}$$

where A_f = free-market acreage of the crop in question; π_n = anticipated short-run profit per acre from production of the crop in question with free-market price; and π_a = anticipated short-run profit per acre from production of competing crops(s).

Profit per acre is defined by price times yield less per-acre production cost, e.g.,

$$\pi_n = P_m \cdot Y_e - C, \tag{2}$$

where P_m = market price; Y_e = expected yield; and C = short-run cost per acre.

When government programs are voluntary, the nonparticipating component of acreage is assumed to follow Eq. (1) on the associated portion of the acreage so that nonparticipating acreage is

$$A_n = (1 - \phi) \cdot A_f(\pi_n, \pi_a), \tag{3}$$

where A_n = nonparticipating acreage; and ϕ = rate of participation in the relevant government program.

The participating acreage is largely determined by program limitations with

$$A_p = B \cdot \phi (1 - \theta) - D(G_a) \tag{4}$$

where B = program base acreage; θ = minimum diversion requirement for participation; D = additional diversion beyond the minimum; and G_a = payment per acre for additional diversion.

The estimating equation for observed total acreage, given the participation level, is obtained by combining (3) and (4),

$$A_t = B \cdot \phi(1 - \theta) - D(G_a) + (1 - \phi) \cdot A_f(\pi_n, \pi_s), \tag{5}$$

where $D(\cdot)$ and $A_f(\cdot)$ follow linear specifications.

Determining the level of participation in this framework is crucial. Each farmer is assumed to participate if perceived profit per acre is greater under participation than under nonparticipation ($\pi^i{}_p > \pi^i{}_n$). Assuming that individual perceived profits differ from an aggregate by an amount characterized by an appropriate random distribution across farmers, the participation rate can be represented by a logistic relationship with

$$\ln \frac{\phi}{1 - \phi} = \phi^*(\pi_n - \pi_p), \tag{6}$$

where π_p = the profit per acre under compliance.

Given the qualitative nature of numerous agricultural policy instruments, a conceptually plausible specification of short-run profit per unit of land (producing plus diverted) on complying farms follows

$$\pi_p = (1 - \theta - \mu)\pi_z + \theta \cdot G_m + \mu \cdot max(G_v, \pi_p), \tag{7}$$

where μ is the maximum proportion of base acreage that can be diverted in addition to minimum diversion, G is the payment per unit of land for minimum diversion (zero if no payment is offered for minimum diversion), G_v is the payment per unit of land for voluntary diversion beyond the minimum, and π_z is the short-run profit per unit of producing land under compliance. The latter term suggests no voluntary additional diversion if $G_v < \pi_z$ and voluntary additional diversion to the maximum if $G_v > \pi_z$.

Conceptually, π_z follows

$$\pi_z = \left[\begin{array}{r} \max(P_t, P_m)Y_p + \max(P_s, P_m) \cdot \max(Y_a - Y_p, 0) \\ + \max(r_m - r_g, 0) \cdot P_s \cdot Y_a - C \end{array}\right], \tag{8}$$

where P_t is the government target price, Y_p is the program yield, P_s is the price support, r_m is the market rate of interest, and r_g is the government subsidized rate of interest on commodity loans under the program (Rausser et al. 1986). Eq. (8) reflects the complicated relationship through which a participating farmer is entitled to at least the target price on his program yield, at least the (lower) support price on all of his production, and gains an additional interest subsidy on a loan against his stored crop (at harvest time) evaluated at the support price. These benefits must be balanced against the opportunity loss of having to divert some land from production reflected by Eq. (7).

Once acreage is determined in this framework, it is simply multiplied by yield and added to carryin to determine crop supply. Of course, the relationships in Eqs. (7) and (8) do not necessarily apply exactly. For example, an uncertain anticipated market price may be discounted by a farmer compared to a target or support price which is known with certainty at the time of acreage decisions. Also, not all farmers place their crop under federal loan to take advantage of the interest subsidy. Nevertheless, intuition and experience implies that Eqs. (7) and (8) apply as reasonable approximations and, furthermore, the approximations apply in a global sense. By comparison, the large number of variables with numerous qualitative relationships involved in these relationships suggests significant problems with objective econometric identification and makes the possibility of obtaining even plausible signs remote with estimation of ad hoc or flexible forms.

To illustrate the difference in performance of the approach of simply adding π_p and G_v to Eq. (1),

$$A_f = A_f(\pi_n, \pi_p, \pi_a, G_v) \tag{9}$$

compared to that in Eqs. (5) and (6), both were used to estimate acreage response of wheat and feed grains in the United States over the period, 1962 to 1982, and

Table 8.1. The performance of structural versus ad hoc models: the case of US wheat and feed grain acreage[a]

Crop	Model definition	Estimation period	Forecast period	Standard error: Within sample	Standard error: Post-sample
	equation			million acres	
Wheat	(9)	1962–1982	1983–1986	4.41	14.90
Wheat	(5)	1962–1982	1983–1986	3.32	6.21
Wheat	(5), (6)	1962–1982	1983–1986	-[b]	9.07
Feed grain	(9)	1962–1982	1983–1987	1.73	6.40
Feed grain	(5)	1962–1982	1983–1987	6.26	6.38
Feed grain	(5), (6)	1962–1982	1983–1987	-[b]	5.50

[a] See the text for equations which define the various models.
[b] No within-sample error is computed since the model is derived by combining the estimated equations corresponding to (5) and (6).
Source: computed.

then to forecast acreage in the 1983–1986 period. The results are given in Table 8.1. The results for Eq. (5) take the participation rate as exogenous whereas the results where the model is specified as Eqs. (5) and (6) include forecasting errors for the participation rate as well.

In the case of feed grains, the ad hoc formulation leads to a much smaller standard error in the sample period than the structural form in Eq. (5), even though the structural form performs better than the ad hoc form in ex ante forecasting of the post-sample period. The model combining Eqs. (5) and (6) obtains an even lower standard error. In the case of wheat, the structural form fits the sample data better than the ad hoc form and performs substantially better in ex ante simulation.

This superior performance of the structural model carries through when errors in forecasting the participation rate are also considered. The reason the structural form can outperform the ad hoc model even in the sample period is that nonlinearities and kinks in response over a wide range of policy parameters put a premium on global properties of the function. The participation rate over the sample period ranges from zero (a kink point) to near 90% in others. As a result, the effects of profits with and without compliance cannot be well represented by a smooth approximating function.

The supply block for each crop is completed by a yield equation and a cost equation. The yield equation represents the effect of yield-increasing technology by a time trend but accounts for the slippage effect of government acreage reducing programs by a diversion effect

$$Y_a = Y_a(t, \theta \cdot \phi \cdot B) \tag{10}$$

where Y_a is the actual yield. This equation represents the phenomena whereby farmers divert the poorest acreage first, thus causing an increase in yields when government acreage restrictions are tight.

A cost equation is included because substantial changes in prices of inputs can occur with major changes in government programs such as contemplated in the policy reforms under consideration. Agricultural acreage equations have been estimated historically as a function of prices and possibly yields with the cost of input prices assumed fixed. Based on a Chicago School view of market adjustments, however, Gardner (1984) argues that the prices of inputs will be bid up until the cost of production is equal to output revenues. This suggests that the per-acre cost of production is not exogenous but rather is responsive to commodity prices. The cost equation used here follows the arguments by Gardner. Thus, the cost per acre is a function of the expected returns per acre,

$$C = C(P_m \cdot Y_e). \tag{11}$$

The empirical results here verify the validity of this view based on observed price variations of the past two decades.

8.4.2 The Crop Demand Structure

Following numerous earlier studies, the demand for crops is broken into food, feed, export, and inventory components for purposes of specification and estimation. The inventory component is further broken into farmer-owned reserve, government-owned, and market components for crops with government programs. The demand system for a given crop is, thus, of the form

$$\begin{aligned}
&Q_i = Q_i\,(P_m, Y_c)\\
&Q_f = Q_f(P_m, P_j)\\
&Q_x = Q_x(P_m, P_{m,-1}, \ldots, E)\\
&Q_r = Q_r(P_m - P_s, Q_{r,-1})\\
&Q_g = Q_g(P_m - P_s, Q_{g,-1}, Q_{r,-1}, D)\\
&Q_m = Q_m(P_m, Q_r, D, S_d)\\
&Q_{r,-1} + Q_{g,-1} + Q_{m,-1} + A_t \cdot Y_a = Q_i + Q_f + Q_x + Q_r + Q_g + Q_m\\
&S_d = Q_{r,t-1} + Q_{g,t-1} + Q_{m,t-1} + A_t \cdot Y_a - Q_x
\end{aligned} \tag{12}$$

including the supply-demand identity where "–1" subscripts refer to 1-year lags and

Q_z = quantity demanded (i = industry or food, f = feed, x = export, r = farmer-owned reserve, g = government stocks, m = market stocks);
P_m = market price;
Y_a = actual average yield;
Y_c = per capita consumer income;
F_j = numbers of various types of livestock on feed;
P_j = prices of various types of livestock meat;
E = trade-weighted exchange rate;
P_s = support price;

D = shift term reflecting the 1983 Payment-in-Kind (PIK) program;
S_d = domestic supply (total supply less exports).

The demand system was not estimated in the form of Eq. (12) because a system that determines price through an identity equation tends to produce erratic price estimates particularly when demands are inelastic. Alternatively, a demand equation in Eq. (12) can be solved for price,

$$P_m = Q_j^{-1} (Q_j, X_j) \tag{13}$$

where X_j represents the other variables included in the equation for Q_j, j = i, f, x, r, g, m. Then the identity can be used to determine Q_j. This approach suffers in practice because the coefficient estimates of exogenous variables in the inverted equation are susceptible to spurious correlations with other factors in the system. This can lead to an unreasonably large contribution of these variables relative to other exogenous variables in the system in determining price predictions in practice. The approach used in this study is to solve the system in (12) for a partial reduced-form price equation which tends to reduce forecasting errors on price.

This partial, reduced form equation can be regarded as a convex combination of equations such as (13) which essentially produces a composite price forecasting equation in the sense of Johnson and Rausser (1982), where the weights are estimated simultaneously with the coefficients of the price equation. For example, one possible price equation would involve inverting the first two equations in (12) to obtain forms as in (13) and then taking a convex combination of the two to obtain

$$P_m = \alpha Q_i^{-1}(Q_j, Y_c) + (1 - \alpha) Q^{-1}{}_f (Q_f, F_j, P_j), \tag{14}$$

where α is the coefficient determining the convex combination. The number of such equations to combine in this manner is roughly determined by the tradeoff between increased forecasting accuracy from combining more forecasting equations and the reduced identification as the total number of variables in the composite forecasting equation increases.

Once a price equation of this type is determined, the standard practice is to use it to replace one of the demand equations, letting that demand be determined by the supply-demand identity. This practice, however, tends to produce erratic predictions for the demand that is dropped. The reason is that all of the errors for quantities generated by other demand equations aggregate into a potentially large error for the quantity determined by the supply-demand identity. In this study an alternative approach is used whereby a demand equation is estimated for each demand as well as the price. Then for purposes of simulation and forecasting, the predicted price and demands are determined by finding the most likely combination of price and demands given estimated standard errors, etc., which satisfies the supply-demand identity.

This procedure can be most easily illustrated in vector notation. Let Q be an n-vector of demands following an n-vector demand equation

$$Q = (P_m, X) + \varepsilon \tag{15}$$

where X is a vector of all exogenous variables in the system and e is a vector of disturbances with $E(\varepsilon) = 0$ and covariance matrix $E(\varepsilon\varepsilon') = \Sigma$. The supply-demand identity is represented by

$$eQ = S, \tag{16}$$

where $e = [1, 1, \ldots 1]$ and S is available supply including carry in. The vector prediction of demands is then derived by maximizing likelihood with respect to the $n + 1$ equations represented by Eqs. (15) and (16). Specifically, where $f(\varepsilon;S)$ represents the probability density function of the random disturbances in Eq. (15), the ε used in Eq. (15) for simulation is the one which maximizes $f(\varepsilon;\Sigma)$ subject to Eqs. (15) and (16) where, of course, the demand equations in (15) and the covariance matrix in f are replaced by estimates.

To capture the qualitative nature of government program effects on the demand side, the farmer-owned reserve and government inventory demand equations are estimated using a qualitative relationship in the difference between market and support price. For example, the farmer-owned reserve demand equation (which applies after the institution of the 1977 Farm Bill) is of the form,

$$Q_r - Q_{r-1} = \begin{cases} a_1(P_m - \alpha P_s) + a_2(P_m - \alpha P_s)Q_{r,-1} & \text{if } P_m - \alpha P_s < 0 \\ a_3(P_m - \alpha P_s)Q_{r,-1} & \text{if } P_m - \alpha P_s > 0. \end{cases} \tag{17}$$

The first right-hand case applies if market price is low relative to the price support level. It reflects how more stocks are placed in the reserve when prices are low, but at a decreasing rate as the farmer-owned reserve "fills up." The second right-hand case reflects how stocks are taken out of the reserve as the market price increases relative to the support price. The current reserve level appears purely multiplicatively in this case to reflect the fact that no stocks can be taken out of the reserve when it becomes depleted. This is one sense in which a plausible qualitative global functional structure is necessary to derive reasonable policy implications for the case where current programs are phased out.

The government inventory demand equation is of the form,

$$Q_g - \beta Q_{g-1} = \begin{cases} b_1(1 - \Phi_{77})(P_m - \alpha P_s) + b_2\Phi_{77}(P_m - \alpha P_s)Q_{r,-1} & \text{if } P_m - \alpha P_s < 0 \\ b_3(P_m - \alpha P_s)Q_{g,-1} & \text{if } P_m - \alpha P_s > 0, \end{cases} \tag{18}$$

where Φ_{77} is an indicator variable for the farmer-owned reserve. The first right-hand case of this equation captures the qualitative relationship whereby stocks under Commodity Credit Corporation (CCC) loan are not turned over to the government until the market price falls to the government support level. The first term applies prior to the 1977 Farm Bill, while the second term applies after institution of the farmer-owned reserve and reflects declining use of traditional CCC channels when the farmer-owned reserve is not full. The second right-hand case reflects the tendency to reduce government stocks as the market price rises above the support price. However, stocks can only be released when they are held. Thus, as in the case of the farmer-owned reserve equation, a purely multiplicative term in carry-in stocks is used to capture the associated global relationship.

8.4.3 The Livestock Supply Structure

The supply of livestock accounts for the dynamic nature of breeding herd adjustment and the long lags in breeding and raising livestock to market weight necessary to capture the dynamics of feed demand. The basic form of the model for beef and pork is as follows. First, a stock equation is included for the size of the national breeding herd of the form,

$$H_i = H_i\,(P_c/P_i\,,\, H_{i,-1}\,,\, r_m), \tag{19}$$

where H_i is herd size (i = 1 for cattle, 2 for pork), P_c is the price of corn, and P_i is the price of meat (i = 1 for beef, 2 for pork). Then, a meat production equation is included of the form,

$$M_i = M_i\,(H_{i,-k}\,, H_i - H_{i,-1}\,, P_c/P_i), \tag{20}$$

where M_i is the production of meat. The term, $H_i - H_{i,-1}$, is included to capture the addition to meat production caused by culling breeding herds, while the term, $H_{i,-k}$, is included to reflect the production cycle from birth to slaughter weight based on the breeding herd with the appropriate lag (k = 2 for beef, 1 for pork). The price term captures the response of feeding intensity to market conditions.

The poultry supply model is a single equation model because of the short production cycle. It has broiler production as a simple function of the corn:broiler price ratio and a time trend to reflect rapid technological change that has occurred over the past two decades.

8.4.4 The Meat Demand Structure

The meat demand system is considered independently of the crop demand systems since meats and grains are not very closely related except as grain prices affect meat supply. Each demand equation is estimated in price-dependent form with

$$P_i/Y = P_i\,(P_j/Y,\, P_o/Y,\, C_i/N), \tag{21}$$

where Y is per capita income, P_j represents prices of other meats (included individually), P_o is a price index for nonfarm prices, C_i is domestic consumption of meat i, and N is population. The meat demand system is completed by net import/export equations of the form,

$$I_i = I_i\,(P_i/W,\, I_{i,-1}\,,\, E), \tag{22}$$

where I_i is net imports (negative for net exports) and E is a trade-weighted exchange rate and identities of the form,

$$M_i + I_i = C_i. \tag{23}$$

8.5 Variable Definition, Model Estimation, and Validation

The data used in estimating the model can be divided into two broad groups – data used in estimating supply, and data used in estimating demand. Data were collected from government publications and contacts with remaining variables computed as transformations of these data. The definitions of variables used for estimating supply of feed grains, wheat, and soybeans are presented (in alphabetical order) in Appendix A, along with sources. The definitions of variables used for estimating demand for feed grains, wheat, soybeans, and both demand and supply for livestock are presented in Appendix B (also in alphabetical order).

The explicit transformations made from basic data to obtain the computed variables used in econometric estimation are also summarized explicitly in Appendix A for the grain supply models and in Appendix B for the variables used in estimation of demand.

The estimated model is reported in detail in Appendix C. The crucial and highly structured government program participation and acreage equations for program crops are characterized by R^2 statistics exceeding 0.97. Yield equations are considerably lower because weather and crop disaster effects are included in the disturbances. Production cost equations have good fits with R^2 statistics from 0.7 to 0.8, based on one independent variable (expected revenue).

The four highly structured, government program-related demand equations for feed grains and wheat have R^2 statistics of 0.71 and 0.78 for farmer-owned reserve demand, respectively, and 0.97 and 0.87 for government demand, respectively.

Model validation is an important step in verifying the usefulness of a policy model. Typical validation procedures here reveal that the model performs acceptably. First, the model can reproduce historically observed data quite well. This validation step was done by simulating the model over the estimation period for each equation in the model (where lagged endogenous variables are evaluated at their predicted rather than actual values). The simulated values of endogenous variables and their actual historical values were plotted together over the respective estimation periods (not reported here to save space). In general, simulated values track historical values quite well. Based on these results and the estimation statistics reported above, the model appears to provide a sufficient basis for simulating the effects of US agricultural policy reform on corn, grain sorghum, and wheat.

8.6 The Base Case

The estimated model was used to consider the effects of implementation of policy reform over a 10-year period. Accordingly, a base case with no policy reform is needed for comparison. The 10-year period chosen for this purpose is 1985–1994. An effort was not made at this stage to estimate the model up to the latest possible

Table 8.2. Base case assumptions for policy instruments, 1986–1984

Target price for corn (TPC)	\$3.03 per bushel
Support price of corn (SPRC)	\$1.92 per bushel
Base acreage for corn (BAC)	81.0 million acres
Base acreage for sorghum (BAGS)	18.7 million acres
Diversion required for feed grains (DRFG)	0.2 (20%)
Diversion payment for corn (DPC)	\$8.21 per acre
Voluntary diversion allowed for feed grains (VDFG)	0 (0%)
Payment for voluntary diversion for corn (VDPC)	\$0.00 per acre
Weighted program yield for feed grains (YLDFGP)	89.97 bushels per acre
Target price for wheat (TPW)	\$4.38 per bushel
Support price for wheat (SPRW)	\$2.40 per bushel
Base acreage for wheat (BAW)	91.7 million acres
Diversion required for wheat (DRW)	0.25 (25%)
Diversion payment for wheat (DPW)	\$3.51 per acre
Voluntary diversion allowed for wheat (VDW)	0.1 (10%)
Payment for voluntary diversion of wheat (VDPW)	\$63.88 per acre
Program yeild for wheat (YLDWP)	31.94 bushels per acre
Subsidized CCC interest rate (ICCC)	6.6438%

date (which would allow the simulation to cover an entirely post-1990 time period corresponding to GATT negotiations).

For purposes of defining the base case, a number of assumptions must be made regarding evolution of the exogenous variables over the simulation period. Government program instruments are assumed to remain constant at their 1986 values from 1986 through 1994 (1985 values are used for 1985). The 1986 values are used rather than 1985 values because there is a more equitable relationship between the wheat and feed grain programs (e.g., both offer a positive but nominal payment for diversion). The specific instrument levels are given in Table 8.2.

Other (nonpolicy) exogenous variables are assumed to continue as in the recent past representing a stable macroeconomy. Other scenarios could be considered, but this allows the discussion to focus on the role of agricultural policy variables alone. Another consideration is the vulnerability of the agricultural economy to macroeconomic instability under various agricultural policy regimes, an issue examined in Chaps. 10 and 11. These considerations might also affect political feasibility considerations if they are perceived by special interest groups. The simple projections of nonpolicy exogenous variables are presented at the end of Appendix C.

Table 8.3. Simulation results for the base case selected years, 1985–1994

Variable	1985	1988	1991	1994
Feed grains program compliance, %	51.89	63.74	65.09	66.82
Wheat program compliance, %	60.00	20.99	66.40	65.98
Corn and sorghum acreage, million acres	97.80	87.97	86.85	86.28
Wheat acreage, million acres	79.20	68.61	79.55	74.29
Soybean acreage, million acres	67.80	57.01	55.28	55.30
Feed grains production, million metric tons	235.41	235.60	245.96	257.58
Wheat production, million metric tons	2,595.00	2,141.47	2,626.95	2,510.22
Soybean production, million metric tons	1,860.97	1,797.41	1,795.45	1,848.94
Feed grains cost of production, dollars per acre	130.86	136.61	148.35	154.50
Wheat cost of production, dollars per acre	53.28	59.84	56.72	64.46
Soybean cost of production, dollars per acre	66.33	64.86	74.73	79.80
Corn price, dollars per bushel	2.81	2.57	2.46	2.30
Sorghum price, dollars per bushel	2.54	2.33	2.23	2.08
Wheat price, dollars per bushel	3.35	3.90	2.96	3.54
Soybean price, dollars per bushel	7.21	6.12	6.83	7.03
Feed use of feed grains, million metric tons	130.60	127.31	128.07	129.69
Industry use of feed grains, million metric tons	33.83	41.58	47.88	52.38
Export of feed grains, million metric tons	45.56	41.51	47.34	51.78
FOR stocks of feed grains, million metric tons	20.94	32.51	35.34	35.89
Government stocks of feed grains, million metric tons	8.70	27.98	42.65	54.01
Market stocks of feed grains, million metric tons	41.19	39.20	40.22	44.56
Feed use of wheat, million bushels	426.97	319.33	417.26	481.60
Other domestic uses of wheat, million bushels	646.55	670.06	693.64	725.09
Export of wheat, million bushels	1,201.98	1,141.68	1,132.16	1,563.50
FOR stocks of wheat, million bushels	699.84	377.01	707.21	704.71
Government stocks of wheat, million bushels	360.46	110.79	586.09	668.77

Market stocks of wheat, million bushels	670.29	432.32	675.13	397.40
Soybean crush, million barrels	1,046.81	1,162.06	1,138.49	1,154.96
Export of soybeans, million bushels	682.08	633.49	694.77	676.07
Stocks of soybeans million bushels	307.78	384.32	329.35	325.45
Beef production, million bushels	24,020.00	23,147.00	22,952.00	22,732.00
Pork production, million bushels	13,160.00	14,636.00	15,235.00	15,674.00
Broiler production, million bushels	13,455.00	15,289.00	16,860.00	18,393.00
Cattle breeding herd size, 1000 head	46,239.00	43,920.00	42,763.00	42,633.00
Hog breeding herd size, 1000 head	5,607.00	6,583.00	7,019.00	7,328.00
Beef price, dollars per pound	2.42	2.50	2.59	2.70
Pork price, dollars per pound	1.90	1.81	1.83	1.87
Broiler price, dollars per pound	0.83	0.85	0.86	0.85
Crop farm market revenue, million dollars	48,135.00	43,184.00	43,878.00	45,172.00
Government payments, million dollars	2,357.00	2,226.00	4,380.00	4,430.00
Crop farm input expense, million dollars	21,516.00	19,821.00	21,527.00	22,532.00
Crop farm profit, million dollars	28,976.00	25,590.00	26,731.00	27,070.00
Livestock producer surplus, million dollars	878.00	828.00	856.00	904.00
Grains consumer surplus, million dollars	62,356.00	72,651.00	80,731.00	91,713.00
Meat consumer surplus, million dollars	340,389.00	369,269.00	394,816.00	417,319.00
Total consumer surplus, million dollars	402,745.00	441,920.00	475,547.00	509,033.00
Net domestic surplus, million dollars	432,599.00	468,338.00	503,133.00	537,007.00
Stockholder surplus, million dollars	32,932.00	29,514.00	41,340.00	26,227.00
Iowa land price, dollars per acre	1,551.64	1,715.36	1,885.33	2,058.53
Oklahoma land price, dollars per acre	726.29	794.52	872.43	950.16
US farmland price, dollars per acre	809.60	889.24	973.34	1,058.10
Total farmland value, million dollars	818,505.00	899,026.00	984,044.00	1,069,742
Export earnings, million dollars	13,981.00	12,525.00	12,686.00	14,968.00

Source: estimated.

The results of simulating the base case are given in Table 8.3 for selected years over the time horizon. Results are reported for each of the major endogenous variables in the model. The results appear quite plausible for a semi-depressed state of the agricultural sector. Program participation levels tend to rise slightly over the 10-year period with market prices declining slightly for corn and sorghum, and fluctuating in reasonable ranges for wheat, soybeans, and meat. Stocks, exports, and domestic use all vary in reasonable ranges according to historical standards. Government and farmer-owned reserve stocks tend to rise slightly consistent with the slightly weakening market conditions. Government payments for the feed grains program rise from $869 million in 1985 to $3.3 billion in 1994 while wheat program payments fall slightly from $1.5 billion in 1985 to $1.1 billion in 1994 (the latter rise to $1.8 billion in 1991). Taken as a whole, these results appear remarkably plausible as a 10-year future of the US agricultural sector with unchanged policy.

8.7 Evaluating the Well Being of Individual Groups Affected by Policy Reform

The evaluation of effects of policy reform involves many dimensions. First, many distinct special interest groups must be considered. Second, many different effects on these individual groups require consideration. The major considerations in this respect involve the temporal distribution of effects and the effects on short-term profits versus asset values and wealth. Some types of policy reform may involve a considerable increase in profits at some future time but cause politically untenable profit levels in the early transition period. Some policies tend to cause a major change in asset values in addition to changes in profits. Policies that attach value to base acreage can be properly evaluated only by estimating effects on land values.

With these considerations in mind, Table 8.3 evaluates the effects of the estimated changes in the endogenous variables in the model by computing a variety of welfare, profit, and asset value effects on several broad interest groups. While the model allows for these effects to be broken down by crop, they are summarized for all crop farmers as a group for purposes of brevity. Crop farmer effects are evaluated by investigating crop market revenue effects, government payment effects (which also reflect government budget exposure), variable crop production expenses, and short-run profits from crop production. These effects are translated into land price effects by state using a land price equation approximating that developed elsewhere by Just (1988). This is done by determining a profit-per-acre effect by crop (including government programs) and then translating those effects into an average profit-per-acre effect by state, using the acreage distribution among crops by state. These profit-per-acre effects by state are then translated into land price effects using the estimated parameters of the land price model developed by Just. While they are only reported here for two representative states, the land price effects can, in principle, be generated for all states and can be used to generate land price effects for the United States as a whole, as reported in Table 8.3. Iowa and Oklahoma were chosen here because Iowa has relatively more of its cropland in wheat than any other state.[2] Thus, they give a reasonable picture on the bounds of land price effects.

Conceptually, the crop-farm profit effects accumulated over a transition period can differ substantially from the asset values accumulated over a transition period. Land prices at the end of the transition period will depend on anticipated profitability beyond the transition period while the profits during the transition period may suffer from costs of adjustment, etc. Both aspects of crop farmer welfare are of interest. One reasonable measure of aggregate crop farmer well-being is to use the discounted sum of profits earned over the transition period plus discounted terminal wealth realized at the end of the transition period. Indeed, land prices at the end of the transition period conceptually reflect the discounted profits that can be earned with the cropland beyond the transition period after adjustments for risk, taxes, etc.

The problem of evaluating effects for livestock producers is similar. Changes in feed and livestock prices cause short-run changes in profits. However, livestock producers also respond by adjusting investment in breeding herds, thereby affecting wealth and future productivity. Effects on investment are a crucial dimension of economic welfare evaluation (Just et al. 1982). In Table 8.3, the short-run profit effects of feed grain, wheat, soybean, and livestock price changes on livestock producers are evaluated by computing the change in consumer surplus associated with feed demand. Where feed is regarded as a necessary input in livestock production and the feed demand is appropriately conditioned on livestock prices, this approach captures all of the short-run profit effects of feed and livestock price changes. In addition, however, the investment of livestock producers must be considered. For this purpose, the breeding herd size for both cattle and hogs is reported in Table 8.3. The value of investment can be calculated using the value-per-head equations for cattle and hogs together with these breeding herd sizes (not shown in Table 8.3).

In addition to these farm welfare effects, a welfare effect on grain consumers is calculated based on the consumer surplus associated with nonfeed disappearance of grains (e.g., industry and food demand). The welfare effect on meat consumers is calculated using consumer surplus associated with meat demand. Total consumer surplus is the sum of grain and meat consumer surplus. Net domestic surplus is calculated as the sum of short-run welfare effects on farmers (both grain and livestock) and consumers (both grain and meat). Stockholder surplus is also calculated as the consumer surplus associated with inventory demand. Finally, in the case of exports, the consumer surplus associated with export demand is not calculated since that is a welfare effect for foreign concerns. However, export

[2] The weights on corn and sorghum land, wheat land, and soybean land are 0.41, 0.0003, and 0.24, respectively, in Iowa; 0.02, 0.24, and 0.001, respectively, in Oklahoma; and 0.10, 0.007, and 0.06, respectively, in the United States as a whole. These weights are based on the 1985 distribution of land among crops.

earnings are estimated since that affects the US trade balance which is an additional concern in US policy considerations.

Turning to the results in Table 8.3 for the base case, the fluctuation in crop farmer profits appears consistent with the case of unchanged policy. Crop revenues tend to decline with a slight rebound at the end of the period while government payments tend to increase. Input expenses increase slightly over the period with profits fluctuating in a reasonable range. Livestock producer surplus fluctuates relatively little but net investment declines substantially (cattle inventory declines by 3.6 million head while breeding hog inventory increases by 1.7 million head). Grain consumer surplus increases as a result of depressed real prices and increasing demand due to population and income growth. Meat consumer surplus grows for the same reason. The latter two effects account for much of the increase in net domestic surplus over the 10-year period. Land prices rise, but only slightly. Increased productivity outweighs the effect of declining prices.

To summarize these profit, welfare, and asset value effects over the entire transition period, Table 8.4 computes the discounted value of the profit and welfare effects as of 1985, assuming a nominal discount rate of 10%. Also, the asset values for land, cattle, and hogs as of 1994 discounted to 1985 are given. Clearly, the asset values for crop and livestock producers are substantial compared to the discounted values of profits and producer surplus. It is also interesting to note that the largest welfare measures are associated with consumers rather than producers.

Table 8.4. Discounted 10-year time stream of welfare effects and terminal asset values of land and livestock (at 10%), base case, 1985–1994

Variable	Predicted (million dollars)
Discounted Welfare Effects	
Crop farm profit	179,721.00
Livestock producer surplus	5,757.00
Grains consumer surplus	499,041.00
Meat consumer surplus	2,529,724.00
Total consumer surplus	3,028,765.00
Net domestic surplus	3,214,243.00
Stockholder surplus	223,595.00
Government payments	20,169.00
Terminal Asset Values	
Total value of cropland	453,675.00
Total value of livestock	9,113.00

The reason is that demand for grains and meat as a whole are quite inelastic so that the area behind the demand curves and above price can be substantially larger than the total market value of grain or meat sales. Thus, the relative magnitudes of the surplus measures are plausible.

8.8 Alternative Policy Reform

A number of policy reform possibilities are under consideration. In this chapter, we examine three of the major alternatives that appear to be most likely in the context of proposal tables by the United States, Canada, and the European Community. Each of the three alternatives includes a phased reduction of current red-light (coupled distortionary) policies. To attain political feasibility, the first case considers direct payments to ill-affected groups. The other two cases consider attaining political feasibility by a phased introduction of two of the most likely green-light policies – land retirement and environmental preservation incentives – with the associated economic effects accomplishing the necessary compensation for political feasibility and sustainability indirectly.

8.8.1 Uncoupled Direct Income Payments

For the case of direct income payments, agricultural programs are assumed to be uncoupled in the truest sense so that all future direct payments are not capitalized into land values or base and farmers have no incentive to increase or maintain production for purposes of being eligible for future benefits. This requires that direct payments are determined on the basis of criteria that cannot be influenced by the farmer once the reform has begun. Thus, aside from affecting current farmer wealth, the effect is equivalent to reducing target and support prices in the model. The direct payments which replace these benefits have no effect on production or market behavior. One such possibility is to determine the payments by the amount of base held at the institution of the reform (e.g., 1985 in our case). With this approach, farmers have no incentive to plant more during the transition period to increase base or to be eligible for future benefits.

8.8.2 Environmental Protection Programs

While environmental protection is a multidimensional issue, it is considered here as an issue of pesticide control. (Nevertheless, similar principles apply to issues of water quality, soil erosion, etc.) Accordingly, the effect of environmental protection is represented by cost and yield effects of pesticide elimination. Cancellation of pesticide registration is one of the foremost issues in environmental protection as it interacts with agriculture. Some of the pesticides that have come under special review (consideration for cancellation) with the Environmental Protection Agency (EPA) have substantial cost and yield effects ranging up to 40% in some cases where no alternative pesticides are available. Given the highly inelastic demand faced by many crops, the possibility exists that farm prices and profits can actually be improved by pesticide elimination. If environmental policies are permitted, then one possibility is to replace part or all of the deficiency payment benefits currently received by farmers under the farm bill by higher profits through selective application of environmental policy or better coordination of environmental policy with agricultural policy (see Rausser and Foster 1991 for related discussion).

8.8.3 Programs to Remove Land from Production

Another set of potential, permitted policies are policies to remove land from production. Such policies also have the potential of being used to compensate farmers for loss of deficiency payments under current farm programs. First, the payment per acre for taking land out of production may be more attractive for farmers if it is higher on average than profits from production or considered to be more stable and reliable than production profits. More importantly, however, for crops that face inelastic demand, retiring land from production causes profits from production to increase because the price effect of reducing production exceeds the production effect.

8.9 Simulation of Policy Reform

These three alternative policy reform regimes are investigated as follows. First, in a "phase out" scenario, a rate of reduction for target and support prices is determined that will just phase out deficiency payments over a 10-year horizon. Then three alternatives are considered for restoring the same level of benefits to crop farmers. This restoration of benefits is considered as a necessary step to make the policy reform politically feasible. In the "direct payments" scenario, the necessary payments can be simply calculated on the basis of the phase out scenario since uncoupled payments will not alter the results for production and market behavior.[3] Next, an "environmental protection" scenario is developed to determine the rate of reduction in yields through pesticide cancellations that would be necessary to restore the same level of benefits to crop farmers as in the base case when direct payments are not used. Third, a "land retirement" scenario is developed to determine the rate of increase in payments for diversion that would be required to restore the same level of benefits to crop farmers as in the base case where neither direct payments nor pesticide cancellations are used. Finally, some additional scenarios are developed where various combinations of the three approaches are used to restore benefits.

---

[3] Here, we assume that farmers are profit maximizers. If farmers are risk averse, then behavior could be altered somewhat just by a change in wealth associated with the direct payments.

8.9.1 The Phase-Out Scenario

Through an iterative search procedure, a rate of reduction for target and support prices was determined that just phases out all deficiency payments by 1994 by beginning the simulation in 1985 as described otherwise for the base case above. The necessary rate of reduction (assuming both target and support prices of both feed grains and wheat are reduced at the same rate) is 5.8% per year. The associated effects on market variables and the welfare of various groups (for comparison with Tables 8.3 and 8.4) are given in Tables 8.5 through 8.9.[4] The "change" columns in these tables report changes from the predictions under the base case scenario in Tables 8.3 and 8.4.

Note that the simulations assume that the policy revisions are put in place late enough in 1985 so that production is unaffected in 1985. Substantial effects begin in the following years with participation levels dropping off by 50–60%, in the case of corn. While participation initially drops substantially in the wheat program, it later returns because of sufficiently depressed market conditions and because diversion begins to become more attractive relative to production under depressed market conditions. Production effects are minor in the early years (about 1.5%), but production eventually rises substantially above the base case because of decreased participation and diversion. These production effects in conjunction with reduced government and farmer-owned reserve demand cause prices to drop below the base case by about 23% for corn and sorghum and by 16% for wheat.

These price impacts and their spill-over effects cause annual crop farm profits to decline by about 36% by the end of the transition period with about half of that loss due directly to a reduction in government payments. The discounted value of crop profits over the 10-year period declines by about 19% because losses are less in the earlier years. Livestock producers gain from the depressed feed prices but most of the gains go to meat consumers in the form of lower meat prices and increased meat consumption.

Overall, the discounted net domestic surplus in Table 8.9 shows a slight decline of less than 1% as a result of phasing out deficiency payments. While the result of a decline may seem contrary to the theoretical implications of eliminating market distortions, two considerations must be borne in mind. First, because of substantial dynamics in the livestock industry, adjustments are not yet complete as of 1994. The high costs of adjustment that are necessarily absorbed by imposing change on an agricultural sector with such sluggish adjustment may prevent realization of a positive welfare effect of removing distortions during even a lengthy transition period. Second, one should bear in mind that the domestic surpluses in this "net domestic surplus" calculation include only grain and livestock producers and grain and livestock consumers. The gains to government and domestic private inventory holders as a result of reduced government storage are not included.

---

[4] Although Table 8.8 shows some minor amounts of government payments at the end of the transition period in 1994, these are payments for diversion rather than deficiency payments. Payments for taking land out of production are under consideration as a permitted policy, so only a phase out of deficiency payments was imposed in this policy scenario.

Table 8.5. Simulation results for the phase-out case, 1985

Variable	Predicted	Change	Percent change
Feed grains program compliance, %	51.89	0.00	0.00
Wheat program compliance, %	60.00	0.00	0.00
Corn and sorghum acreage, million acres	97.80	0.00	0.00
Wheat acreage, million acres	79.20	0.00	0.00
Soybean acreage, million acres	67.80	0.00	0.00
Feed grains production, million metric tons	235.41	–0.00	–0.00
Wheat production, million metric tons	2,595.00	0.00	0.00
Soybean production, million metric tons	1,860.97	–0.00	–0.00
Feed grains cost of production, dollars per acre	129.90	–0.96	–0.73
Wheat cost of production, dollars per acre	52.47	–0.82	–1.53
Soybean cost of production, dollars per acre	66.33	0.00	0.00
Corn price, dollars per bushel	2.76	–0.05	–1.70
Sorghum price, dollars per bushel	2.50	–0.04	–1.70
Wheat price, dollars per bushel	3.22	–0.12	–3.71
Soybean price, dollars per bushel	7.21	0.00	0.00
Feed use of feed grains, million metric tons	132.00	1.40	1.07
Industry use of feed grains, million metric tons	34.25	0.42	1.25
Export of feed grains, million metric tons	46.08	0.52	1.14
FOR stock of feed grains, million metric tons	18.67	–2.27	–10.84
Government stocks of feed grains, million metric tons	5.93	–2.77	–31.81
Market stocks of feed grains, million metric tons	43.88	2.69	6.54
Feed use of wheat, million bushels	439.13	12.15	2.85
Other domestic uses of wheat, million bushels	649.84	3.28	0.51
Export of wheat, million bushels	1,232.10	30.12	2.51
FOR stocks of wheat, million bushels	683.08	–16.77	–2.40
Government stocks of wheat, million bushels	316.40	–44.07	–12.23
Market stocks of wheat, million bushels	685.57	15.28	2.28
Soybean crush, million bushels	1,046.81	–0.00	–0.00
Export of soybeans, million bushels	682.08	–0.00	–0.00
Stocks of soybeans, million bushels	307.78	–0.00	–0.00
Beef production, million bushels	24,020.00	–0.00	–0.00
Pork production, million bushels	13,160.00	0.00	0.00
Broiler production, million bushels	13,455.00	–0.00	–0.00
Cattle breeding herd size, 1000 head	46,239.00	0.00	0.00
Hog breeding herd size, 1000 head	5,607.00	0.00	0.00
Beef price, dollars per pound	2.42	–0.00	–0.00
Pork price, dollars per pound	1.90	0.00	0.00
Broiler price, dollars per pound	0.83	0.00	0.00
Crop farm market revenue, million dollars	47,371.00	–764.00	–1.59

Table 8.5 (continued)

Variable	Predicted	Change	Percent change
Government payments, million dollars	2,668.00	311.00	13.21
Crop farm input expense, million dollars	21,357.00	–159.00	–0.74
Crop farm profit, million dollars	28,682.00	–294.00	–1.01
Livestock producer surplus, million dollars	896.00	19.00	2.16
Grains consumer surplus, million dollars	62,903.00	547.00	0.88
Meat consumer surplus, million dollars	340,389.00	0.00	0.00
Total consumer surplus, million dollars	403,292.00	547.00	0.14
Net domestic surplus, million dollars	432,870.00	272.00	0.06
Stockholder surplus, million dollars	33,293.00	991.00	3.01
Iowa land price, dollars per acre	1,547.21	–4.43	–0.29
Oklahoma land price, dollars per acre	721.26	–5.03	–0.69
US farmland price, dollars per acre	807.02	–2.58	–0.32
Total farmland value, millions dollars	815,898.00	–2,607.00	–0.32
Export earnings, million dollars	13,900.00	–81.00	–0.58

Source: estimated.

Table 8.6. Simulation results for the phase-out case, 1988

Variable	Predicted	Change	Percent change
Feed grains program compliance, %	53.20	–10.54	–16.53
Wheat program compliance, %	5.33	–15.66	–74.60
Corn and sorghum acreage, million acres	89.90	1.93	2.20
Wheat acreage, million acres	68.24	–0.37	–0.54
Soybean acreage, million acres	57.21	0.21	0.37
Feed grains production, million metric tons	239.15	3.55	1.51
Wheat production, million metric tons	2,105.58	–35.89	–1.68
Soybean production, million metric tons	1,797.46	0.05	0.00
Feed grains cost of production, dollars per acre	129.80	–6.81	–4.98
Wheat cost of production, dollars per acre	59.20	–0.64	–1.06
Soybean cost of production, dollars per acre	64.29	–0.57	–0.87
Corn price, dollars per bushel	2.27	–0.31	–11.90
Sorghum price, dollars per bushel	2.05	–0.28	–11.90
Wheat price, dollars per bushel	3.81	–0.09	–2.19
Soybean price, dollars per bushel	6.03	–0.09	–1.41
Feed use of feed grains, million metric tons	132.56	5.24	4.12
Industry use of feed grains, million metric tons	43.07	1.48	3.57
Export of feed grains, million metric tons	44.70	3.20	7.70
FOR stocks of feed grains, million metric tons	26.80	–5.71	–17.57

Table 8.6 (continued)

Variable	Predicted	Change	Percent change
Government stocks of feed grains, million metric tons	9.32	−18.66	−66.70
Market stocks of feed grains, million metric tons	50.09	10.89	27.77
Feed use of wheat, million bushels	284.26	−35.07	−10.98
Other domestic use of wheat, million bushels	667.77	−2.29	−0.34
Exports of wheat, million bushels	1,150.94	9.26	0.81
FOR stocks of wheat, million bushels	207.13	−169.89	−45.06
Government stocks of wheat, million bushels	93.20	−17.59	−15.87
Market stocks of wheat, million bushels	490.77	58.45	13.52
Soybean crush, million bushels	1,160.80	−1.27	−0.11
Export of soybeans, million bushels	633.65	−0.16	0.02
Stocks of soybeans, million bushels	383.81	−0.50	−0.13
Beef production, million bushels	22,835.00	−312.00	−1.35
Pork production, million bushels	14,904.00	268.00	1.83
Broiler production, million bushels	15,403.00	114.00	0.74
Cattle breeding herd size, 1000 head	45,173.00	1,253.00	2.85
Hog breeding herd size, 1000 head	6,757.00	174.00	2.65
Beef price, dollars per pound	2.52	0.02	0.77
Pork price, dollars per pound	1.77	−0.04	−2.07
Broiler price, dollars per pound	0.83	−0.01	−1.58
Crop farm market revenue, million dollars	40,190.00	−2,994.00	−6.93
Government payments, million dollars	1,024.00	1,202.00	−54.00
Crop farm input expense, million dollars	19,388.00	−433.00	−2.18
Crop farm profit, million dollars	21,826.00	−3,764.00	−14.71
Livestock producer surplus, million dollars	892.00	63.00	7.65
Grains consumer surplus, million dollars	72,242.00	−409.00	−0.56
Meat consumer surplus, million dollars	369,619.00	351.00	0.09
Total consumer surplus, million dollars	441,861.00	−59.00	−0.01
Net domestic surplus, million dollars	464,579.00	−3,759.00	−0.80
Stockholder surplus, million dollars	32,301.00	2,787.00	9.44
Iowa land price, dollars per acre	1,651.45	−63.92	−3.73
Oklahoma land price, dollars per acre	782.65	−11.87	−1.49
US farmland price, dollars per acre	871.03	−18.22	−2.05
Total farmland value, million dollars	880,608.00	−18,418.00	−2.05
Export earnings, million dollars	12,195.00	−331.00	−2.64

Source: estimated.

Table 8.7. Simulation results for the phase-out case, 1991

Variable	Predicted	Change	Percent change
Feed grains program compliance, %	24.47	–40.61	–62.40
Wheat program compliance, %	64.40	–2.00	–3.02
Corn and sorghum acreage, million acres	92.63	5.79	6.66
Wheat acreage, million acres	84.53	4.98	6.26
Soybean acreage, million acres	56.08	0.80	1.45
Feed grains production, million metric tons	256.28	10.32	4.20
Wheat production, million metric tons	2,787.41	160.46	6.11
Soybean production, million metric tons	1,796.89	1.44	0.08
Feed grains cost of production, dollars per acre	135.94	–12.41	–8.37
Wheat cost of production, dollars per acre	51.10	–5.62	–9.91
Soybean cost of production, dollars per acre	73.41	–1.32	–1.77
Corn price, dollars per bushel	1.97	–0.49	–19.79
Sorghum price, dollars per bushel	1.79	–0.44	–19.79
Wheat price, dollars per bushel	2.14	–0.82	–27.59
Soybean price, dollars per bushel	6.65	–0.18	–2.62
Feed use of feed grains, million metric tons	134.70	6.63	5.17
Industry use of feed grains, million metric tons	49.65	1.77	3.70
Export of feed grains, million metric tons	53.84	6.50	13.73
FOR stocks of feed grains, million metric tons	27.81	–7.54	–21.32
Government stocks of feed grains, million metric tons	3.35	–39.30	–92.14
Market stocks of feed grains, million metric tons	55.66	15.44	38.38
Feed use of wheat, million bushels	478.96	61.70	14.79
Other domestic uses of wheat, million bushels	699.08	5.44	0.78
Export of wheat, million bushels	1,283.46	151.30	13.36
FOR stocks of wheat, million bushels	580.14	–127.07	–17.97
Government stocks of wheat, million bushels	142.54	–443.54	–75.68
Market stocks of wheat, million bushels	903.68	228.55	33.85
Soybean crush, million bushels	1,133.72	–4.77	–0.42
Export of soybeans, million bushels	704.79	10.03	1.44
Stocks of soybeans, million bushels	326.20	–3.15	–0.96
Beef production, million bushels	22,470.00	–482.00	–2.10
Pork production, million bushels	15,784.00	550.00	3.61
Broiler production, million bushels	17,051.00	191.00	1.14
Cattle breeding herd size, 1000 head	46,582.00	3,819.00	8.93
Hog breeding herd size, 1000 head	7,389.00	370.00	5.27
Beef price, dollars per pound	2.61	0.03	1.11
Pork price, dollars per pound	1.75	–0.08	–4.28
Broiler price, dollars per pound	0.83	–0.02	–2.90
Crop farm market revenue, million dollars	37,849.00	–6,029.00	–13.74
Government payments, million dollars	1,446.00	–2,934.00	–66.98

Table 8.7 (continued)

Variable	Predicted	Change	Percent change
Crop farm input expense, million dollars	21,029.00	–498.00	–2.31
Crop farm profit, million dollars	18,266.00	–8,464.00	–31.67
Livestock producer surplus, million dollars	931.00	76.00	8.82
Grains consumer surplus, million dollars	81,731.00	1,000.00	1.24
Meat consumer surplus, million dollars	395,912.00	1,096.00	0.28
Total consumer surplus, million dollars	477,643.00	2,096.00	0.44
Net domestic surplus, million dollars	496,840.00	–6,293.00	–1.25
Stockholder surplus, million dollars	61,307.00	19,967.00	48.30
Iowa land price, dollars per acre	1,724.54	–160.79	–8.53
Oklahoma land price, dollars per acre	827.50	–44.94	–5.15
US farmland price, dollars per acre	922.74	–50.59	–5.20
Total farmland value, million dollars	932,894.00	–51,150.00	–5.20
Export earnings, million dollars	11,625.00	–1,061.00	–8.36

Source: estimated.

Table 8.8. Simulation results for the phase-out case, 1994

Variable	Predicted	Change	Percent
Feed grains program compliance, %	29.53	–37.30	–55.81
Wheat program compliance, %	59.84	–6.14	–9.31
Corn and sorghum acreage, million acres	90.24	3.96	4.58
Wheat acreage, million acres	81.48	7.19	9.68
Soybean acreage, million acres	56.48	1.19	2.14
Feed grains production, million metric tons	264.20	6.62	2.57
Wheat production, million metric tons	2,741.77	231.55	9.22
Soybean production, million metric tons	1,865.81	16.87	0.91
Feed grains cost of production, dollars per acre	140.19	–14.31	–9.26
Wheat cost of production, dollars per acre	60.46	–3.99	–6.20
Soybean cost of production, dollars per acre	77.50	–2.30	–2.88
Corn price, dollars per bushel	1.78	–0.52	–22.61
Sorghum price, dollars per bushel	1.61	–0.47	–22.61
Wheat price, dollars per bushel	2.98	–0.56	–15.72
Soybean price, dollars per bushel	6.74	–0.29	–4.15
Feed use of feed grains, million metric tons	134.41	4.73	3.64
Industry use of feed grains, million metric tons	53.51	1.13	2.15
Export of feed grains, million metric tons	59.00	7.22	13.95
FOR stocks of feed grains, million metric tons	9.87	–26.02	–72.49
Government stocks of feed grains, million metric tons	0.65	–53.36	–98.79

Table 8.8 (continued)

Variable	Predicted	Change	Percent
Market stocks of feed grains, million metric tons	73.57	29.01	65.11
Feed use of wheat, million bushels	511.31	29.71	6.17
Other domestic uses of wheat, million bushels	729.29	4.20	0.58
Export of wheat, million bushels	1,861.37	297.88	19.05
FOR stocks of wheat, million bushels	293.48	–411.24	–58.36
Government stocks of wheat, million bushels	83.19	–585.58	–87.56
Market stocks of wheat, million bushels	652.27	254.87	64.13
Soybean crush, million bushels	1,148.36	–6.60	–0.57
Export of soybeans, million bushels	698.70	22.63	3.35
Stocks of soybeans, million bushels	321.25	–4.21	–1.29
Beef production, million bushels	22,177.00	–555.00	–2.44
Pork production, million bushels	16,444.00	770.00	4.91
Broiler production, million bushels	18,640.00	247.00	1.34
Cattle breeding herd size, 1000 head	49,816.00	7,183.00	16.85
Hog breeding herd size, 1000 head	7,854.00	526.00	7.18
Beef price, dollars per pound	2.73	0.03	1.10
Pork price, dollars per pound	1.75	–0.11	–6.11
Broiler price, dollars per pound	0.82	–0.03	–4.10
Crop farm market revenue, million dollars	39,238.00	–5,933.00	–13.14
Government payments, million dollars	97.00	–4,333.00	–97.81
Crop farm input expense, million dollars	21,955.00	–577.00	–2.56
Crop farm profit, million dollars	17,380.00	–9,690.00	–35.79
Livestock producer surplus, million dollars	947.00	43.00	4.73
Grains consumer surplus, million dollars	92,481.00	767.00	0.84
Meat consumer surplus, million dollars	419,322.00	2,003.00	0.48
Total consumer surplus, million dollars	511,803.00	2,770.00	0.54
Net domestic surplus, million dollars	530,130.00	–6,877.00	–1.28
Stockholder surplus, million dollars	42,151.00	15,924.00	60.72
Iowa land price, dollars per acre	1,782.61	–275.91	–13.40
Oklahoma land price, dollars per acre	872.03	–78.13	–8.22
US farmland price, dollars per acre	970.93	–87.17	–8.24
Total farmland value, million dollars	981,609.00	–88,133.00	–8.24
Export earnings, million dollars	14,389.00	–579.00	–3.87

Source: estimated.

Table 8.9. Discounted 10-year time stream of welfare effects and terminal asset values of land and livestock (at 10%), phase-out case, 1985–1994

Variable	Predicted	Change	Percent change
	millions of dollars		
Discounted Welfare Effects			
Crop farm profit	144,932.00	–34,780.00	–19.36
Livestock producer surplus	6,132.00	375.00	6.51
Grains consumer surplus	503,675.00	4,634.00	0.93
Meat consumer surplus	2,534,462.00	4,738.00	0.19
Total consumer surplus	3,038,137.00	9,372.00	0.31
Net domestic surplus	3,189,200.00	–25,043.00	0.78
Stockholder surplus	283,498.00	59,903.00	26.79
Government payments	8,239.00	–11,930.00	–59.15
Terminal Asset Values			
Total value of crop land	416,298.00	–37,377.00	–8.24
Total value of livestock	10,745.00	1,632.00	17.91
Overall net welfare effect	11,044.00		

Their inclusion leads to a net result consistent with theory (a net gain of $46.8 billion).

An additional dimension of the picture suggested by Table 8.9, however, is the wealth effects at the end of the transition period. Cropland owners lose $37.4 billion in discounted wealth as a result of the curtailment of deficiency payments. While cropland value theoretically does not affect economic efficiency under profit maximization (assuming competitive capital markets, etc.), this is a welfare effect that cannot be ignored in considering the political feasibility of policy reform. Similarly, livestock producers gain about $1.6 billion in wealth associated with holding breeding herds as a result of increasing profitability with declining feed prices.

To make this policy reform palatable for crop farmers would thus require direct payments sufficient to compensate for both lost profits during the 10-year transition period (discounted value of $34.8 billion in 1985) and the lost wealth at the end of the transition period (discounted value of $37.4 billion in 1985) for a total of $72.2 billion in 1985 discounted value. Of course, a reasonable approach would be to distribute these direct payments in a series of annual payments over the adjustment period rather than make them as a lump sum at the beginning of the period. Otherwise, the crop farmers would have an incentive to lobby for a return of deficiency payments after having received the lump sum distributions. For example, a payment of $11.74 billion per year for 10 years would be equivalent to $72.2 billion in 1985 at a 10% discount rate. If these payments were made in a truly uncoupled fashion, say, on the basis of base acreage held in 1985, then this would amount to a payment of $61 per acre of base held in corn, sorghum, and wheat per year for 10 years. Since all other groups gain from curtailing deficiency payments, no other compensation would be required to make the policy reform politically sustainable.

8.9.2 The Environmental Protection Scenario

Given the 5.8% rate of reduction in target and support prices in the phase-out scenario, an iterative procedure was used to determine the rate of yield decrease per year which would be required as a result of pesticide cancellation to just restore crop farmer discounted profits to the base case level in lieu of direct payments. This rate of reduction in yields was approximately 6.1%. The reader should bear in mind, however, that this is a rate of reduction from the yields that would occur in the base case as a result of normal technological development. Normal technological development has resulted in growth in yields per year of approximately 3.7, 1.7, and 2.3% for feed grains, wheat, and soybeans, respectively. Thus, this scenario would correspond to an actual fall in yields of about 2.3% per year for feed grains and 4.3% for wheat.

The results for this case are presented in Tables 8.10 through 8.14 and correspond to Tables 8.3 and 8.4 for the base case and Tables 8.5 through 8.9 for the phase-out case. This case leads to a much greater decline in program participation because of the rising prices caused by yield reductions. The corn price rises to $2.91 in 1988, $4.07 in 1991, and $5.05 in 1994. Wheat price rises to $4.96 in

Table 8.10. Simulation results for the environmental enhancement case, 1985

Variable	Predicted	Change	Percent change
Feed grains program compliance, %	51.89	0.00	0.00
Wheat program compliance, %	60.00	0.00	0.00
Corn and sorghum acreage, million acres	97.80	0.00	0.00
Wheat acreage, million acres	79.20	0.00	0.00
Soybean acreage, million acres	67.80	0.00	0.00
Feed grains production, million metric tons	235.41	−0.00	−0.00
Wheat production, million metric tons	2,595.00	0.00	0.00
Soybean production, million metric tons	1,860.97	−0.00	−0.00
Feed grains cost of production, dollars per acre	126.51	−4.35	−3.32
Wheat cost of production, dollars per acre	51.17	−2.11	−3.96
Soybean cost of production, dollars per acre	64.02	−2.31	−3.48
Corn price, dollars per bushel	2.76	−0.05	−1.70
Sorghum price, dollars per bushel	2.50	−0.04	−1.70
Wheat price, dollars per bushel	3.22	−0.12	−3.71
Soybean price, dollars per bushel	7.21	0.00	0.00
Feed use of feed grains, million metric tons	132.00	1.40	1.07
Industry use of feed grains, million metric tons	34.25	0.42	1.25
Export of feed grains, million metric tons	46.08	0.52	1.14
FOR stocks of feed grains, million metric tons	18.67	−2.27	−10.84
Government stocks of feed grains, million metric tons	5.93	−2.77	−31.81
Market stocks of feed grains, million metric tons	43.88	2.69	6.54
Feed use of wheat, million bushels	439.13	12.15	2.85
Other domestic uses of wheat, million bushels	649.84	3.28	0.51
Export of wheat, million bushels	1,232.10	30.12	2.51
FOR stocks of wheat, million bushels	683.08	−16.77	−2.40
Government stocks of wheat, million bushels	316.40	−44.07	−12.23
Market stocks of wheat, million bushels	685.57	15.28	2.28
Soybean crush, million bushels	1,046.81	−0.00	−0.00
Export of soybeans, million bushels	682.08	−0.00	−0.00
Stocks of soybeans, million bushels	307.78	−0.00	−0.00
Beef production, million bushels	24,020.00	−0.00	−0.00
Pork production, million bushels	13,160.00	0.00	0.00
Broiler production, million bushels	13,455.00	−0.00	−0.00
Cattle breeding herd size, 1000 head	46,239.00	0.00	0.00
Hog breeding herd size, 1000 head	5,607.00	0.00	0.00
Beef price, dollars per pound	2.42	−0.00	−0.00
Pork price, dollars per pound	1.90	0.00	0.00
Broiler price, dollars per pound	0.83	0.00	0.00
Crop farm market revenue, million dollars	47,371.00	−764.00	−1.59
Government payments, million dollars	2,668.00	311.00	13.21

Table 8.10 (continued)

Variable	Predicted	Change	Percent change
Crop farm input expense, million dollars	20,767.00	–749.00	–3.48
Crop farm profit, million dollars	29,272.00	–296.00	–1.02
Livestock producer surplus, million dollars	896.00	19.00	2.16
Grains consumer surplus, million dollars	62,903.00	547.00	0.88
Meat consumer surplus, million dollars	340,389.00	0.00	0.00
Total consumer surplus, million dollars	403,292.00	547.00	0.14
Net domestic surplus, million dollars	433,461.00	862.00	0.20
Stockholder surplus, million dollar	33,923.00	991.00	3.01
Iowa land price, dollars per acre	1,549.19	–2.46	–0.16
Oklahoma land price, dollars per acre	721.64	–4.65	–0.64
US farmland price, dollars per acre	807.60	–2.00	–0.25
Total farmland value, millions dollars	816,482.00	–2,022.00	–0.25
Export earnings, million dollars	13,900.00	–81.00	–0.58

Source: estimated.

Table 8.11. Simulation results for the environmental enhancement case, 1988

Variable	Predicted	Change	Percent change
Feed grains program compliance, %	10.79	–52.95	–83.07
Wheat program compliance, %	5.42	–15.58	–74.20
Corn and sorghum acreage, million acres	98.18	10.21	11.61
Wheat acreage, million acres	69.77	1.16	1.69
Soybean acreage, million acres	56.98	–0.03	–0.05
Feed grains production, million metric tons	213.57	–22.03	–9.35
Wheat production, million metric tons	1,783.09	–358.38	–16.74
Soybean production, million metric tons	1,461.01	–336.40	–18.72
Feed grains cost of production, dollars per acre	129.44	–7.18	–5.25
Wheat cost of production, dollars per acre	59.53	–0.31	–0.51
Soybean cost of production, dollars per acre	66.07	1.22	1.88
Corn price, dollars per bushel	2.91	0.34	13.07
Sorghum price, dollars per bushel	2.63	0.30	13.07
Wheat price, dollars per bushel	4.96	1.07	27.37
Soybean price, dollars per bushel	8.20	2.08	34.02
Feed use of feed grains, million metric tons	119.68	–7.63	–5.99
Industry use of feed grains, million metric tons	39.36	–2.22	–5.35
Export of feed grains, million metric tons	38.67	–2.84	–6.84
FOR stocks of feed grains, million metric tons	00.0	–32.51	–100.00

Table 8.11 (continued)

Variable	Predicted	Change	Percent change
Government stocks of feed grains, million metric tons	1.94	–26.05	–93.08
Market stocks of feed grains, million metric tons	56.24	17.03	43.45
Feed use of wheat, million bushels	235.47	–83.86	–26.26
Other domestic uses of wheat, million bushels	653.82	–16.24	–2.42
Exports of wheat, million bushels	970.45	–171.23	–15.00
FOR stocks of wheat, million bushels	36.86	–340.16	–90.22
Government stocks of wheat, million bushels	87.83	–22.96	–20.72
Market stocks of wheat, million bushels	356.57	–75.75	–17.52
Soybean crush, million bushels	994.17	–167.89	–14.45
Export of soybeans, million bushels	524.48	–109.01	–17.21
Stocks of soybeans, million bushels	211.09	–173.22	–45.07
Beef production, million bushels	23,361.00	214.00	0.92
Pork production, million bushels	14,512.00	–124.00	–0.85
Broiler production, million bushels	15,186.00	–102.00	–0.67
Cattle breeding herd size, 1000 head	43,694.00	–226.00	–0.52
Hog breeding herd size, 1000 head	6,516.00	–67.00	–1.02
Beef price, dollars per pound	2.50	–0.01	–0.35
Pork price, dollars per pound	1.83	0.02	1.23
Broiler price, dollars per pound	0.86	0.01	1.39
Crop farm market revenue, million dollars	45,268.00	2,083.00	4.82
Government payments, million dollars	22.00	–2,204.00	–99.00
Crop farm input expense, million dollars	20,627.00	806.00	4.07
Crop farm profit, million dollars	24,663.00	–927.00	–3.62
Livestock producer surplus, million dollars	735.00	93.00	–11.19
Grains consumer surplus, million dollars	66,491.00	6,160.00	–8.48
Meat consumer surplus, million dollars	368,929.00	–339.00	–0.09
Total consumer surplus, million dollars	435,421.00	–6,499.00	–1.47
Net domestic surplus, million dollars	460,819.00	–7,519.00	–1.61
Stockholder surplus, million dollars	12,709.00	–16,805.00	–56.94
Iowa land price, dollars per acre	1,664.13	–51.24	–2.99
Oklahoma land price, dollars per acre	786.17	–8.35	–1.05
US farmland price, dollars per acre	875.20	–14.05	–1.58
Total farmland value, million dollars	884,823.00	–14,203.00	–1.58
Export earnings, million dollars	13,542.00	–1,017.00	8.12

Source: estimated.

Table 8.12. Simulation results for the environmental enhancement case, 1991

Variable	Predicted	Change	Percent change
Feed grains program compliance, %	5.91	–59.17	–90.91
Wheat program compliance, %	34.26	–32.14	–48.41
Corn and sorghum acreage, million acres	95.41	8.57	9.86
Wheat acreage, million acres	91.07	11.52	14.48
Soybean acreage, million acres	56.22	0.95	1.71
Feed grains production, million metric tons	184.79	–61.16	–24.87
Wheat production, million metric tons	2,017.33	–609.62	–23.21
Soybean production, million metric tons	1,228.09	–567.35	–31.60
Feed grains cost of production, dollars per acre	150.76	2.40	1.62
Wheat cost of production, dollars per acre	55.09	–1.63	–2.87
Soybean cost of production, dollars per acre	72.58	–2.15	–2.88
Corn price, dollars per bushel	4.07	1.61	65.30
Sorghum price, dollars per bushel	3.69	1.46	65.30
Wheat price, dollars per bushel	4.30	1.34	45.14
Soybean price, dollars per bushel	10.21	3.38	49.45
Feed use of feed grains, million metric tons	101.86	–26.22	–20.47
Industry use of feed grains, million metric tons	40.61	–7.27	–15.18
Export of feed grains, million metric tons	33.53	–13.81	–29.17
FOR stocks of feed grains, million metric tons	0.00	–35.34	–100.00
Government stocks of feed grains, million metric tons	0.23	–42.42	–99.47
Market stocks of feed grains, million metric tons	33.17	–7.05	–17.52
Feed use of wheat, million bushels	412.48	–4.78	–1.14
Other domestic uses of wheat, million bushels	686.24	–7.40	–1.07
Export of wheat, million bushels	766.74	–365.42	–32.28
FOR stocks of wheat, million bushels	36.82	–670.39	–94.79
Government stocks of wheat, million bushels	23.68	–562.40	–95.96
Market stocks of wheat, million bushels	782.48	107.35	15.90
Soybean crush, million bushels	911.17	–227.31	–19.97
Export of soybeans, million bushels	359.53	–335.24	–48.25
Stocks of soybeans, million bushels	92.47	–236.88	–71.92
Beef production, million bushels	24,326.00	1,374.00	5.99
Pork production, million bushels	14,152.00	–1,083.00	–7.11
Broiler production, million bushels	16,338.00	–522.00	–3.09
Cattle breeding herd size, 1000 head	37,678.00	–5,085.00	–11.89
Hog breeding herd size, 1000 head	6,322.00	–687.00	–9.79
Beef price, dollars per pound	2.50	–0.09	–3.42
Pork price, dollars per pound	1.99	0.16	8.84
Broiler price, dollars per pound	0.92	0.06	7.35
Crop farm market revenue, million dollars	50,809.00	6,931.00	15.80
Government payments, million dollars	37.00	–4,343.00	–99.15

Table 8.12 (continued)

Variable	Predicted	Change	Percent change
Crop farm input expense, million dollars	23,482.00	1,955.00	9.08
Crop farm profit, million dollars	27,364.00	633.00	2.37
Livestock producer surplus, million dollars	557.00	–299.00	–34.93
Grains consumer surplus, million dollars	74.626.00	6,106.00	–7.56
Meat consumer surplus, million dollars	393,368.00	–1,448.00	–0.37
Total consumer surplus, million dollars	467,994.00	–7,553.00	–1.59
Net domestic surplus, million dollars	495,915.00	–7,219.00	–1.43
Stockholder surplus, million dollars	35,544.00	–5,796.00	–14.02
Iowa land price, dollars per acre	1,804.94	–80.39	–4.26
Oklahoma land price, dollars per acre	841.26	–31.18	–3.57
US farmland price, dollars per acre	945.67	–27.67	–2.84
Total farmland value, million dollars	956,068.00	–27,976.00	–2.84
Export earnings, million dollars	12,337.00	–349.00	–2.75

Source: estimated.

Table 8.13. Simulation results for the environmental enhancement case, 1994

Variable	Predicted	Change	Percent change
Feed grains program compliance, %	4.01	–62.82	–94.00
Wheat program compliance, %	22.20	–43.79	–66.36
Corn and sorghum acreage, million acres	96.85	10.56	12.24
Wheat acreage, million acres	76.53	2.24	3.01
Soybean acreage, million acres	55.62	0.32	0.59
Feed grains production, million metric tons	166.32	–91.25	–35.43
Wheat production, million metric tons	1,425.83	–1,084.39	–43.20
Soybean production, million metric tons	1,035.21	–813.74	–44.01
Feed grains cost of production, dollars per acre	162.76	8.26	5.34
Wheat cost of production, dollars per acre	65.02	0.56	0.88
Soybean cost of production, dollars per acre	74.27	–5.53	–6.92
Corn price, dollars per bushel	5.05	2.75	119.71
Sorghum price, dollars per bushel	4.57	2.49	119.71
Wheat price, dollars per bushel	6.96	3.42	96.57
Soybean price, dollars per bushel	11.74	4.71	66.98
Feed use of feed grains, million metric tons	90.61	–39.07	–30.13
Industry use of feed grains, million metric tons	41.73	–10.66	–20.34
Export of feed grains, million metric tons	19.26	–32.53	–62.81
FOR stocks of feed grains, million metric tons	0.00	–35.89	–100.00

Table 8.13 (continued)

Variable	Predicted	Change	Percent change
Government stocks of feed grains, million metric tons	0.02	–54.00	–99.97
Market stocks of feed grains, million metric tons	18.46	–26.10	–58.57
Feed use of wheat, million bushels	378.31	–103.29	–21.45
Other domestic uses of wheat, million bushels	681.20	–43.89	–6.05
Export of wheat, million bushels	626.57	–936.92	–59.92
FOR stocks of wheat, million bushels	36.75	–667.96	–94.78
Government stocks of wheat, million bushels	5.65	–663.12	–99.16
Market stocks of wheat, million bushels	312.35	–85.05	–21.40
Soybean crush, million bushels	884.40	–270.56	–23.43
Export of soybeans, million bushels	162.96	–513.11	–75.90
Stocks of soybeans, million bushels	40.98	–284.47	–87.41
Beef production, million bushels	25,781.00	3,049.00	13.41
Pork production, million bushels	13,034.00	–2,640.00	–16.84
Broiler production, million bushels	17,366.00	–1,027.00	–5.58
Cattle breeding herd size, 1000 head	23,038.00	–19,595.00	–45.96
Hog breeding herd size, 1000 head	5,569.00	–1,758.00	–24.00
Beef price, dollars per pound	2.46	–0.23	–8.68
Pork price, dollars per pound	2.25	0.38	20.44
Broiler price, dollars per pound	0.89	0.13	15.24
Crop farm market revenue, million dollars	55,111.00	9,939.00	22.00
Government payments, million dollars	24.00	–4,406.00	–99.45
Crop farm input expense, million dollars	24,870.00	2,338.00	10.37
Crop farm profit, million dollars	30,266.00	3,196.00	11.81
Livestock producer surplus, million dollars	468.00	–436.00	–48.24
Grains consumer surplus, million dollars	76,643.00	–15,070.00	–16.43
Meat consumer surplus, million dollars	415,070.00	–2,249.00	–0.54
Total consumer surplus, million dollars	491,714.00	–17,319.00	–3.40
Net domestic surplus, million dollars	522,448.00	–14,559.00	–2.71
Stockholder surplus, million dollars	6,164.00	–20,063.00	–76.50
Iowa land price, dollars per acre	1,970.04	–88.48	–4.30
Oklahoma land price, dollars per acre	902.69	–47.47	–5.00
US farmland price, dollars per acre	1,023.30	–34.81	–3.29
Total farmland value, million dollars	1,034,553.00	–35,189.00	–3.29
Export earnings, million dollars	10,097.00	–4,872.00	–32.55

Source: estimated.

Table 8.14. Discounted 10-year time stream of welfare effects and terminal asset values of land and livestock (at 10%), environmental enhancement case, 1985–1994

Variable	Predicted	Change	Percent change
	millions of dollars		
Discounted Welfare Effects			
Crop farm profit	179,721.00	–0.00	–0.00
Livestock producer surplus	4,743.00	–1,014.00	–17.62
Grains consumer surplus	464,066.00	–34,975.00	–7.01
Meat consumer surplus	2,525,039.00	–4,685.00	–0.19
Total consumer surplus	2,989,105.00	39,660.00	–1.31
Net domestic surplus	3,173,569.00	–40,675.00	–1.27
Stockholder surplus	161,390.00	–62,205.00	–27.82
Government payments	3,199.00	–16,970.00	–84.14
Terminal Asset Values			
Total value of crop land	438,751.00	–14,924.00	–3.29
Total value of livestock	4,524.00	–4,589.00	–50.36
Net welfare effect	–105,423.00		

1988, \$4.30 in 1991, and \$6.96 in 1994. Government- and farmer-owned reserves approach zero much more rapidly and production declines by 35 to 45% over the 10-year period. Feed use declines by about 20%, and pork and broiler meat prices rise by 15 to 20%. Export earnings drop over 32%.

Turning to the evaluation of welfare effects in Table 8.14, discounted crop farm profits are just restored to the level of the base case. This restoration of benefits, however, comes at great expense to purchasers of grain. Grain consumers lose about \$35 billion and meat consumers lose another \$4.7 billion. Livestock producers lose about \$1 billion in producer surplus (short-run profits) but lose another \$10.8 billion in herd value. Only the government gains in this scenario by saving \$17 billion in government payments.

A major factor not considered in Table 8.14, however, is the environmental benefits of the associated pesticide cancellations. To get some idea of the magnitude of these benefits, note that Ott (1990) has found that 80% of consumers are willing to pay at least 10% more for pesticide-free food products. The annual consumer expenditure on cereals, meats, dairy products, and eggs (which embody the crops considered herein) is roughly \$200 billion, which implies an annual benefit of \$16 billion or a discounted value over 10 years of \$98 billion (with a 10% discount rate). In addition, Andrew Manelli of the Agricultural Policy Branch of the EPA has calculated that the additional cost of water purification caused by pesticide/chemical use is \$50 per person. If the population using water affected by pesticide/chemical use on these crops is 100 million, then the cost saving would be $5 billion annually. If this benefit does not begin until 10 years after pesticide/chemical use is cut, then the discounted value associated with water purification would be about $19 billion. These two benefits, along with the additional benefits associated with worker safety and ecological side effects, could conceivably exceed the net welfare loss of $105.4 billion in Table 8.4.

8.9.3 The Land Retirement Scenario

Given the 5.8% rate of reduction in target and support prices in the phase-out scenario, an iterative procedure was used to determine the accompanying rate of increase in per-acre diversion payments which would be required to increase program participation to achieve a level of land diversion which would just restore crop farmer profits. The restoration of crop farm profits in this case results from production declines as more land is diverted or "retired" from production. For the purposes of this scenario, land retirement decisions are assumed to be made annually consistent with the way acreage reduction programs have operated rather than on a 10-year basis as under the conservation reserve program.

The rate of increase in diversion payments required for this purpose is 70.6% per year. While this rate of increase seems absurd, one must bear in mind that diversion payments in recent years have been nominal with payments for voluntary diversion more than an order of magnitude greater. This rate of increase results in a diversion payment for wheat of $17.42 per acre in 1988, $86.53 in 1991, and $429.65 in 1994. Only the last few years reach levels which clearly involve heavy subsidies beyond normal profit rates.

The results for this case are presented in Tables 8.15 through 8.19 and correspond to Tables 8.3 and 8.4 for the base case, Tables 8.5 through 8.9 for the phase-out case, and Tables 8.10 through 8.14 for the environmental protection case. Unlike the other cases, this case generally leads to an increase in program participation over the period as might be expected with the increase in diversion payments. Prices and production, however, are relatively unaffected by the base case. Government- and farmer-owned reserve stocks tend to be depleted as a result of eliminating deficiency payments and price supports. Market stocks are increased somewhat to make up for the farmer-owned reserve effect. All other market variables generally change by less than 2 or 3% by 1994 from the base case. Basically, a reduction in deficiency payments is simply replaced by an increase in diversion payments.

Turning to the evaluation of welfare effects in Table 8.19, discounted crop farm profits are just restored to the level of the base case, while cropland value changes marginally. In this case, however, the restoration of crop farmer profit is achieved without a loss for other market groups. Livestock producers gain a small amount in producer surplus and about $2 billion in herd value; grain and meat consumers gain about $3 billion each (only 0.6 and 0.12%, respectively) as a result of slightly lower feed grain and meat prices. Private stockholders gain the most (about $67 billion) as a result of replacing the government stockholding activity. The only important expense of this case is an increase in government outlays of $14 billion due to the high diversion payments.

Table 8.15. Simulation results for the land retirement case, 1985

Variable	Predicted	Change	Percent change
Feed grains program compliance, %	51.89	0.00	0.00
Wheat program compliance, %	60.00	0.00	0.00
Corn and sorghum acreage, million acres	97.80	0.00	0.00
Wheat acreage, million acres	79.20	0.00	0.00
Soybean acreage, million acres	67.80	0.00	0.00
Feed grains production, million metric tons	235.41	–0.00	–0.00
Wheat production, million metric tons	2,595.00	0.00	0.00
Soybean production, million metric tons	1,860.97	–0.00	–0.00
Feed grains cost of production, dollars per acre	129.90	–0.96	–0.73
Wheat cost of production, dollars per acre	52.47	–0.82	–1.53
Soybean cost of production, dollars per acre	66.33	0.00	0.00
Corn price, dollars per bushel	2.76	–0.05	–1.70
Sorghum price, dollars per bushel	2.50	–0.04	–1.70
Wheat price, dollars per bushel	3.22	–0.12	–3.71
Soybean price, dollars per bushel	7.21	0.00	0.00
Feed use of feed grains, million metric tons	132.00	1.40	1.07
Industry use of feed grains, million metric tons	34.25	0.42	1.25
Export of feed grains, million metric tons	46.08	0.52	1.14
FOR stock of feed grains, million metric tons	18.67	–2.27	–10.84
Government stocks of feed grains, million metric tons	5.93	–2.77	–31.81
Market stocks of feed grains, million metric tons	43.88	2.69	6.54
Feed use of wheat, million bushels	439.13	12.15	2.85
Other domestic uses of wheat, million bushels	649.84	3.28	0.51
Export of wheat, million bushels	1,232.10	30.12	2.51
FOR stocks of wheat, million bushels	683.08	–16.77	–2.40
Government stocks of wheat, million bushels	316.40	–44.07	–12.23
Market stocks of wheat, million bushels	685.57	15.28	2.28
Soybean crush, million bushels	1,046.81	–0.00	–0.00
Export of soybeans, million bushels	682.08	–0.00	–0.00
Stocks of soybeans, million bushels	307.78	–0.00	–0.00
Beef production, million bushels	24,020.00	–0.00	–0.00
Pork production, million bushels	13,160.00	0.00	0.00
Broiler production, million bushels	13,455.00	–0.00	–0.00
Cattle breeding herd size, 1000 head	46,239.00	0.00	0.00
Hog breeding herd size, 1000 head	5,607.00	0.00	0.00
Beef price, dollars per pound	2.42	–0.00	–0.00
Pork price, dollars per pound	1.90	0.00	0.00
Broiler price, dollars per pound	0.83	0.00	0.00
Crop farm market revenue, million dollars	47,371.00	–764.00	–1.59
Government payments, million dollars	2,668.00	311.00	13.21

Table 8.15 (continued)

Variable	Predicted	Change	Percent change
Crop farm input expense, million dollars	21,357.00	–159.00	–0.74
Crop farm profit, million dollars	28,682.00	–294.00	–1.01
Livestock producer surplus, million dollars	896.00	19.00	2.16
Grains consumer surplus, million dollars	62,903.00	547.00	0.88
Meat consumer surplus, million dollars	340,389.00	0.00	0.00
Total consumer surplus, million dollars	403,292.00	547.00	0.14
Net domestic surplus, million dollars	432,870.00	272.00	0.06
Stockholder surplus, million dollars	33,923.00	991.00	3.01
Iowa land price, dollars per acre	1,547.21	–4.43	–0.29
Oklahoma land price, dollars per acre	721.26	–5.03	–0.69
US farmland price, dollars per acre	807.02	–2.58	–0.32
Total farmland value, millions dollars	815,898.00	–2,607.00	–0.32
Export earnings, million dollars	13,900.00	–81.00	–0.58

Source: estimated.

Table 8.16. Simulation results for the environmental enhancement case, 1988

Variable	Predicted	Change	Percent change
Feed grains program compliance, %	61.81	–1.93	–3.03
Wheat program compliance, %	10.46	–10.54	–50.18
Corn and sorghum acreage, million acres	88.18	0.21	0.24
Wheat acreage, million acres	68.47	–0.14	–0.20
Soybean acreage, million acres	57.19	0.19	0.33
Feed grains production, million metric tons	235.89	0.29	0.12
Wheat production, million metric tons	2,120.75	–20.72	–0.97
Soybean production, million metric tons	1,802.08	4.67	0.26
Feed grains cost of production, dollars per acre	130.91	–5.71	–4.18
Wheat cost of production, dollars per acre	59.10	–0.73	–1.23
Soybean cost of production, dollars per acre	64.20	–0.65	–1.01
Corn price, dollars per bushel	2.31	–0.26	–10.02
Sorghum price, dollars per bushel	2.09	–0.23	–10.02
Wheat price, dollars per bushel	3.79	–0.10	–2.67
Soybean price, dollars per bushel	6.01	–0.11	–1.76
Feed use of feed grains, million metric tons	132.00	4.69	3.69
Industry use of feed grains, million metric tons	42.93	1.35	3.24
Export of feed grains, million metric tons	44.59	3.08	7.42
FOR stocks of feed grains, million metric tons	26.12	–6.39	–19.65

Table 8.16 (continued)

Variable	Predicted	Change	Percent change
Government stocks of feed grains, million metric tons	7.62	–20.37	–72.78
Market stocks of feed grains, million metric tons	49.32	10.12	25.81
Feed use of wheat, million bushels	288.22	–31.11	–9.74
Other domestic uses of wheat, million bushels	668.20	–1.86	–0.28
Export of wheat, million bushels	1,156.27	14.58	1.28
FOR stocks of wheat, million bushels	206.92	–170.10	–45.12
Government stocks of wheat, million bushels	93.30	–17.49	–15.79
Market stocks of wheat, million bushels	504.44	72.12	16.68
Soybean crush, million bushels	1,162.83	0.77	0.07
Export of soybeans, million bushels	634.51	1.02	0.16
Stocks of soybeans, million bushels	385.88	1.56	0.41
Beef production, million bushels	22,846.00	–301.00	–1.30
Pork production, million bushels	14,894.00	258.00	1.77
Broiler production, million bushels	15,399.00	110.00	0.72
Cattle breeding herd size, 1000 head	45,126.00	1,207.00	2.75
Hog breeding herd size, 1000 head	6,750.00	168.00	–2.55
Beef price, dollars per pound	2.52	0.02	0.74
Pork price, dollars per pound	1.77	–0.04	–2.00
Broiler price, dollars per pound	0.83	–0.01	–1.53
Crop farm market revenue, million dollars	40,355.00	–2,829.00	–6.55
Government payments, million dollars	1,436.00	–790.00	–35.49
Crop farm input expense, million dollars	19,262.00	–559.00	–2.82
Crop farm profit, million dollars	22,529.00	3,061.00	–11.96
Livestock producer surplus, million dollars	884.00	56.00	6.78
Grains consumer surplus, million dollars	72,358.00	–293.00	–0.40
Meat consumer surplus, million dollars	369,609.00	340.00	0.09
Total consumer surplus, million dollars	441,967.00	47.00	0.01
Net domestic surplus, million dollars	465,381.00	–2,957.00	–0.63
Stockholder surplus, million dollars	33,211.00	3,697.00	12.53
Iowa land price, dollars per acre	1,657.39	–57.97	–3.38
Oklahoma land price, dollars per acre	783.11	–11.42	–1.44
US farmland price, dollars per acre	872.52	–16.73	–1.88
Total farmland value, million dollars	882,114.00	–16,912.00	–1.88
Export earnings, million dollars	12,260.00	–266.00	–2.12

Source: estimated.

Table 8.17. Simulation results for the environmental enhancement case, 1991

Variable	Predicted	Change	Percent change
Feed grains program compliance, %	66.79	1.70	2.61
Wheat program compliance, %	71.27	4.86	7.33
Corn and sorghum acreage, million acres	85.94	–0.91	–1.04
Wheat acreage, million acres	81.93	2.38	2.99
Soybean acreage, million acres	55.81	0.53	0.96
Feed grains production, million metric tons	243.77	–2.19	–0.89
Wheat production, million metric tons	2,714.51	87.56	3.33
Soybean production, million metric tons	1,813.77	18.33	1.02
Feed grains cost of production, dollars per acre	142.45	–5.90	–3.98
Wheat cost of production, dollars per acre	52.50	–4.22	–7.44
Soybean cost of production, dollars per acre	73.42	–1.31	–1.76
Corn price, dollars per bushel	2.22	–0.24	–9.72
Sorghum price, dollars per bushel	2.01	–0.22	–9.72
Wheat price, dollars per bushel	2.34	–0.63	–21.14
Soybean price, dollars per bushel	6.62	–0.22	–3.16
Feed use of feed grains, million metric tons	129.74	1.67	1.31
Industry use of feed grains, million metric tons	48.27	0.39	0.81
Export of feed grains, million metric tons	51.32	3.98	8.41
FOR stocks of feed grains, million metric tons	1.80	–33.54	–94.91
Government stocks of feed grains, million metric tons	1.46	–41.18	–96.57
Market stocks of feed grains, million metric tons	70.03	29.81	74.12
Feed use of wheat, million bushels	503.31	86.06	20.62
Other domestic uses of wheat, million bushels	700.09	6.45	0.93
Export of wheat, million bushels	1,299.64	167.48	14.79
FOR stocks of wheat, million bushels	445.95	–261.25	–36.94
Government stocks of wheat, million bushels	83.83	–502.25	–85.70
Market stocks of wheat, million bushels	928.44	253.31	37.52
Soybean crush, million bushels	1,142.00	3.51	0.31
Export of soybeans, million bushels	708.99	14.22	2.05
Stocks of soybeans, million bushels	334.20	4.85	1.47
Beef production, million bushels	22,641.00	–311.00	–1.36
Pork production, million bushels	15,623.00	389.00	2.55
Broiler production, million bushels	16,982.00	122.00	0.73
Cattle breeding herd size, 1000 head	45,842.00	3,080.00	7.20
Hog breeding herd size, 1000 head	7,285.00	266.00	3.79
Beef price, dollars per pound	2.60	0.02	0.72
Pork price, dollars per pound	1.77	–0.05	–2.98
Broiler price, dollars per pound	0.84	–0.02	–1.89
Crop farm market revenue, million dollars	39,664.00	–4,214.00	–9.60
Government payments, million dollars	5,117.00	737.00	16.83

Table 8.17 (continued)

Variable	Predicted	Change	Percent change
Crop farm input expense, million dollars	20,641.00	–886.00	–4.12
Crop farm profit, million dollars	24,140.00	–2,590.00	–9.69
Livestock producer surplus, million dollars	868.00	12.00	1.45
Grains consumer surplus, million dollars	82,082.00	1,351.00	1.67
Meat consumer surplus, million dollars	395,590.00	774.00	0.20
Total consumer surplus, million dollars	477,672.00	2,125.00	0.45
Net domestic surplus, million dollars	502,680.00	–453.00	–0.09
Stockholder surplus, million dollars	64,724.00	23,384.00	56.57
Iowa land price, dollars per acre	1,770.48	–114.85	–6.09
Oklahoma land price, dollars per acre	839.22	–33.22	–3.81
US farmland price, dollars per acre	936.83	–36.50	–3.75
Total farmland value, million dollars	947,138.00	–36,906.00	–3.75
Export earnings, million dollars	12,215.00	–471.00	–3.71

Source: estimated.

Table 8.18. Simulation results for the land retirement case, 1994

Variable	Predicted	Change	Percent change
Feed grains program compliance, %	90.29	23.47	35.12
Wheat program compliance, %	87.67	21.69	32.78
Corn and sorghum acreage, million acres	81.98	–4.30	–4.99
Wheat acreage, million acres	75.68	1.39	1.87
Soybean acreage, million acres	54.83	–0.47	–0.84
Feed grains production, million metric tons	248.12	–9.46	–3.67
Wheat production, million metric tons	2,594.43	84.21	3.35
Soybean production, million metric tons	1,847.38	–1.67	–0.09
Feed grains cost of production, dollars per acre	153.16	–1.34	–0.87
Wheat cost of production, dollars per acre	66.29	1.83	2.84
Soybean cost of production, dollars per acre	80.00	0.20	0.25
Corn price, dollars per bushel	2.24	–0.06	–2.68
Sorghum price, dollars per bushel	2.02	0.06	–2.68
Wheat price, dollars per bushel	3.78	0.24	6.75
Soybean price, dollars per bushel	7.04	0.01	0.16
Feed use of feed grains, million metric tons	126.04	–3.65	–2.81
Industry use of feed grains, million metric tons	51.21	–1.18	2.25
Export of feed grains, million metric tons	52.99	1.21	2.34
FOR stocks of feed grains, million metric tons	0.00	–35.89	–100.00

Table 8.18 (continued)

Variable	Predicted	Change	Percent change
Government stocks of feed grains, million metric tons	0.25	–53.76	–99.53
Market stocks of feed grains, million metric tons	73.23	28.67	64.35
Feed use of wheat, million bushels	436.14	–45.46	–9.44
Other domestic uses of wheat, million bushels	715.16	–9.93	–1.37
Export of wheat, million bushels	1,657.39	93.89	6.01
FOR stocks of wheat, million bushels	262.22	–442.50	–62.79
Government stocks of wheat, million bushels	24.58	–644.19	–96.32
Market stocks of wheat, million bushels	610.36	212.96	53.59
Soybean crush, million bushels	1,143.46	–11.50	–1.00
Export of soybeans, million bushels	695.01	18.94	2.80
Stocks of soybeans, million bushels	314.63	–10.82	–3.33
Beef production, million bushels	22,607.00	–125.00	–0.55
Pork production, million bushels	15,965.00	291.00	1.86
Broiler production, million bushels	18,453.00	60.00	0.32
Cattle breeding herd size, 1000 head	47,022.00	4,389.00	10.29
Hog breeding herd size, 1000 head	7,537.00	210.00	2.86
Beef price, dollars per pound	2.70	0.01	0.20
Pork price, dollars per pound	1.83	–0.04	–2.19
Broiler price, dollars per pound	0.84	–0.01	–1.12
Crop farm market revenue, million dollars	44,643.00	–529.00	–1.17
Government payments, million dollars	26,873.00	22,442.00	506.57
Crop farm input expense, million dollars	21,959.00	–573.00	–2.55
Crop farm profit, million dollars	49,557.00	22,487.00	83.07
Livestock producer surplus, million dollars	846.00	–59.00	–6.47
Grains consumer surplus, million dollars	89,309.00	–2,404.00	–2.62
Meat consumer surplus, million dollars	418,110.00	791.00	0.19
Total consumer surplus, million dollars	507,419.00	–1,613.00	–0.32
Net domestic surplus, million dollars	557,822.00	20,815.00	3.88
Stockholder surplus, million dollars	38,301.00	12,074.00	46.04
Iowa land price, dollars per acre	2,030.21	–28.32	–1.38
Oklahoma land price, dollars per acre	947.14	–3.02	–0.32
US farmland price, dollars per acre	1,050.65	–7.46	–0.70
Total farmland value, million dollars	1,062,203.00	–7,540.00	–0.70
Export earnings, million dollars	15,819.00	850.00	5.68

Source: estimated.

Table 8.19. Discounted 10-year time stream of welfare effects and terminal asset values of land and livestock (at 10%), land retirement case, 1985–1994

Variable	Predicted	Change	Percent change
	millions of dollars		
Discounted Welfare Effects			
Crop farm profit	179,721.00	–0.00	–0.00
Livestock producer surplus	5,919.00	162.00	2.82
Grains consumer surplus	502,029.00	2,988.00	0.60
Meat consumer surplus	2,532,764.00	3,040.00	0.12
Total consumer surplus	3,034,793.00	6,028.00	0.20
Net domestic surplus	3,220,433.00	6,190.00	0.19
Stockholder surplus	290,907.00	67,312.00	30.10
Government payments	34,184.00	14,014.00	69.49
Terminal Asset Values			
Total value of crop land	450,478.00	–3,198.00	–0.70
Total value of livestock	10,051.00	938.00	10.29
Net welfare effect	57,227.00		

8.9.4 Combinations of Policies

While the case of land retirement appears to suggest a politically feasible improvement over continuation of past policies if permitted policies include land retirement programs, several reservations should be borne in mind. First, the level of diversion payments in the last few years of this case are perhaps implausibly high. A better rate of change in payments without such rapid growth at the end of the period (linear as opposed to geometric growth) may be more plausible. Further refinement of this work should focus on dynamic plausibility considerations. Second, most of the net increase in benefits from this case accrue to private stockholders. However, much of the increase in private stockholding comes as a result of replacing farmer-owned reserve stockholding. The extent of welfare gain for stockholders would appear differently if a welfare benefit were ascribed to farmer-owned reserve stockholding. Further refinement of this work should develop an appropriate measure for the welfare of holders of farmer-owned reserve stocks and these welfare effects should be incorporated into the overall evaluation of policy reform alternatives.

Finally, various combinations of permitted policies need to be considered. This is done briefly in Table 8.20 where various combinations of growth rates in diversion payments and reduction rates in yields due to pesticide cancellations are used to generate discounted welfare effects comparable to Tables 8.4, 8.9, 8.14, and 8.19. In each case, the potential use of direct payments can also be considered to make up the difference between the crop farmer profits in Table 8.20 and the $179,721 billion in crop farmer profits attained with the base case. In effect, Table 8.20 represents a coarse grid search for an optimal policy where the multidimensional aspects of distribution are considered.

Generally, the results show that crop farmers are better off with higher diversion payment growth rates and higher rates of yield reduction due to pesticide cancellation while consumers and livestock producers are better off with lower rates of both. Interestingly, stockholders are better off with higher diversion payment growth rates when no pesticide cancellations are used but are better off with low diversion payment growth rates when high rates of yield reductions occur through pesticide cancellations. The government budget is obviously better off with lower diversion payment growth rates but also with higher rates of yield reduction which tends to strengthen market prices and reduce deficiency payments faster. The net welfare effects over all groups considered in Table 8.20 are higher with higher diversion payment growth and worse with higher rates of yield reduction. Again, it should be noted here that any gains in environmental quality are not incorporated in the net welfare effects. If these gains are high enough, then welfare effects may also be improving in the rate of yield reduction.

The objective of this study was to investigate policy reform which makes welfare-improving policies politically sustainable. In this sense, the upper left-hand cell of Table 8.20 corresponds to the case where deficiency payments are phased out and direct payments are used to enforce political sustainability (e.g., the loss from $179.721 billion to $144.932 billion is discounted crop farmer profits and the loss from $453.675 billion to $416.298 in discounted terminal cropland value which would be replaced with direct payments is reflected in the "Net Welfare Effect"). This policy is welfare improving in terms of the standard criterion of net efficiency because it gains $11.044 billion overall. It is politically sustainable as long as the losing group (crop farmers) are compensated with direct payments. However, the results in Table 8.20 also suggest that better policies are possible when land retirement policies are permitted so that compensation can be made indirectly through other policies. The case with the highest growth rate of diversion payments (upper right-hand cell) attains a net welfare gain of $42.968 billion. With no yield reduction, the policy reform which replaces deficiency payments with diversion payments improves overall welfare more than policy reform which replaces deficiency payments with direct uncoupled payments.

Based on Table 8.20, it is clear that a set of values of the two rates can be developed for which discounted farmer profits are held constant at the base case level. This is done in Table 8.21. That is, for several selected values of the rate-of-yield reduction, a rate of increase in diversion payment is calculated which just restores crop farmer benefits to the base case without direct payments. As these re-

Table 8.20. Discounted 10-year time stream of welfare effects (at 10%) and terminal asset values of land and livestock for selected policy alternatives (in millions of dollars), base case, 1985–1994

	Diversion payment growth rate			
Welfare or asset value effect	0.0	0.2	0.4	0.6
	Yield reduction rate: 0.00			
Discounted Welfare Effects				
Crop farm profit	144,932.00	147,458.00	153,925.00	166,518.00
Livestock producer surplus	6,132.00	6,080.00	5,995.00	5,945.00
Grains consumer surplus	503,675.00	503,516.00	502,984.00	502,341.00
Meat consumer surplus	2,534,462.00	2,534,120.00	2,533,400.00	2,532,955.00
Total consumer surplus	3,038,137.00	3,037,636.00	3,036,384.00	3,035,296.00
Net domestic surplus	3,189,200.00	3,191,175.00	3,196,304.00	3,207,758.00
Stockholder surplus	283,498.00	286,150.00	290,161.00	290,875.00
Government payments	8,239.00	9,054.00	12,233.00	22,351.00
Terminal Asset Values				
Total value of crop land	416,298.00	418,734.00	424,822.00	436,996.00
Total value of livestock	10,745.00	10,616.00	10,322.00	10,148.00
Overall net welfare effect	11,044.00	17,163.00	28,917.00	42,968.00
	Yield reduction rate: 0.02			
Discounted Welfare Effects				
Crop farm profit	157,752.00	159,945.00	168,579.00	182,438.00
Livestock producer surplus	5,665.00	5,634.00	5,526.00	5,442.00
Grains consumer surplus	488,711.00	488,561.00	488,234.00	487,661.00

Meat consumer surplus	2,530,501.00	2,530,222.00	2,529,410.00	2,528,699.00
Total consumer surplus	3,019,212.00	3,018,783.00	3,017,644.00	3,016,360.00
Net domestic surplus	3,182,630.00	3,184,362.00	3,191,749.00	3,204,240.00
Stockholder surplus	253,036.00	251,914.00	251,307.00	247,411.00
Government payments	4,664.00	5,287.00	8,724.00	18,735.00
Terminal Asset Values				
Total value of crop land	425,755.00	427,735.00	436,608.00	449,889.00
Total value of livestock	8,971.00	8,827.00	8,278.00	7,886.00
Overall net welfare effect	–14,731.00	–12,907.00	–1,241.00	10,233.00
	Yield reduction rate: 0.04			
Discounted Welfare Effects				
Crop farm profit	170,890.00	172,438.00	179,824.00	194,019.00
Livestock producer surplus	5,185.00	5,164.00	5,068.00	4,976.00
Grains consumer surplus	475,854.00	475,737.00	475,578.00	475,103.00
Meat consumer surplus	2,527,021.00	2,526,872.00	2,526,469.00	2,526,035.00
Total consumer surplus	3,002,875.00	3,002,609.00	3,002,047.00	3,001,137.00
Net domestic surplus	3,178,950.00	3,180,212.00	3,186,940.00	3,200,132.00
Stockholder surplus	200,101.00	197,961.00	196,713.00	194,872.00
Government payments	3,437.00	3,948.00	7,327.00	17,685.00
Terminal Asset Values				
Total value of crop land	434,273.00	435,612.00	443,846.00	457,688.00
Total value of livestock	6,774.00	6,674.00	6,186.00	5,661.00
Overall net welfare effect	–63,798.00	63,948.00	–54,099.00	39,710.00

Table 8.20 (continued)

Welfare or asset value effect	Diversion payment growth rate 0.0	0.2	0.4	0.6
	Yield reduction rate: 0.04			
Discounted Welfare Effects				
Crop farm profit	179,481.00	180,753.00	186,931.00	200,326.00
Livestock producer surplus	4,760.00	4,742.00	4,663.00	4,576.00
Grains consumer surplus	464,527.00	464,458.00	464,367.00	464,064.00
Meat consumer surplus	2,525,077.00	2,524,983.00	2,524,893.00	2,524,921.00
Total consumer surplus	2,989,603.00	2,989,441.00	2,989,260.00	2,989,984.00
Net domestic surplus	3,173,844.00	3,174,936.00	3,180,853.00	3,193,886.00
Stockholder surplus	162,700.00	161,243.00	159,709.00	159,128.00
Government payments	3,202.00	3,678.00	6,975.00	17,288.00
Terminal Asset Values				
Total value of crop land	438,658.00	439,786.00	447,163.00	460,640.00
Total value of livestock	4,615.00	4,535.00	4,126.00	3,609.00
Overall net welfare effect	−103,843.00	−103,635.00	−95,582.00	−80,482.00

Table 8.21. Discounted 10-year time stream of welfare effects (at 10%) and terminal asset values of land and livestock for selected policy combinations which restore farm profits (in millions of dollars), 1985–1994

	Yield reduction rate			
	0.0	0.2	0.4	0.6
	Diversion payment growth rate			
Welfare or asset value effect	0.7059	0.5699	0.3979	0.06016
Discounted Welfare Effects				
Crop farm profit	179,721.00	179,721.00	179,721.00	179,720.00
Livestock producer surplus	5,919.00	5,452.00	5,069.00	4,756.00
Grains consumer surplus	502,029.00	487,734.00	475,582.00	464,509.00
Meat consumer surplus	2,534,764.00	2,528,779.00	2,526,473.00	2,525,052.00
Total consumer surplus	3,034,793.00	3,016,512.00	3,002,055.00	2,989,561.00
Net domestic surplus	3,220,433.00	3,201,686.00	3,186,845.00	3,174,038.00
Stockholder surplus	290,907.00	248,171.00	196,720.00	162,382.00
Government payments	34,184.00	16,446.00	7,268.00	3,283.00
Terminal Asset Values				
Total value of crop land	450,978.00	447,198.00	443,742.00	438,840.00
Total value of livestock	10,051.00	7,921.00	6,193.00	4,681.00
Overall net welfare effect	69,347.00	8,072.00	–54,226.00	–103,880.00

sults show, all groups are generally better off as reliance on environmental protection is replaced by reliance on land retirement holding the discounted profits of crop farmers constant again ignoring environmental benefits of pesticide cancellation. However, comparing the results of Tables 8.20 and 8.21, it is clear that the largest overall welfare gain is attained by switching to a policy reform relying solely on land retirement rather than combinations involving either direct payments or environmental protection. Of course, if reliable estimates of the environmental benefits of pesticide cancellation were available and could be included, then the environmental protection alternative or some combination involving it may be preferred.

8.10 Conclusion

This chapter has developed a complete model of the corn sorghum, wheat, soybean, and livestock sectors of the US agricultural economy for the purpose of investigating policy reform under the GATT. The model generates plausible estimates of policy effects over a 10-year forecasting horizon and is used to investigate several specific policy scenarios. First, a rate of reduction in target and support prices was determined to just drive deficiency payments to zero during a 10-year transition period. It was found that about $72 billion in direct payments to crop farmers would be required to make this reform politically feasible. However, an additional $83.2 billion in net benefits are gained by other groups (including an $11.9 billion reduction in deficiency payments). Alternatively, an environmental protection approach to restoring crop farmer benefits was investigated. Here, it was found that the environmental benefits of pesticide reduction achieved by a 21 to 36% reduction in yields over a 10-year period would have to be worth about $105 billion to be as well off as in the case of no policy reform. In this case, crop farmer profits are maintained without large government payments, which is politically more attractive. Finally, a case of increased payment for acreage reduction ("retirement") was considered as a means of restoring crop farmer benefits. Here, it was found that all groups except government gain so the policy reform is uniformly politically feasible. However, this policy involves an increase of $14 billion in government outlays.

This study represents a first attempt to develop a model sufficient for analyzing the temporal feasibility and sustainability of policy reform. Additional policy reform alternatives need to be considered such as disaster programs, food, air, marketing programs, and stockpiling. Also, various mixes and combinations of policies need to be considered in evaluating policy reform alternatives. Finally, differentiation in policy reform parameters among commodities needs to be considered. While much remains to be done, the framework and the modeling techniques of this study offer the basic principles by which these refinements can be accomplished.

References

Chambers RG, Foster WE (1983) Participation in the farmer-owned reserve program: a discrete choice model. Am J Agric Econ 65:120–124

Chambers RG, Just RE (1981) Effects of exchange rates on US agriculture: a dynamic analysis. Am J Agric Econ 63:32–46

Gardner BG (1984) Policy options for grains. In: Rausser GC, Farrell KR (eds) Alternative agricultural and food policies and the 1985 Farm Bill. Giannini Foundation of Agricultural Economics, University of California, Berkeley, pp 81–92

Gardner BG (1988) Gains and losses in the wheat program. Working Paper 88–11, Department of Agricultural and Resource Economics, University of Maryland, College Park

Johnson SR, Rausser GC (1982) Composite forecasting in commodity systems. In: Rausser GC (ed) New directions in econometric modeling and forecasting in US agriculture, Chap 21. Elsevier North-Holland, New York

Just RE (1981) Farmer-owned grain reserve program needs modification to improve effectiveness: theoretical and empirical considerations in agricultural buffer stock policy under the Food and Agriculture Act of 1977. Prepared for Report to Congress, US General Accounting Office, US Government Printing Office, Washington, DC

Just RE (1987) Making economic welfare analysis useful in the policy process: implications of the public choice literature. Am J Agric Econ 70:488–453

Just RE (1988) The role of monetary, fiscal, and agricultural variables in farmland prices. Working Paper, University of Maryland, College Park

Just RE (1990) Modelling the interactive effect of alternative sets of policies on agricultural prices. In: Winters LA, Sapsford D (eds) Primary commodity prices: economic models and economic policy. Center for Economic Policy Research, London

Just RE, Hueth DK, Schmitz A (1982) Applied welfare economics amd public policy. Prentice-Hall, Englewood Cliffs

Just RE, Zilberman D, Rausser GC (1989) Building sustainable coalitions for welfare improving policies. Prepared for the Office of the US Trade Representative, Washington, DC, Dec

Just RE, Zilberman D, Rausser GC (1990) A framework for evaluating specific policy reform. Prepared for the Office of the US Trade Representative, Washington, DC, Feb

Lichtenberg E, Zilberman D (1988) Efficient regulation of environmental health risks. Q J Econ 103, 1:167–178

Lins Q (1988) Gains and losses from the corn program. Unpublished Working Paper, Economic Research Service, US Department of Agriculture, Washington, DC

Love HA (1987) Flexible public policy: the case of the United States wheat sector. Ph D Diss, University of California, Berkeley

Ott SL (1990) Supermarket shoppers' pesticide concerns and willingness to purchase certified pesticide residue-free fresh produce. Agribusiness: Int J 6:593–602

Rausser GC (1985) Macroeconomics and US agricultural policy. In: Gardner BL (ed) US agricultural policy: 1985 farm legislation. American Enterprise Institute for Public Policy Research, Washington, DC, pp 207–252

Rausser GC (1992) Predatory versus productive governments: the case of US agricultural policy. J Econ Perspect 6: 133–158

Rausser GC, Foster WE (1992) The evolution and coordination of US commodity and resource policies. In: Just RE, Bockstael N (eds) Commodity and resource policies in agricultural systems. Agricultural management and economics. Springer, Berlin Heidelberg New York, pp 17–45

Rausser GC, Just RE (1982) Principles of policy modeling. In: Rausser GC (ed) New directions in econometric modeling and forecasting in US agriculture. Elsevier North-Holland, New York, pp 763–800

Rausser GC, Chalfant JA, Love HA, Stamoulis KG (1986) Macroeconoic linkages, taxes, and subsidies in the US agricultural sector. Am J Agric Econ 68:399–412

Appendix A

The definitions of variables in the supply data (in alphabetical order and including sources) are as follows:

AC	=	Acreage of corn in million acres [*Feed: Outlook & Situation Report*]
ACGS	=	Acreage of corn and grain sorghum (computed from AC and AGS)
ACGSN	=	Corn and grain sorghum acreage on noncomplying farms (computed from ACGS, BACGS, COMPFG, and DRFG)
AGS	=	Acreage of grain sorghum in million acres [*Feed: Outlook & Situation Report*]
AS	=	Acreage of soybeans in million acres [*Oil Crops: Outlook & Situation Report*]
AW	=	Acreage of wheat in million acres [*Wheat: Outlook & Situation Report*]
AWN	=	Acreage of wheat on noncomplying farms (computed from AW, BAW, COMPW, and DRW)
BAC	=	Base acreage of corn and grain sorghum [*Feed: Outlook & Situation Report*, text of various issues]
BACGS	=	Base acreage for corn and grain sorghum (computed from BAC and BAGS)
BAGS	=	Base acreage of grain sorghum in million acres [*Feed: Outlook & Situation Report*, text of various issues]
BAW	=	Base acreage of wheat in million acres [*Wheat: Outlook & Situation Report*, text of various issues]
RVDP	=	Real inducement for additional voluntary diversion (computed from VDPC, COMPFG, and GNPD)
COMPFG	=	Program participation rate for corn and grain sorghum in percent of acreage (US Agricultural Stabilization and Conservation Service)
COMPW	=	Program participation rate for wheat in percent of acreage (US Agricultural Stabilization and Conservation Service)
COSTC	=	Variable costs per acre of corn in dollars (includes seed, chemicals, and labor) [USDA, ESCS, Paul Gallagher, *Costs of Producing Selected Crops in the US*]
COSTCGS	=	Variable costs per acre for corn and grain sorghum in dollars per acre (computed from COSTC, AC, COSTGS, AGS, and ACGS)
COSTGS	=	Variable costs per acre of sorghum in dollars (includes seed, chemicals, and labor) [USDA, ESCS, Paul Gallagher, *Costs of Producing Selected Crops in the US*]

COSTS	=	Variable costs per acre of soybeans in dollars (includes seed, chemicals, and labor) [USDA, ESCS, Paul Gallagher, *Costs of Producing Selected Crops in the US*]
COSTW	=	Variable costs per acre of wheat in dollars (includes seed, chemicals, and labor) [USDA, ESCS, Paul Gallagher, *Costs of Producing Selected Crops in the US*]
D71	=	Dummy variable, 1 if 1971, 0 if not
D72	=	Dummy variable, 1 if 1972, 0 if not
D73	=	Dummy variable, 1 if 1973, 0 if not
DIVFG	=	Minimum feed grain diversion acreage (computed from COMPFG, DRFG, and BACGS)
DIVW	=	Minimum wheat diversion acreage (computed from COMPW, DRW, and BAW)
DPC	=	Diversion payment for corn (paid diversion) in dollars per acre [*Feed: Outlook & Situation Report*, text of various issues]
DPW	=	Diversion payment for wheat (paid diversion) in dollars per acre [*Wheat: Outlook & Situation Report*, text of various issues]
DRFG	=	Diversion requirement of feed grains in percent of base acreage [*Feed: Outlook & Situation Report*, text of various issues]
DRW	=	Diversion requirement of wheat in percent of base acreage [*Wheat: Outlook & Situation Report*, text of various issues]
EYLDCGS	=	Expected yield for corn and grain sorghum (computed from lagged values of YLDCGS)
EYLDS	=	Expected yield for soybeans (computed from lagged values of YLDS)
EYLDW	=	Expected yield for wheat (computed from lagged values of YLDW)
FGVOL	=	Maximum return on acreage eligible for additional voluntary participation in the feed grain program (computed from VDPC and RETFGP)
ICCC	=	Interest rate charged on CCC nonrecourse loans in percent [*Wheat: Outlook & Situation Report* and *Feed: Outlook & Situation Report*]
LCOMPFG	=	Linearized logistic representation of the feed grain program participation rate (computed from COMPFG)
LCOMPW	=	Linearized logistic representation of the wheat program participation rate (computed from COMPW)
LRPROFF	=	Lagged real profit of feed grain production effect on current nonparticipating feed grain acreage (computed from PROFFGN and COMPFG)
LRPROFFS	=	Lagged real profit of soybean production effect on current nonparticipating feed grain acreage (computed from PROFS and COMPFG)

LRPROFFW = Lagged real profit of feed grain production effect on current nonparticipating wheat acreage (computed from PROFFG and COMPW)

LRPROFS = Lagged real profit of soybean production effect on current soybean acreage (computed from PROFS)

LRPROFSF = Lagged real profit of soybean production effect on current nonparticipating feed grain acreage (computed from PROFS and COMPFG)

LRPROFW = Lagged real profit of feed grain production effect on current nonparticipating wheat acreage (computed from PROFFG and COMPW)

MAXRATE = Interest rate subsidy on stocks under CCC loan (computed from RS and ICCC)

MAXYLDFG = Expected corn/grain sorghum yield eligible for price support (computed from EYLDCGS and YLDFGP)

MAXYLDW = Expected wheat yield eligible for price support (computed from EYLDW and YLDWP)

MSPC = Effective farm price for corn/grain sorghum including price support payments (computed from SPRC and PAFC)

MSPW = Effective farm price for wheat includng price support payments (computed from SPRW and PAFW)

MTPC = Effective farm price for corn/grain sorghum including deficiency payments (computed from TPC and PAFC)

MTPW = Effective farm price for wheat including deficiency payments (computed from TPW and PAFW)

NOPROGFG = Dummy variable: 1, if a feed grain program is in effect (restricted acreage or diversion is required to receive support); 0, if not

NOPROGW = Dummy variable: 1, if a wheat program is in effect; 0, if not

PAFC = Corn, price at farm, US average in dollars per bushel, October–September [*Feed: Outlook & Situation Report*]

PAFS = Soybeans, price at farm, US average in dollars per bushel, October–September [*Oil Crops: Outlook & Situation Report*]

PAFSG = Grain sorghum, price at farm, US average in dollars per bushel, October–September [*Feed: Outlook & Situation Report*]

PAFW = Wheat, price at farm, US average, in dollars per bushel, June–May [*Wheat: Outlook & Situation Report*]

PROFDIFF = Profit differential for participation versus nonparticipation in the feed grain program (computed from PROFFGC, PROFFGN, DRFG, VDFG, COSTCGS, and GNPD)

PROFDIFW = Profit differential for participation versus nonparticipation in the wheat program (computed from PROFWC, PROFWN, DRW, VDW, COSTW, and GNPD)

PROFFG = Average real quasirent per acre among all farms for feed grains (computed from COMPFG, PROFFGN, and PROFFGC)

PROFFGC = Average real quasirent per acre on farms participating in the feed grain program (computed from DRFG, VDFG, RETFGP, FGVOL, DPC, and GNPD)

PROFFGN = Real quasirent per acre for corn/grain sorghum without participation (computed from PAFC, EYLDCGS, COSTCGS, and GNPD)

PROFS = Real quasirent per acre for soybeans (computed from PAFS1, EYLDS, COSTS, and GNPD)

PROFW = Average real quasirent per acre among all farms for wheat (computed from COMPW, PROFWN, and PROFWC)

PROFWC = Average real quasirent per acre on farms participating in the wheat program (computed from DRW, VDW, RETWP, WVOL, DPW, and GNPD)

PROFWN = Real quasirent per acre for wheat without participation (computed from PAFW, EYLDW, COSTW, and GNPD)

RCOSTCGS = Real variable costs per acre for corn and grain sorghum (computed from COSTCGS and GNPD)

RCOSTS = Real variable costs per acre for soybeans (computed from COSTS and GNPD)

RCOSTW = Real variable costs per acre for wheat (computed from COSTW and GNPD)

RETFGP = Quasirent per acre for corn/grain sorghum acreage in production under participation (computed from MTPC, YLDFGP, MSPC, MAXYLDFG, MAXRATE, SPRC, EYLDCGS, and COSTCGS)

RETWP = Quasirent per acre for wheat acreage in production under participation (computed from MTPW, YLDWP, MSPW, MAXYLDW, MAXRATE, SPRW, EYLDW, and COSTW)

RPAFC = Real price of corn at the farm level (computed from PAFC)

RPAFS = Real price of soybeans at the farm level (computed from PAFS)

RPAFW = Real price of wheat at the farm level (computed from PAFW)

SPRC = Support price of corn in dollars per bushel [*Feed: Outlook & Situation Report*]

SPRW = Support price of wheat in dollars per bushel [*Wheat: Outlook & Situation Report*]

TPC = Target price of corn in dollars per bushel (support price and additional support payment prior to 1973) [*US Corn Industry*]

TPW = Target price of wheat in dollars per bushel (support price and additional support payment prior to 1973) [*Wheat: Outlook & Situation Report*]

VDFG = Additional voluntary paid diversion for feed grain in percent of base acreage [*Feed: Outlook & Situation Report*, text of various issues]

VDW	=	Additional voluntary paid diversion for wheat in percent of base acreage [*Wheat: Outlook & Situation Report*, text of various issues]
VDPC	=	Additional voluntary diversion payment for corn in dollars per acre [*Feed: Outlook & Situation Report*, text of various issues]
VDPW	=	Additional voluntary diversion payment for wheat in dollars per acre [*Wheat: Outlook & Situation Report*, text of various issues]
WVOL	=	Maximum return on acreage eligible for additional voluntary participation in the wheat program (computed from VDPW and RETWP)
YEAR	=	Two-digit year
YLDCGS	=	Yield per planted acre of corn and grain sorghum in bushels per acre [calculated from acreage and production in *Feed: Outlook & Situation Report*]
YLDFGP	=	Program yield of feed grains (corn and grain sorghum) in bushels per acre [*Feed: Outlook & Situation Report*, text of various issues]
YLDS	=	Yield per planted acre of soybeans in bushels per acre [from acreage and production in *Oil Crops: Outlook & Situation Report*]
YLDW	=	Yield per planted acre of wheat in bushels per acre [from acreage and production in *Wheat: Outlook & Situation Report*]
YLDWP	=	Program yield of wheat in bushels per acre [*Wheat: Outlook & Situation Report*, text of varous issues]

The transformations made from basic data to obtain the computed variables used in econometric estimation of supply are defined as follows:

RPAFC	=	PAFC/GNPD
RPAFS	=	PAFS/GNPD
MAXRATE	=	$0.01 \cdot \text{RS}0.01 \cdot \text{ICC}$; if MAXRATE < 0, then MAXRATE = 0
ACGS	=	AC + AGS
BACGS	=	BAC + BAGS
EYLDCGS	=	[YLDCGS (T – 1) + YLDCGS (T – 2) + ... + YLDCGS (T – 5)]/3
EYLDS	=	[YLDS (T – 1) + YLDS (T – 2) + ... + YLDS (T – 5)]/3
MAXYLDFG	=	EYLDCGS – YLDFGP; if MAXYLDFG < 0, then MAXYLDFG = 0
MTPC	=	TPC; if MTPC < PAFC, then MTPC = PAFC
MSPC	=	SPRC; if MSPC < PAFC, then MSPC = PAFC
COSTCGS	=	$\text{COSTC} \cdot (\text{AC/ACGS}) + \text{COSTGS} \cdot (\text{AGS/ACGS})$
RCOSTCGS	=	COSTCGS/GNPD
RCOSTS	=	COSTS/GNPD
PROFFGN	=	$(\text{PAFC} \cdot \text{EYLDCGS} - \text{COSTCGS})/\text{GNPD}$

RETFGP	=	MTPC · YLDFGP + MSPC · MAXYLDFG + MAXRATE · SPRC · EYLDCGS = COSTCGS
FGVOL	=	VDPC; if FGVOL < RETFGP, then FGVOL = RETFGP
PROFFGC	=	[(1 – DRFG – VDFG) · RETFGP + VDFG · FGVOL + DPC · DRFG]/GNPD
PROFS	=	(PAFS · EYLDS – COSTS)/GNPD
PROFFG	=	(1 – COMPFG) · PROFFGN + COMPFG · PROFFGC
LCOMPFG	=	LOG[(0.001 + COMPFG)/(1 – CPMPFG)]
DIVFG	=	COMPFG · DRFG · BACGS
LRPROFF	=	[PROFFGN (T – 1) + PROFFGN (T – 2) + ... + PROFFGN (T – 5)] · (1 – COMPFG)/5
PROFDIFF	=	PROFFGC – PROFFGN – (DRFG + VDFG) · [COSTCGS/GNPD – COSTCGS (T –1)]/GNPD (T – 1)]
LRPROFFS	=	[PROFFG (T – 1) + PROFFG (T – 2) + PROFFG (T – 3) + PROFFG (T – 4) + PROFFG (T – 5)]/5
LRPROFS	=	[PROFS (T – 1) + PROFS (T – 2) + ... + PROFS (T – 5)]/5
LRPROFSF	=	LRPROFS · (1 – COMPFG)
ACGSN	=	ACGS – BACGS · COMPFG · (1 – DRFG)
RVDP	=	COMPFG · VDPC/GNPD
RPAFW	=	PAFW/GNPD
EYLDW	=	[YLDW (T – 1) + YLDW (T – 2) + ... + YLDW (T–5)]/5
MAXYLDW	=	EYLDW – YLDWP; if MAXYLDW < 0, then MAXYLDW = 0
MTPW	=	TPW; if MTPW < PAFW, then MTPW = PAFW
MSPW	=	SPRW; if MSPW < PAFW, then MSPW = PAFW
RCOSTW	=	COSTW/GNPD
PROFWN	=	(PAFW · EYLDW – COSTW)/GNPD
RETWP	=	MTPW · YLDWP + MSPW · MAXYLDW + MAXRATE · SPRW · EYLDW – COSTW
WVOL	=	VDPW; if WVOL < RETWP, then WVOL = RETWP
PROFWC	=	[(1 – DRW – VDW) · RETWP + VDW · WVOL + DPW · DRW]/GNPD
PROFW	=	(1 – COMPW) · PROFWN + COMPW · PROFWC
LCOMPW	=	LOG[(0.001 + COMPW)/(1 – COMPW)]
LRPROFW	=	[PROFWN (T – 1) + PROFWN (T – 2) + ... + PROFWN (T – 5)] · (1 – COMPW)/5
LRPROFFW	=	[PROFFG (T – 1) + PROFFG (T – 2) + ... + PROFFG (T – 5)] · (1 – COMPW)/5
AWN	=	AW – BAW · COMPW · (1 – DRW)
DIVW	=	DRW · BAW · COMPW
PROFDIFW	=	PROFWC – PROFWN – (DRW + VDW) · [COSTW/GNPD – COSTW (T – 1)/GNDP (T – 1)]

Appendix B

The definitions of variables used for estimating demand (in alphabetical order and with sources) are as follows:

BRHOGKE	=	Breeding hog inventory for 10 states, 1000 head, September 1 [*Livestock & Poultry: Outlook & Situation Report*]
COMPFG	=	Program participation of corn and grain sorghum in percent of acreage (US Agricultural Stabilization and Conservation Service)
CRUSH	=	US crushings of soybeans, million bushels
DDFG	=	US total domestic use of feed grains (corn, sorghum, oats, barley) million metric tons [*Feed: Outlook & Situation Report*]
DDW	=	US total domestic use of wheat, million bushels [*Wheat: Outlook & Situation Report*]
DFW	=	US domestic food use of wheat, million bushels [*Wheat: Outlook & Situation Report*]
DFWCAP	=	US per-capita domestic food use of wheat, million bushels (calculated from DFW and POP)
DINDFG	=	US feed grain use by industry (computed from DDFG and DLVKFG)
DLVKFG	=	US feed and residual of feed grains (corn, sorghum, oats, barley), million metric tons [*Feed: Outlook & Situation Report*]
DLVKW	=	US feed and residual of wheat, million bushels [*Wheat: Outlook & Situation Report*]
DMYPIK	=	Dummy variable for Payment-in-Kind Program
ERPCPW	=	Exponential of the ratio of corn price to wheat price (computed from RPAFC and RPAFW)
EXR	=	Exchange rate, trade weighted index in dollars per unit of foreign currency, 1972 = 1.00
GNPD	=	Gross National Product price deflator
KCCCFGE	=	Ending feed grain stocks under CCC loan, million metric tons (US Agricultural Stabilization and Conservation Service)
KCCCWE	=	Ending wheat stocks under CCC loan, million bushels (US Agricultural Stabilization and Conservation Service)
KFORFGE	=	US ending farmer-owned reserve stocks of feed grains, million metric tons (US Agricultural Stabilization and Conservation Service)
KFORWE	=	US ending farmer-owned reserve stocks of wheat, million bushels (US Agricultural Stabilization and Conservation Service)
KGOVFGE	=	Ending government-owned feed grain stocks (total CCC inventory), million metric tons (US Agricultural Stabilization and Conservation Service) [*Feed: Outlook & Situation Report*]

KGOVWE = Ending government-owned wheat stocks (total CCC inventory), million bushels (US Agricultural Stabilization and Conservation Service)
KMKTFGE = US ending market inventory of feed grain (computed from KPRIFGE and KCCCFGE)
KMKTWE = US ending market inventory of wheat (computed from KPRIWE and KCCCWE)
KPRIFGE = US free ending stocks of feed grains, million metric tons US Agricultural Stabilization and Conservation Service)
KPRIWE = US free ending stocks of wheat, million bushels (US Agricultural Stabilization and Conservation Service)
KPRISBE = US free ending soybean stocks, million bushels
LPCX = Lagged real border price of corn (computed from PAFC, EXR, and GNPD)
LPSX = Lagged real border price of soybeans (computed from PAFS, EXR, and GNPD)
LPWX = Lagged real border price of wheat (computed from PAFW, EXR, and GNPD)
N = Total US population in millions
MBEEF = Net US imports of beef (computed from PRDBEEF, PCDBEEF, and N)
MPORK = Net US imports of pork (computed from PRDPORK, PCDPORK, and N)
PAFC = Corn, price at farm, US average in dollars per bushel, October–September [*Feed: Outlook & Situation Report*]
PAFS = Soybeans, price at farm, US average in dollars per bushel, October–September [*Oil Crops: Outlook & Situation Report*]
PAFSG = Grain sorghum, price at farm, US average in dollars per bushel, October–September [*Feed: Outlook & Situation Report*]
PAFW = Wheat, price at farm, US average in dollars per bushel, June–May [*Wheat: Outlook & Situation Report*]
PBEEF = Average retail price of choice beef in cents per pound [*Livestock & Poultry: Outlook & Situation Report*]
PBR = Average retail price of broilers in four regions, in cents per pound [*Livestock & Poultry: Outlook & Situation Report*]
PCDBEEF = Per-capita disappearance of carcass weight of beef in pounds [*Livestock & Poultry: Outlook & Situation Report*, Supply & Utility Table]
PCDBR = Per-capita civilian disappearance of young chickens in pounds [*Livestock & Poultry: Outlook & Situation Report*]
PCDPORK = Per-capita disappearance of carcass weight of pork in pounds [*Livestock & Poultry: Outlook & Situation Report*]
PCX = Border price of corn (computed from PAFC and EXR)
PF = US Gross National Product price deflator for nonfarm total sales, 1982 dollars

POP	=	US noninstitutional population over 16 years in millions
PPORK	=	Average retail price pork in cents per pound [*Livestock & Poultry: Outlook & Situation Report*]
PRDBEEF	=	Commercial production of beef, million pounds [*Livestock & Poultry: Outlook & Situation Report*]
PRDBR	=	Total production of young chicken, million pounds [*Livestock & Poultry: Outlook & Situation Report*]
PRDFG	=	US production of feed grains (corn, sorghum, oats, barley), million metric tons [*Feed: Outlook & Situation Report*]
PRDOTH	=	US production of feed grains other than corn and sorghum, million metric tons (computed from PRDFG, ACGS, and YLDCGS)
PRDPORK	=	Commercial production of pork, million pounds [*Livestock & Poultry: Outlook & Situation Report*]
PRDSB	=	US production of soybeans, million bushels (computed from KPRISBE, CRUSH, and XSB)
PRDW	=	US production of wheat, million bushels [*Wheat: Outlook & Situation Report*]
PSX	=	Border price of soybeans (computed from PAFS and EXR)
PWX	=	Border price of wheat (computed from PAFW and EXR)
RELOTH	=	Ratio of price of other goods to consumer income (computed from RPF and RPYD)
RELPBEEF	=	Ratio of beef price to consumer income (computed from RPBEEF and RPYD)
RELPBR	=	Ratio of broiler price to consumer income (computed from RPBR and RPYD)
RELPORK	=	Ratio of pork price to consumer income (computed from RPPORK and RPYD)
RPAFC	=	Real US corn price (computed from PAFC and GNPD)
RPAFCMT	=	Ratio of corn price to the price of meat (computed from PAFC and PMEAT)
RPAFS	=	Real US soybean price (computed from PAFS and GNPD)
RPAFW	=	Real US wheat price (computed from PAFW and GNPD)
RPBEEF	=	Real price of beef (computed from PBEEF and GNPD)
RPBR	=	Real price of broilers (computed from PBR and GNPD)
RPCPB	=	Ratio of corn price to the price of beef (computed from PAFC and PBEEF)
RPCPBR	=	Ratio of corn price to the price of broilers (computed from PAFC and PBR)
RPCPP	=	Ratio of corn price to the price of pork (computed from PAFC and PPORK)
RPF	=	Real price of nonfood products (computed from PF and GNPD)
RPMEAT	=	Real price of meat (computed from PMEAT and GNPD)
RPPORK	=	Real price of pork (computed from PPORK and GNPD)

RPYD	=	Real disposable income per capita (computed from YD, GNPD, and POP)
RRS	=	Real interest rate (computed from RS and GNPD)
RS	=	Three-month US Treasury bill rate (percentage points)
RSPFORC	=	Real farmer-owned reserve support price for corn (computed from SPFORC and GNPD)
RSPFORW	=	Real farmer-owned reserve support price for wheat (computed from SPFORW and GNPD)
SPFORC	=	Support price for farmer-owned reserve corn (CCC support price prior to institution of the farmer-owned reserve) in dollars per bushel [*Feed: Outlook & Situation Report*]
SPFORW	=	Support price for farmer-owned reserve wheat (CCC support price prior to institution of the farmer-owned reserve) in dollars per bushel [*Wheat: Outlook & Situation Report*]
SUPDOMFG	=	Domestic supply of feed grains (computed from PRDFG, KMKTFGE, KFORFGE, KGOVFGE, and XFG)
SUPDOMSB	=	Domestic supply of soybeans (computed from PRDSB, KPRISBE, and XSB)
SUPDOMW	=	Domestic supply of wheat (computed from PRDW, KMKTWE, KFORWE, KGOVWE, and XW)
SUPC	=	Indicator variable for effectiveness of corn price support (computed from RSPFORC and RPAFC)
SUPW	=	Indicator variable for effectiveness of wheat price support (computed from RSPFORW and RPAFW)
TCOWKE	=	Cows and heifers that have calved (cow inventory) in the United States, 1000 head, average quarterly ending inventory [*Livestock & Poultry: Outlook & Situation Report*]
TPBEEF	=	Real border price of beef (computed from RPBEEF and EXR)
TPBR	=	Real border price of broilers (computed from RPBR and EXR)
TPPORK	=	Real border price of pork (computed from RPPORK and EXR)
VALCAT	=	Value per head for cattle, in dollars [*Agricultural Statistics*]
VALHOG	=	Value per head for hogs, in dollars [*Agricultural Statistics*]
XBR	=	Net US exports of broilers (computed from PRDBR, PCDBR, and N)
XFG	=	US exports of feed grains (corn, sorghum, oats, barley), million metric tons [*Feed: Outlook & Situation Report*]
XSB	=	US exports of soybeans, million bushels [*Oil Crops: Outlook & Situation Report*]
XW	=	US exports of wheat, million bushels [*Wheat: Outlook & Situation Report*]
YD	=	US disposable income in billion dollars [*Statistical Abstract of the United States*]
YEAR	=	Two-digit year

The transformations made from basic data to obtain the computed variables used in econometric estimation of demand are defined as follows:

RPYD	=	YD/(GNPD · POP)
RRS	=	RS – [GNPD – GNPD (T – 1)]/GNPD (T – 1) · 100
PCX	=	PAFC/(EXR · GNPD)
LPCX	=	[PCX (T – 1) + PCX (T – 2) + ... + PCX (T – 5)]/5
SUPDOMFG	=	KMKTFGE (T – 1) + PRDFG – XFG
KMKTFGE	=	KPRIFGE + KCCCFGE
DINDFG	=	DDFG – DLVKFG
RPAFC	=	PAFC/GNPD
RSPFORC	=	SPFORC/GNPD
SUPC	=	1; if RSPFORC – 0.85 · RPAFC < 0, then SUPC = 0
FORINC	=	(RSPFORC – 0.85 · RPAFC) · SUPC
FOROUTC	=	(RPAFC · 0.85 – RSPFORC) · KFORFGE (T – 1) · (1 – SUPC)
FORGOVC	=	(RSPFORC – 0.85 – RPAFC) · KFORFGE (T – 1) · SUPC
GOVINC	=	FORINC; if YEAR > 77, then GOVINC = 0
GOVOUTC	=	(RPAFC · 0.85 – RSPFORC) · KGOVFGE (T – 1) · (1 – SUPC)
PRDOTH	=	PRDFG – 0.0254 · ACGS · YLDCGS
PWX	=	PAFW/(EXR · GNPD)
LPWX	=	[PWX (T – 1) + PWX (T – 2) + ... + PWX (T – 5)]/5
KMKTWE	=	KPRIWE + KCCCWE
DLVKW	=	DDW – DFW
RPAFW	=	PAFW/GNPD
DFWCAP	=	DFW/POP
RSPFORW	=	SPFORW/GNPD
SUPW	=	1; if RSPFORW – 0.85 · RPAFW < 0, then SUPW = 0
FORINW	=	(RSPFORW – 0.85 · RPAFW) · SUPW
FOROUTW	=	(RPAFW · 0.85 – RSPFORW) · [KFORWE (T – 1)/100]^4 · (1 – SUPW)
FORGOVW	=	(RSPFORW – 0.85 · RPAFW) · KFORWE (T – 1) · SUPW
GOVINW	=	FORINW; if YEAR > 77, then GOVINW = 0
GOVOUTW	=	(RPAFW · 0.85 – FSPFORW) · KGOVWE (T – 1) · (1 – SUPW)
SUPDOMW	=	PRDW + KMKTWE (T – 1) – XW
ERPCPW	=	EXP(RPAFC/RPAFW)
PRDSB	=	KPRISBE – KPRISBE (T – 1) + XSB + CRUSH
PSX	=	PAFS/(EXR · GNPD)
LPSX	=	[PSX (T – 1) + PSX (T – 2) + ... + PSX (T – 5)]/5
SUPDOMSB	=	KPRISBE (T – 1) + PRDSB – XSB
RPAFS	=	PAFS/GNPD
RPCPB	=	PAFC/PBEEF
RPCPP	=	PAFC/PPORK
RPCPBR	=	PAFC/PBR

PMEAT	=	0.37 · PBEEF + 0.12 · PBR + 0.51 · PPORK
RPMEAT	=	PMEAT/GNPD
RPAFCMT	=	PAFC/PMEAT
RPBEEF	=	PBEEF/GNPD
RPPORK	=	PPORK/GNPD
RPBR	=	PBR/GNPD
RPF	=	PF/GNPD
RELPBEEF	=	RPBEEF/RPYD

Appendix C

Using the variables defined in Appendix A and Appendix B, and the model structure defined in the text, the following equations are estimated.

Feed Grain Supply Block

The feed grain supply block consists of a logistic equation that explains program participation, an equation that explains nonparticipating feed grain acreage and variation from program acreage (base acreage less minimum diversion requirements) on participating farms, an equation that represents feed grain yield, an equation that explains how per-acre costs of feed grain production respond to feed grain prices. The participation equation follows (6) with a dummy variable added to represent years when diversion was not required to receive feed grain program benefits. The acreage equation follows (5) with soybeans as the competing crop (the predominant competitive relationship in the corn belt). The yield equation is a simple time trend modified to represent response of yields to diversion which presumably removes poorer acreage from production first. The cost equation specifies cost of production as a function of returns per acre following the arguments of gardener where the prices of inputs are bid up to exhaust rents.

Estimated Equation for Feed Grain Program Participation (LCOMPFG):

$$\underset{}{\text{LCOMPFG}} = \underset{(0.38)}{-6.90} + \underset{(0.38)}{6.30}\,[\text{PROFDIFF} \cdot (1 - \text{NOPROGFG})].$$

$R^2 = 0.971$ $\bar{R}^2 = 0.968$ D.W. = 2.64 Sample: 1977–1987

Estimated Equation for Nonparticipating Corn and Sorghum Acreage (ACGSN):

$$\text{ACGSN} = \underset{(15.5)}{90.8}\,(1 - \text{COMPFG}) + \underset{(0.881)}{0.631}\,\text{LRPROFF} - \underset{(0.888)}{0.157}\,\text{LRPROFSFS} - \underset{(0.196)}{0.429}\,\text{RVDP}$$

$R^2 = 0.977$ $\bar{R}^2 = 0.972$ D.W. = 1.67 Sample: 1968–1986

Estimated Equation for Corn and Sorghum Yield (YLDCGS):

$$\text{YLDCGS} = \underset{(27.5)}{-76.5} + \underset{(0.35)}{1.96}\,\text{YEAR} + \underset{(0.258)}{0.253}\,\text{DIFVG}$$

$R^2 = 0.650$ $\bar{R}^2 = 0.609$ D.W. = 2.27 Sample: 1968–1987

Estimated Equation for Real Production Costs for Corn and Sorghum (RCOSTCGS):

$$\text{RCOSTCGS} = \underset{(2.3)}{16.8} + \underset{(0.042)}{0.248}\,\text{RPAFC} \cdot \text{EYLDCGS}$$

$R^2 = 0.665$ $\bar{R}^2 = 0.646$ D.W. = 0.986 Sample: 1968–1987

Wheat Supply Block

The wheat supply block similarly consists of a logistic equation that explains program participation, an equation that explains program nonparticipating wheat acreage and variation from program acreage (base acreage less minimum diversion requirements) on participating farms, an equation for wheat yield, and an equation explaining per-acre costs of wheat production. The participation equation follows (6) with a dummy variable added to represent years when diversion was not required to receive wheat program benefits. The acreage equation follows (5) with sorghum and/or corn considered as the competing crop (a predominant competitive relationship in the Southern and Central Great Plains). The yield equation is a simple time trend modified to represent response of yields diversion. The cost equation specifies cost of production as a function of returns per acre following Gardener's arguments.

Estimated Equation for Wheat Program Participation (LCOMPW):

$$\underset{}{\text{LCOMPW}} = \underset{(0.27)}{-6.87} + \underset{(0.29)}{6.80}\ [\text{PROFDIFW} \cdot (1 - \text{NOPROGW})].$$

$R^2 = 0.986$ $\bar{R}^2 = 0.984$ D.W. = 2.29 Sample: 1977–1986

Estimated Equation for Nonparticipating Wheat Acreage (AWN):

$$\text{AWN} = \underset{(12.0)}{101.0}\ (1 - \text{COMPW}) + \underset{(1.71)}{4.41}\ \text{LRPROFW} - \underset{(1.48)}{3.70}\ \text{LRPROFFW}$$

$$+ \underset{(0.028)}{0.104}\ \text{BAW} \cdot \text{COMPW} + \underset{(4.4)}{43.8}\ \text{D71} + \underset{(4.4)}{46.1}\ \text{D72} + \underset{(4.4)}{52.5}\ \text{D73}$$

$R^2 = 0.980$ $\bar{R}^2 = 0.970$ D.W. = 2.22 Sample: 1968–1986

Estimated Equation for Wheat Program Yield (YLDW):

$$\text{YLDW} = \underset{(5.17)}{8.23} + \underset{(0.068)}{0.259}\ \text{YEAR} + \underset{(0.0511)}{0.994}\ \text{DIVW}$$

$R^2 = 0.609$ $\bar{R}^2 = 0.560$ D.W. = 0.139 Sample: 1968–1986

Estimated Equation for Production Costs for Wheat (RCOSTW):

$$\text{RCOSTW} = \underset{(0.70)}{7.08} + \underset{(0.023)}{0.209}\ \text{RPARW} \cdot \text{EYLDW}$$

$R^2 = 0.825$ $\bar{R}^2 = 0.815$ D.W. = 0.894 Sample: 1968–1986

Soybean Supply Block

The soybean supply block has a structure similar to feed grains and wheat except that no partcipation is included since there has been no voluntary program. Hence, the acreage equaiton follows the free market form in (1). The yield equation follows a simple time trend with variations in response to feed grain diversion (which presumably removes poorer acreage from soybean as well as corn production). The cost equation is similar in form to the feed grain and wheat cases.

Estimated Equation for Soybean Acreage (AS):

$$\underset{}{AS} = \underset{(2.6)}{44.5} + \underset{(0.264)}{0.575}\ \text{LRPROFS} - \underset{(0.371)}{0.196}\ \text{LRPROFFS}$$

$R^2 = 0.663$ $\bar{R}^2 = 0.624$ D.W. = 0.625 Sample: 1968–1987

Estimated Equation for Soybean Yield (YLDS):

$$\text{YLDS} = \underset{(6.91)}{3.68} + \underset{(0.088)}{0.312}\ \text{YEAR} + \underset{(0.0649)}{0.0537}\ \text{DIVGF}$$

$R^2 = 0.429$ $\bar{R}^2 = 0.362$ D.W. = 0.243 Sample: 1968–1987

Estimated Equation for Real Production Costs for Soybean (RCOSTS):

$$\text{RCOSTS} = \underset{(1.30)}{6.43} + \underset{(0.245)}{0.189}\ \text{RPAFS} \cdot \text{EYLDS}$$

$R^2 = 0.768$ $\bar{R}^2 = 0.756$ D.W. = 1.71 Sample: 1968–1987

Feed Grain Demand Block

For purposes of estimation, the demand for feed grains is broken into the demand for feed, industry, exports, farmer-owned reserve, government-owned stocks, and free market stocks (including stocks held under CCC loan). Feed demands depends on the relationship of the price of corn to the price of beef, pork, and poultry. The specification of export and inventory equations follows the previous dicussion.

Estimated Equation for Feed Demand for Feed Grains (DLVKFG):

$$\text{DLVKFG} = \underset{(10.0)}{158.0} - \underset{(627.0)}{1{,}909.0}\ \text{RPAFCMT}$$

$R^2 = 0.382$ $\bar{R}^2 = 0.341$ D.W. = 1.69 Sample: 1971–1987

Estimated Equation for Industry Demand for Feed Grains (DINDFG):

$$\text{DINDFG} = \text{DINDFG}\ (T - 1) - \underset{(22.0)}{94.6} - \underset{(4.7)}{11.1}\ \text{RPAFC} + \underset{(6.0)}{39.6}\ \text{RPYD}$$

$R^2 = 0.901$ $\bar{R}^2 = 0.887$ D.W. = 1.21 Sample: 1971–1987

Estimated Equation for Feed Grain Exports (XFG):

$$\text{XFG} = \underset{(12.4)}{19.6} - \underset{(20.0)}{64.7}\ \text{LPCX} + \underset{(17.7)}{90.1}\ \text{EXR}$$

$R^2 = 0.745$ $\bar{R}^2 = 0.688$ D.W. = 0.265 Sample: 1976–1987

Estimated Equation for Farmerr-Owned Reserve Demand for Feed Grains (KFORFGE):

$$\text{KFORFGE} = \text{KFORFGE}(T-1) + \underset{(62.0)}{182.0}\ \text{FORINC} - \underset{(8.0)}{21.6}\ \text{FOROUTC} - \underset{(2.76)}{5.05}\ \text{FORGOVE}$$

$R^2 = 0.711$ $\bar{R}^2 = 0.628$ D.W. = 0.945 Sample: 1978–1987

Estimated Equation for Government Demand for Feed Grains (KGOVFGE):

$$\text{KGOVFGE} = \underset{(0.37)}{6.50}\ \text{FORGOVC} + \underset{(52.2)}{65.7}\ \text{GOVINC} - \underset{(0.990)}{0.379}\ \text{GOVOUTC}$$

$$+ \underset{(0.082)}{0.585}\ \text{KGOVFGE}(T-1) - \underset{(9.6)}{41.6}\ \text{DMYPIK}$$

$R^2 = 0.972$ $\bar{R}^2 = 0.962$ D.W. = 0.801 Sample: 1971–1987

Estimated Equation for Free Market Stocks for Feed Grains (KMKTFGE):

$$\text{KMKTFGE} = \underset{(21.8)}{31.3} - \underset{(15.6)}{41.7}\ \text{RPAFC} - \underset{(0.229)}{0.734}\ \text{KFORFGE} + \underset{(0.067)}{0.249}\ \text{SUPDOMFG}$$

$R^2 = 0.805$ $\bar{R}^2 = 0.760$ D.W. = 2.05 Sample: 1971–1987

Estimated Equation for the Real Price of Corn (RPAFC):

$$\text{RPAFC} = \underset{(0.3862)}{0.0617} + \underset{(0.0063)}{0.0258}\ \text{RPMEAT} - \underset{(0.00078)}{0.00276}\ \text{SUPDOMFG}$$

$R^2 = 0.767$ $\bar{R}^2 = 0.734$ D.W. = 1.73 Sample: 1971–1987

Estimated Equation for the Price of Sorghum (PAFSG):

$$\text{PAFSG} = \underset{(0.01)}{1.62}\ \text{PAFC}$$

$R^2 = 0.976$ $R^2 = 0.976$ D.W. = 2.03 Sample: 1971–1987

Wheat Demand Block

The demand for wheat is similarly broken into the demnad for feed, food, exports, farmer-owned reserve, government-owned stocks, and free market stocks (includiung stocks held under CCC loan). Feed demands depends on broiler numbers since poultry is primary user of feed wheat. In addition, the use of wheat for feed depnds crtically on the ratio of corn price to wheat price. This term is included nonlinerly to capture the globar relationship whereby feed use of wheat increase asymptotically as wheat price declines relative to corn price. The specification of export and inventory equations follows the previous discussion.

Estimated Equation for the Feed Demand for Wheat (DLVKW):

$$\text{DLVKW} = \underset{(226.0)}{-694.0} + \underset{(91.0)}{284.0}\ \text{ERPCPW}\ (T-1) + \underset{(0.0087)}{0.0324}\ \text{PRDBR}$$

$R^2 = 0.664$ $\bar{R}^2 = 0.609$ D.W. = 1.11 Sample: 1971–1986

Estimated Equation for the Per-Capita Food Demand for Wheat (DFWCAP):

$$\text{DFWCAP} = \underset{(0.55)}{3.46} - \underset{(0.0680)}{0.0964}\ \text{RPAFW} + \underset{(0.1641)}{0.0653}\ \text{RPYD}$$

$R^2 = 0.187$ $\bar{R}^2 = 0.061$ D.W. = 1.77 Sample: 1971–1986

Estimated Equation for the Export Demand for Wheat (XW):

$$\text{XW} = \underset{(451.0)}{1{,}592.0} - \underset{(424.0)}{1{,}617.0}\ \text{LPWX} + \underset{(616.0)}{1{,}502.0}\ \text{EXR}$$

$R^2 = 0.646$ $\bar{R}^2 = 0.558$ D.W. = 1.93 Sample: 1976–1986

Estimated Equation for Farmer-Owned Reserve Demand for Wheat (KFORWE):

$$\text{KFORWE} = \text{KFORWE}\ (T-1) + \underset{(1{,}393.0)}{6{,}940.0}\ \text{FORINW} - \underset{(1.77)}{1.24}\ \text{FOROUTW} - \underset{(2.03)}{9.21}\ \text{FORGOVW}$$

$R^2 = 0.782$ $\bar{R}^2 = 0.709$ D.W. = 1.63 Sample: 1978–1986

Estimated Equation for Government Demand for Wheat (KGOVWE):

$$\text{KGOVWE} = \underset{(1.17)}{3.70}\ \text{FORGOVW} + \underset{(1{,}830.0)}{2{,}857.0}\ \text{GOVINW} - \underset{(0.452)}{0.104}\ \text{GOVOUTW}$$

$$+ \underset{(0.237)}{0.678}\ \text{KGOVWE} - \underset{(573.0)}{1{,}922.0}\ \text{DMYPIK}$$

$R^2 = 0.870$ $\bar{R}^2 = 0.882$ D.W. = 1.94 Sample: 1971–1986

Estimated Equation for the Free Market Stock Demand for Wheat (KMKTWE):

$$\text{KMKTWE} = \underset{(338.0)}{-112.0} - \underset{(142.5)}{46.7}\ \text{RPAFW} - \underset{(0.134)}{0.397}\ \text{KFORWE} + \underset{(0.153)}{0.568}\ \text{SUPDOMW}$$

$R^2 = 0.713$ $\bar{R}^2 = 0.642$ D.W. = 2.68 Sample: 1971–1986

Estimated Equation for the Real Price of Wheat (RPAFW):

$$\text{RPAFW} = \underset{(1.0237)}{0.0803} + \underset{(0.0159)}{0.0335}\ \text{RPMEAT} - \underset{(0.000246)}{0.000455}\ \text{SUPDOMW}$$

$R^2 = 0.594$ $\bar{R}^2 = 0.531$ D.W. = 1.49 Sample: 1971–1986

Estimated Equation for the Real Price Differential Wheat and Corn (RPAFW–RPAFC):

$$\text{RPAFW} - \text{RPAFC} = -0.508 + 0.0153\ \text{RPMEAT} - 0.000205\ \text{SUPDOMW}$$
(06.81) (0.0097) (0.000169)

$$+\ 0.0162\ \text{XW} + 0.000676\ \text{SUPDOMFG}$$
(0.000134) (0.001375)

$R^2 = 0.499$ $\bar{R}^2 = 0.317$ D.W. = 1.95 Sample: 1971–1986

Soybean Demand Block

The soybean demand block contains equations for exports, crushing, inventory, and price. The structure of the export equation is essentially the same as for feed grains. Crushing demand is determined by meat prices reflecting feed demand for soybean meal and consumer income reflecting demand for soybean oil. The inventory and price equations follow the same structure as in the cases of feed grains and wheat aside from the role of government programs in inventory determination.

Estimated Equation for Soybean Export Demand (XSB):

$$\text{XSB} = 714.0 - 533.0\ \text{LPSX} + 1{,}131.0\ \text{EXR}$$
(246.0) (225.0) (520.0)

$R^2 = 0.385$ $\bar{R}^2 = 0.249$ D.W. = 2.14 Sample: 1976–1986

Estimated Equation for Soybean Crushings (CRUSH):

$$\text{CRUSH} = -1{,}367.0 + 570.0\ \text{RPYD} + 20.8\ \text{RPMEAT} - 269.0\ \text{RPAFS}$$
(909.0) (207.0) (11.6) (110.0)

$R^2 = 0.633$ $\bar{R}^2$ 0.549 D.W. = 1.19 Sample: 1971–1987

Estimated Equation for Free Market Stocks of Soybeans (KPRISBE):

$$\text{KPRISBE} = -209.0 - 18.0\ \text{RPAFS} + 0.411\ \text{SUPDOMSB}$$
(138.0) (34.1) (0.068)

$R^2 = 0.890$ $\bar{R}^2$ 0.874 D.W. = 1.45 Sample: 1971–1987

Estimated Equation for the Real Price of Soybeans (RPAFS):

$$\text{RPAFS} = -0.586 - 0.000732\ \text{SUPDOMSB} + 0.0705\ \text{RPMEAT}$$
(1.208) (0.000363) (0.0188)

$R^2 = 0.754$ $\bar{R}^2$ 0.719 D.W. = 2.33 Sample: 1971–1987

Meat Supply Block

The structure of the meat supply block follows the earlier generic discussion with breeding herd and production equations included for cattle, hogs, and poultry. The beef supply component of the meat supply block includes the following estimated equations.

Estimated Equation for Breeding Cow Inventory (TCOWKE):

$$\text{TCOWKE} = \text{TCOWKE}(T-1) + \underset{(2{,}202.0)}{7{,}365.0} - \underset{(162.0)}{421.0}\ \text{RRS} - \underset{(157{,}100.0)}{556{,}300.0}\ \text{RPCPB}(T-1)$$

$R^2 = 0.494$ $\bar{R}^2$ 0.416 D.W. = 0.987 Sample: 1972–1987

Estimated Equation for Beef Production (PRDBEEF):

$$\text{PRDBEEF} = \underset{(846.0)}{22{,}248.0} + \underset{(0.0174)}{0.0128}\ \text{TCOWKE}(T-2) - \underset{(0.128)}{0.540}\ [\text{TCOWKE} - \text{TCOWKE}(T-1)]$$

$R^2 = 0.634$ $\bar{R}^2$ 0.578 D.W. = 1.06 Sample: 1972–1987

The pork supply component of the meat supply block includes the following estimated equations.

Estimated Equation for Breeding Hog Inventory (BRHOGKE):

$$\text{BRHOGKE} = \underset{(1{,}209.0)}{3{,}211.0} + \underset{(0.175)}{0.711}\ \text{BRHOGKE}(T-1) - \underset{(37{,}160.0)}{80{,}290.0}\ \text{RPCPP}(T-1) = \underset{(42.8)}{76.5}\ \text{RRS}$$

$R^2 = 0.489$ $\bar{R}^2$ 0.361 D.W. = 1.15 Sample: 1972–1987

Estimated Equation for Pork Production (PRDPORK):

$$\text{PRDPORK} = \underset{(3{,}020.0)}{13{,}530.0} + \underset{(0.416)}{0.682}\ \text{BRHOGKE}(T-1) - \underset{(82{,}500.0)}{211{,}2100.0}\ \text{RPCPP}(T-1)$$
$$+ \underset{(0.583)}{0.878}\ [\text{BRHOGKE} - \text{BRHOGKE}(T-1)]$$

$R^2 = 0.489$ $\bar{R}^2$ 0.361 D.W. = 1.15 Sample: 1972–1987

The poultry component of the meat supply block includes the following estimated equations.

Estimated Equation for Poultry Production (PRDBR):

$$\text{PRDBR} = \underset{(1{,}403.0)}{-26{,}120.0} - \underset{(12{,}480.0)}{46{,}910.0}\ \text{RPCPBR} + \underset{(16.0)}{487.0}\ \text{YEAR}$$

$R^2 = 0.986$ $\bar{R}^2$ 0.984 D.W. = 1.67 Sample: 1972–1987

Meat Demand Block

The meat demand block includes three sets of estimated equations. The domestic meat demand component of the meat demand block forms a simultaneous multivariate system where each demand equation is represented in a price-dependant form with each demand depending on the prices of the other two meat types and the price of all other goods. Consumer income is included and homogeneity is imposed by expressing all prices relative to consumer income. The second set of equations depicts trade in meat. While US trade in meats is not insignificant, these equations are needed to close the system. They relate trade in meat to the border price of meat and the exchange rate. The third set of equations translates meat prices for beef and pork into value per head for breeding cattle and hogs, respectively. These equations are used later to estimate changes in investment in breeding herds.

Estimated Equation for the Real Price of Beef (RPBEEF):

$$RPBEEF = \underset{(25.9)}{25.1} + \underset{(1.65)}{5.62}\ RELPBR - \underset{(1.149)}{0.301}\ RELPPORK + \underset{(419.0)}{1{,}028.0}\ RELOTH - \underset{(0.129)}{0.456}\ PCDBEEF$$

$R^2 = 0.812$ R^2 0.743 D.W. = 1.60 Sample: 1972–1987

Estimated Equation for the Real Price of Pork (RPPORK):

$$RPPORK = \underset{(14.1)}{4.27} + \underset{(0.58)}{3.27}\ RELPBR + \underset{(0.409)}{0.533}\ RELPBEEF = \underset{(237.0)}{103.0}\ RELOTH - \underset{(0.079)}{0.420}\ PCDPORK$$

$R^2 = 0.946$ R^2 0.926 D.W. = 2.91 Sample: 1972–1987

Estimated Equation for the Real Price of Broilers (RPBR):

$$RPBR = \underset{(8.0)}{103.0} + \underset{(0.083)}{0.358}\ RELPBEEF + \underset{(0.128)}{0.352}\ RELPPORK - \underset{(74.0)}{888.0}\ RELOTH - \underset{(0.50)}{0.496}\ PCDBR$$

$R^2 = 0.989$ R^2 0.985 D.W. = 2.27 Sample: 1972–1987

The meat trade component of the meat demand block consists of these equations explaining the net imports (exports) of beef, pork, and poultry.

Estimated Equation for Beef Imports (MBEEF):

$$MBEEF = \underset{(635.0)}{-426.0} + \underset{(637.0)}{2{,}647.0}\ EXR$$

$R^2 = 0.590$ R^2 0.556 D.W. = 2.01 Sample: 1974–1987

Estimated Equation for Pork Imports (MPORK):

$$MPORK = \underset{(639.0)}{1{,}955.0} - \underset{(380.0)}{1{,}822.0}\ EXR + \underset{(8.90)}{9.03}\ TPPORK$$

$R^2 = 0.700$ R^2 0.653 D.W. = 0.861 Sample:1972–1987

Estimated Equation for Broiler Exports (XBR):

$$XBR = \underset{(498.0)}{2{,}788.0} - \underset{(15.6)}{67.7}\ TPBR - \underset{(336.0)}{866.0}\ EXR - \underset{(165.0)}{385.0}\ D83$$

$R^2 = 0.670$ R^2 0.587 D.W. = 1.52 Sample: 1972–1987

The equations for breeding-herd value express value per head as a simple linear function of the respective meat price.

$$\underset{}{\text{VALCAT}} = \underset{(53.2)}{-82.4} + \underset{(0.26)}{2.12}\ \text{PBEEF}$$

$R^2 = 0.821$ $\bar{R}^2$ 0.809 D.W. = 1.20 Sample: 1972–1987

$$\text{VALHOGCAT} = \underset{(17.9)}{22.6} + \underset{(0.126)}{0.319}\ \text{PORK}$$

$R^2 = 0.331$ $\bar{R}^2$ 0.279 D.W. = 3.16 Sample: 1972–1986

Identities

The grain model is closed by acreage (total acreage is equal to participating plus nonparticipating acreage), production (production is equal to the product of acreage and yield), and supply-demand identity equations for each of the grain commodities.

*Identities for Feed Grains**

ACGS = ACGSN + BACGS · COMPFG · (1 – DRFG)
PRDFG = 0.0254 · ACGS · YLDCGS + PRDOTH
PRDFG + KMKTFGE (T – 1) + KFORFGE (T – 1) + KGOVFGE (T – 1)
= DINDFG + DLVKFG + XFG + KMKTRGE + KFORFGE + KGOVFGE

Identities for Wheat

AW = AWN + BAW · COMPW · (1 – DRW)
PRDW = AW · YLDW
PRDW + KMKTWE (T – 1) + KFORWE (T – 1) + KGOVWE (T – 1)
= DFWCAP · POP + DLVKW + XW + KMKTWE + KFORWE + KGOVWE

Identities for Soybeans

PRDSB = AS · YLDS
PRDSB + KPRISBE (T – 1) = CRUSH + XSB + KPRISBE

The meat model is closed by supply-demand identity equations for each of the meats which equate production plus net imports to domestic consumption (per-capita consumption multiplied by population).

* Note that the production of feed grains other than corn and grain sorghum has been nearly constant at 18 million metric tons per year. Nevertheless, feed grain production system is closed by estimated PRDOTH as a linear function of corn and grain sorghum production. The estimated equation is PRDOTH = 18.0 + 0.00178 ACGS · YLDCGS. Note that the coefficient, 0.0254, converts bushels to metric tons for purposes of relating supply to demand.

Identity for Beef

PRDBEEF + MBEEF = PCDBEEF · POP

Identity for Pork

PRDPORK + MPORK = PCDPORK · POP

Identity for Broilers

PRDBR – XBR = PCDBR · POP

Projection Equations for Nonpolicy Exogenous Variables

For purposes of simulation, the nonpolicy exogenous variables are projected according to simple assumptions that represent a stable macroeconomy. The nominal rate of interest is assumed to contiue thoughtout the entire period at its 1985 level of 7.95%. The gross national product price deflator, consumer income, population, and prices of nonagricultural goods are assumed to increase by a constant amount each year as characterized by the recent past.

GNPD = -7.08 + 0.135
(0.33) (0.004)

$R^2 = 0.997$ $\bar{R}^2$ 0.997 D.W. = 1.48 Sample: 1983–1987

YD = -12,719.0 + 182.0 YEAR
(712.0) (8.0)

$R^2 = 0.994$ $\bar{R}^2$ 0.992 D.W. = 2.03 Sample: 1983–1987

POP = -0.650 + 2.13 YEAR
(3.065) (0.04)

$R^2 = 0.999$ $\bar{R}^2$ 0.999 D.W. = 1.80 Sample: 1983–1987

N = 80.0 + 0.876 POP
(3.9) (0.023)

$R^2 = 0.990$ $\bar{R}^2$ 0.990 D.W. = 1.85 Sample: 1972–1987

PF = -1.55 + 0.0299 YEAR
(0.07) (0.0008)

$R^2 = 0.998$ $\bar{R}^2$ 0.997 D.W. = 1.93 Sample: 1983–1987

Chapter 9

The Determination of Technology and Commodity Policy in the US Dairy Industry

Harry de Gorter, David J. Nielson, and Gordon C. Rausser

9.1 Introduction

United States dairy policy includes both predatory and productive components. The milk price support program is designed to transfer income to dairy farmers while research and extension expenditures are designed to increase social welfare. The purpose of this chapter is to provide an empirical example of the theory developed in Chapter 4.

It has long been recognized that government research and extension policies have been significant contributors to technological advance in agriculture (Evenson and Kislev 1975; Evenson et al. 1979). The advent of technological change in agriculture and its policy implications was first noted by Schultz (1945, 1953). Some, like Cochrane (1958) in characterizing his famous technological treadmill thesis, argued for price supports in order to compensate farmers for the adverse effects of research on farmers' welfare. Indeed, commentators like Thurow (1971) and Schlesinger (1984) argue that public good provision in agriculture is one of the few major economic success stories of government intervention in the history of the United States. Nevertheless, one of the most stylized facts on government policy intervention in agriculture is the pervasive level and overwhelming evidence of underinvestment in public research (Ruttan 1982). Concomitant to this notion, economists have alleged that governments "overinvest" in commodity policies because of the associated deadweight losses generated.

This chapter examines the way in which predatory and productive policies are determined in the context of the US dairy sector. The paper develops an analytical and empirical model of the endogenous determination of research expenditures and price supports in the dairy industry. The results also provide an explanation for the persistent underinvestment in agricultural research and for the apparent contradiction between the two public policies that both increase output and yet individually have opposite effects on social welfare.

9.2 Analytical Model

In this section, a stylized model of the US dairy industry is developed. This model is used to analyze the joint determination of government expenditures on cost-reducing public research and on price support activities. The analysis is performed under the assumption of competitive market conditions using a deterministic, static, and partial-equilibrium representation of the dairy industry.

In the United States, the price of milk at the farm level is supported through a standing offer by the Commodity Credit Corporation (CCC) to purchase several of the manufactured dairy products at pre-announced price levels.[1] These purchases guarantee that the market price of Class II milk does not fall below the support price. Class II milk refers to that milk which is purchased for use in producing a variety of manufactured dairy products such as cheeses and butter. Class I milk used for fluid milk consumption is purchased at a fixed price above the Class II support price. The difference between these two prices is called the Class I differential and is assumed to be set exogenously in the analysis to follow. Farmers receive a blend price which is calculated as a weighted average of the Class I and the Class II prices.

In the notation adopted here, Class I and Class II prices are linked by a fixed differential, α,

$$P_I = P_{II} + \alpha. \tag{1}$$

In equilibrium, the total supply of milk is equal to the total demand for milk. The total demand is made up of the three components D_I, D_{II}, and NR; D_I and D_{II} are the Class I and Class II levels of milk demand, respectively, and NR represents the milk equivalent units of CCC purchases of manufactured dairy products, often referred to as net removals of the government.[2] Net removals (NR) are determined endogenously according to the following expression:

$$NR = S - D_I - D_{II}. \tag{2}$$

The blend price received by farmers (P_b) is given by[3]

$$P_b = \frac{P_I D_I + P_{II}(D_{II} - NR)}{S} \tag{3}$$

The government choice variable for the commodity policy is the Class II price support, P_{II}. The fluid milk price faced by consumers and the blend price received by farmers are uniquely related to the choice variable, P_{II}, as described in Eqs. (1), (2), and (3).

Government costs, G, due to the CCC price support program are given by:

$$G = \partial P_{II}(S - D_I - D_{II}) \tag{4}$$

---

[1] The CCC has made positive net purchases each year since 1955. However, market prices often are above Class II support prices during a year, since CCC purchases are seasonal. The Class II support price, P_{II}, was found to be 99.9% of the market price on average for the time period 1955–1988. Hence, we ignore the distinction between P_{II} and the market price for manufacturing products in this study.

[2] For a review on the mechanics of the US dairy program, see Ippolito and Masson (1978).

[3] Imports of manufacturing milk products in the United States are very low and stable throughout the historical time period under investigation. Hence, they are ignored throughout the analysis, without consequence for our results.

where $\delta = 0.85$. The value of δ indicates that only 15% of CCC removal costs are recovered from sales (Tauer and Kaiser 1990). This number is an average figure for the past 30 years and indicates that much of the CCC resale revenues are offset by the processing, storage, and handling costs incurred by the CCC after the product has been purchased at the support price.

The government also intervenes in the dairy industry in providing publicly funded research which is designed to increase productive efficiency at the farm level. Improvements in productive efficiency have the effect of shifting outward and downward the supply schedule for milk. Public research expenditures are denoted here by E. Total expenditures on both productive and predatory policies are provided and financed by the government with consumers as taxpayers being coerced to pay for the entire budgetary costs.

In order to understand the ways in which these two policy instruments, P_{II} and E, affect the various characteristics of the market for milk, it is necessary to consider the behavior of the individual actors who comprise the demand and supply side of the market. Suppose n identical consumers choose their optimal level of Class I and Class II milk consumption by maximizing on a preference function represented by a twice differentiable, additive and separable utility function $U(q^d_j, Z) = U(q_j) + Z$ where $j = \text{I, II}$; q is the amount of milk consumption; Z is the numeraire good $= M_0 - \Sigma_j p_j q^d_j - E/n - \delta P_{II}[S - \Sigma_j q^d_j]$; and M_0 is endowment income. Each consumer and producer takes both P_{II} (and hence P_I and P_b) and E as given in making their consumption and production choices. At optimality, $U_q(q^d) - \lambda p = 0$ where λ is the Lagrange multiplier of the budget constraint. Assuming that the consumption behavior of individual consumers has no effect on market prices, the following are optimal consumption schedules $q^d_j = U^{-1}{}_q(\lambda^* p_j)$ where λ^* is the constant marginal utility of income. The industry demand schedule for each class of milk is the summation over n consumers,

$$\sum_{i=1}^{n} \sum_{j} q_{ij}^{d}(P_j) = D_j(P_j).$$

Suppose m producers choose their individual level of production, q^s, so as to maximize profits $\Pi = P_b q - C(q^s, E)$ where $C(q^s, E)$ represents individual i's cost function. It is assumed that $C(q^s, E)$ is everywhere increasing in q^s and decreasing in E; i.e., $\partial C/\partial q^s = C_q > 0$ and $\partial C/\partial E = C_E < 0$. It is further assumed that E decreases marginal units; i.e., $\partial^2 C/\partial q^2 \partial E = C_{QE} < 0$.

Profit maximization results in $P_b = C^s{}_q(q^s, E)$, which can be used to solve for the individual's supply schedule, $q^s(P_b, E) = c^{s-1}{}_q(P_b, E)$. The industry supply schedule, $S(P_b, E)$ is the summation over m producers,

$$\sum_{k=1}^{m} q_k^{s}(P_b, E) = S(P_b, E).$$

The government's choice problem is to choose some optimal levels of P_{II} and E. Because of the indirect interactive effects between the two instruments (i.e., the price supports affect the efficiency and income distribution of research, and vice-versa), a rational and informed government will choose the level of the two instruments jointly. Suppose that the government's objective function is characterized by the indirect welfare function, $V(P_I, P_{II}, P_b, P_{-j}, Y)$, where P_{-j} is the vector of all other prices in the economy and Y reflects aggregate income[4]. We assume that the government's objectives can be characterized by the preference function $V(V_1, V_2)$. This function is assumed to be linear and additive in its arguments which consist of a weighted sum of aggregate consumer income (net of optimal consumption expenditures) and optimal producer profits. Thus, the government's decision problem is to choose P_b and E so as to satisfy the following:

$$\max_{P_b, E} V = w_1 V_1(P^*, M^*) + w_2 V_2(\Pi_2), \tag{5}$$

where w_1, w_2 are political weights (that sum to unity)[5] assigned to the welfare of consumers and producers – respectively,

$$M^* = M_0 - E - P_{II^*}\{S^*[P_b(P_{II}, E), E] - D_I(P_I) - D_{II}(P_{II})\} \text{ and}$$
$$\Pi^* = [P_b(P_{II}, E)]S[P_b(P_{II}, E), E] - C\{S[P_b(P_{II}, E), E], E\}.$$

The superscript star represents equilibrium levels.

The necessary first-order conditions for a maximum, using Roy's identity, can be written as

$$w_1\left[-D_I - D_{II} + \frac{\partial M}{\partial P_{II}}\right] + w_2 \frac{\partial \Pi}{\partial P_{II}} = 0 \tag{6}$$

and

$$w_1 M_E + w_2 \Pi_E = 0 \tag{7}$$

where $V_y = V_m = V_\Pi$ = marginal utility of income. The net effect of the price supports on taxpayers income is described by

$$\frac{\partial M}{\partial P_{II}} = -\delta\left[NR + P_{II} \cdot \left(\frac{\partial S}{\partial P_{II}} - \frac{-\partial D_I}{\partial P_I} - \frac{-\partial D_{II}}{\partial P_{II}}\right)\right] < 0. \tag{8}$$

The effect is negative and the negative impact is stronger – the larger are the values of δ, the levels of net removals, the Class II price support, and the absolute values of both the supply or the demand elasticities.

The effect of price supports on the level of producer profits is described in the equation,

---

[4] This formulation assumes negligible income effects with expenditures on price supports and research, i.e., it is assumed that income is independent of changes in the price vector. This is equivalent to asserting that the same goods do not serve as both inputs and outputs (Varian 1984, p. 276).

[5] These implicit weights placed on producer and consumer welfare by government reflect the manner in which changes in economic welfare affect political support. The underlying conceptual framework that generates these political weights may be found in Rausser and Foster (1990).

$$\frac{\partial \Pi}{\partial P_{II}} = S\left(\frac{\partial P_b}{\partial P_{II}}\right) \geq 0. \tag{9}$$

This expression depends critically on the change in the blend price resulting from a change in Class II support prices:

$$\frac{\partial P_b}{\partial P_{II}} = \left[\frac{S + \frac{\partial D_I}{\partial P_I} \cdot (P_I - P_{II})}{S^2 + \frac{\partial S}{\partial P_b} \cdot (P_I - P_{II}) \cdot D_I}\right] S. \tag{10}$$

The effect of research expenditures on taxpayer income is given by

$$M = -\delta P_{II}\left(\frac{dS}{dE}\right) - 1 < 0. \tag{11}$$

Eq. (10) demonstrates that the cost of research to consumers is affected by the presence of a price support. The price support has an impact upon taxpayer costs, not only through CCC purchases as reflected in the values of δ and the Class II price support but also as a result of the responsiveness of supply to research expenditures.

9.2.1 Impact of Research On Producer Profits

The impact of research on producer profits is given by

$$\Pi_E^* = S\left(\frac{\partial P_b}{\partial E}\right) - C_E \lesseqgtr 0, \tag{12}$$

where

$$\frac{dS}{dE} = \frac{\partial S}{\partial P_b}\frac{\partial P_b}{\partial E} + \frac{\partial S}{\partial E} \tag{13}$$

and

$$\frac{\partial P_b}{\partial E} = \frac{-dS}{dE}\left(\frac{D_I(P_I - P_{II})}{S^2}\right). \tag{14}$$

Substitution of (12) into (13) yields

$$\frac{\partial P_b}{\partial E} = \frac{-S_E D_I \cdot (P_I - P_{II})}{S^2 + \frac{\partial S}{\partial P_b} \cdot (P_I - P_{II})}. \tag{15}$$

The effect of research expenditures on the blend price of milk received by farmers is indirect in that it operates through changes in the relative weights in the blend price formula. Because research expenditures ultimately have the effect of shifting out the industry supply schedule, they increase the amount of milk produced at any given price. A change in production at given price levels affects the total consumption of Class II milk via increased *NR* but leaves the demand for Class I milk unchanged. Thus, it is the change in the relative percentage of the milk which is used for Class I and Class II purposes which alters the weights in the blend price formula shown in Eq. (2) and which results in an indirect change in the blend price, even if the Class I and Class II prices remain unchanged.

9.2.2 The Joint Determination of Research and Price Supports

Expressions (8) through (15) emphasize the way in which each policy affects the consumer, producers, and taxpayers. Farmers always gain from the provision of Class II price supports. They may, however, lose from publicly funded research expenditures, even if a price support is present. In their role as taxpayers, consumers lose from the provision of either policy. In the case of public research expenditures, the losses which consumers feel as taxpayers can easily be offset by gains in their role as consumers. These gains to consumers are generated by the price reductions which accompany the productivity enhancing results of research.

Expressions (5) and (6) characterize the way in which the welfare of consumers and producers are balanced against each other in the government's choice of P_{II} and E. Eq. (5) indicates that the level of the price support must be chosen such that the weighted marginal cost to consumers and taxpayers of increasing the price support, $[w_1(-D_I-D_{II} + \partial M/\partial P_{II})]$, is just equivalent to the weighted marginal benefit of the subsidy to producers $[w_2\partial\Pi/\partial P_{II}]$. Eq. (6) dictates that E be chosen such that the weighted marginal cost to consumers of additional research expenditures (i.e., w_1M_E) is just equated with the weighted marginal benefit to producers $(w_2\Pi_E)$.

Given a fixed level of research expenditures, E, Eq. (5) can be used to solve for the level of P_{II} which is consistent with an optimal solution to the government's choice problem, viz.,

$$P_{II} = \left[(D_I + D_{II})\left(1 - \frac{1}{\delta}\right) + S\left(\frac{\gamma}{\delta}\frac{dP_b}{dP_{II}} - 1\right)\right] / \frac{\partial NR}{\partial P_{II}}. \tag{5a}$$

The price support is "effective" – i.e., expression (5a) becomes positive – if $\gamma > 1$. Given that $\delta < 1$, the price support becomes ineffective if demand exceeds supply at the political optimal level of P_{II}. Expression (5a) shows that Class II support prices are positively related with the relative political weight given to farmers γ and the response of net removals to the support price, $\partial NR/\partial P_{II}$.

The politically optimal level of research expenditures can be determined by substituting appropriate terms into Eq. (6),

$$\gamma\left(S\frac{\partial P_b}{\partial E} - C_E\right)\delta P_{II}\left(\frac{dS}{dE}\right) = 1. \tag{6a}$$

If research expenditures are evaluated assuming equal welfare weights (i.e., $\gamma = 1$) and that no price supports exist, then the optimal condition is $-C_E = 1$. This characterizes an efficient allocation of resources, and a rationally chosen allocation, if one ignores the existence of price supports and the political component of government policy. However, the optimal level of research must reflect the effects of research on price support costs as well, as shown in Eq. (6a).

It is clear from the necessary conditions for a maximum in the government's choice problem that the way in which each of the two policies affects the choice of the other is dependent upon, among other things, the value of the relative weights as reflected in γ, and whether farmers are helped or injured at the margin by the effects of additional research. Casual observation of the government involvement in the dairy industry suggests that producer income is weighted more heavily than that of consumers/taxpayers for the purpose of determining dairy policy (i.e., $\gamma > 1$). We now turn our attention to the way in which producer incomes are affected by research-induced shifts in their cost functions.

9.2.3 When Are Research and Price Supports Complementary Policy Instruments?

Having outlined the nature of the government's decision problem in setting each of the two policy instruments jointly, we examine whether the provision of price support activity tends to result in more or less research being provided than would otherwise be the case. It was shown in the previous chapter that it depends on the parameters of the market. It is possible that the government is able to use the price support instrument to compensate farmers for losses which they experience as a result of research-induced technical change. As shown in Chapter 8, it can occur when government places a greater weight upon producer income than it places upon consumer/taxpayer income (i.e., when $\gamma > 1$) and when farmers' welfare is injured as a result of the research program. In such cases, the existence of price support policy allows a greater level of research to be provided than would otherwise be the case.

It is possible to examine whether the chosen levels of the two policies are positively related to each other in the neighborhood of the optimal pair (E^*, P_{II}^*) by totally differentiating expression (6a) after utilizing (5a) to substitute for the chosen level of P_{II}. This allows the slope of the choice of E as a function of P_{II} to be determined in the neighborhood of the chosen pair (P_{II}^*, E^*). It is convenient to work with simplified expressions (5a') and (6a') in examining the sign of the slope at this reaction function. Expressions (5a') and (6a') are derived from (5a) and (6a) under the simplifying assumption that only one demand curve is faced by producers. This allows for ease of presentation without affecting the features of interest. Operating under this simplifying assumption, expressions (5a') and (6a') take the following form:

$$P = \frac{(\gamma - 1)S}{S_p - D_p} \tag{5a'}$$

$$-C_E = \frac{1 + PS_e}{\gamma}. \tag{6a'}$$

The slope, $dE(P_{II})/dP_{II}$, is described by $(d^2V/dEdP_{II})/(d^2V/dE^2)$:

$$\frac{dE(P_{II})}{dP_{II}} = -\left\{\left[-\frac{\left(1 + \frac{S_{PE}}{S_P S_E}\right)}{S_p - D_p} + \left(\frac{P_M(S_{pp} - D_{pp})}{S_p - D_p}\right)\frac{1}{\eta^s(S_p - D_p)} + \left(\frac{1}{1-\gamma}\right)\frac{P}{S\eta^s}\right]\right\} / \frac{d^2V}{dE^2}.$$

The sign of $dE(P_{II})/dP_{II}$ is guaranteed to be positive if the first term in brackets in Eq. (15) is sufficiently negative and the second term is sufficiently positive. The first term in brackets is negative only if S_{pe} is negative, and the second term in brackets is positive only if $S_{pp} - D_{pp}$ is positive. Note that it is not a necessary condition that either term be of a particular value or sign for expression (15) to be positive. As for the third term in brackets, we assume that $\gamma > 1$, so that it is strictly positive.

The value and sign of S_{pe} depends upon the particular form of the response of the supply schedule to research induced technical change. A pivot-like shift in the supply curve, representing a large impact of the technical improvement on variable costs relative to fixed costs, can cause $dE(P)/dP$ to be negative. On the other hand, if the shift in the supply schedule is a parallel one (as assumed in Alston et al. 1988), the first term in brackets takes a value of zero and the likelihood of complementarity between E and P_{II} is increased. The value and sign of $(S_{pp} - D_{pp})$ represents the degree of concavity of net removals with respect to the support price.[6] Note in the earlier discussion that producer welfare at the margin is likely to be positive with a more inelastic supply curve. Hence, $dE(P_{II})/dP_{II}$ is more likely positive under conditions when farmers are hurt by public research expenditures. Accordingly, governments are more willing to jointly set P_{II} and E as complementary instruments when farmers lose from research. Moreover, if the supply and demand functions are linear, $(S_{pp} - D_{pp})$ = zero, and a parallel shift in the supply curve due to research (as in Alston 1988), guarantees that dE/dP_{II} is positive.

9.3 Empirical Evidence

Are public research expenditures higher in the presence of price supports than they would be otherwise? Are both consumers/taxpayers and farmers better off with the existence of price supports than they would be in their absence? If the an-

[6] For the case of a constant elasticity supply curve with an elasticity less than one, the $(S_{pp} - D_{pp})$ term is more likely negative, enhancing the likelihood that $dE(P_{II})/dP_{II}$ is positive.

swers to these questions are yes, the price-support policy and public-research expenditures appear to be employed as complements and the availability of the price-support policy may allow a potential Pareto improvement to be achieved, relative to the outcome which would be otherwise obtained.

In this section we describe the relevant characteristics of the milk supply industry and the empirical model employed here to capture these features. The remaining discussion focuses on the results of an empirical model of the US dairy industry.

9.3.1 Public Research and Milk Supply

As in other agricultural sectors, the real level of public expenditures on agricultural research has increased steadily over time in the US dairy sector (Fig. 9.1). Meanwhile, productivity in the production of milk has increased dramatically. Two measures of productivity, labor per unit of output and milk produced per cow, are also reported in Fig. 9.2. These two figures demonstrate the dramatic improvement in productivity in historical milk production. This increase in productivity has resulted in a sharp reduction over the same period of time in the number of milk cows as well as in the number of operating dairy farms in the United States. These developments in agriculture are not limited to the livestock sector but also characterize developments in the crop sectors (Griliches 1958).

The estimated effects of public research expenditures on milk supply and on each of the inputs – grain concentrates, hay, labor, cows, and capital – are reported in Table 9.1. The econometric model employed in generating these estimates is specified in Nielson (1989). The model of aggregate milk production was developed from micro-level foundations. Individual firms were hypothesized to operate under a fixed coefficient technology in labor, capital and cows. They were specified to be able to substitute between hay and grain in the composition of the ra-

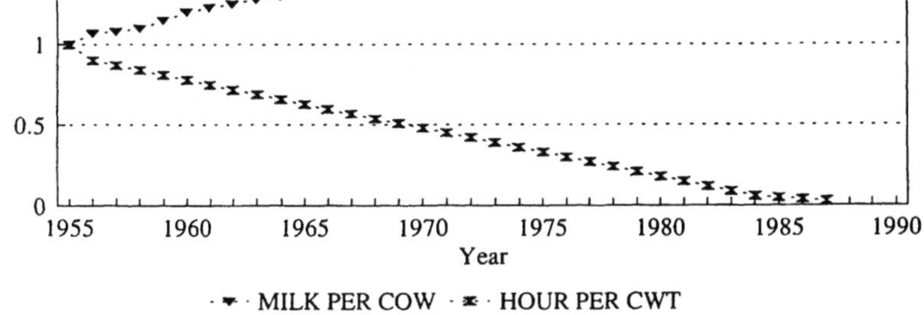

Fig. 9.1. Public research expenditures on US dairying, 1955–1988 (measured in constant 1988 dollars)

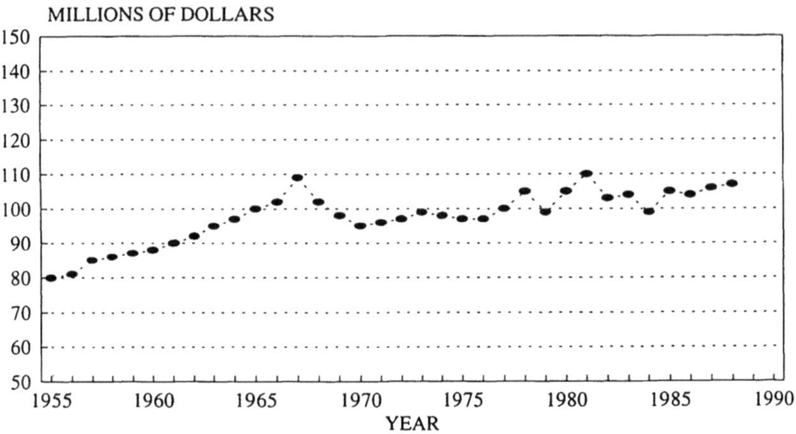

Fig. 9.2. Productivity improvements in US dairying, 1955–1988

tions of the cows. This substitution between the feeds is captured in a standard Cobb-Douglas specification for the yield of milk per cow as a function of the amount of each of the feeds. Thus producers employ a fixed coefficient technology in some of the inputs while having the ability to substitute between the feeds. The model was constructed so as to incorporate heterogeneity between producers. Because producers are specified to differ in the relative efficiencies in which they produce milk, they also differ in their profitability. As a consequence, entry and exit of individual producers of differing abilities is observed to correspond to changes in the economic environment.

The model is constructed such that the aggregate expressions for output and for the input demands are exactly consistent with the theory of the profit maximizing behavior of the individual heterogeneous firms. At the aggregate level, substitution between each of the inputs is possible in response to changes in the economic environment. This is in contrast to the fixed coefficient technology which prevails between some of the inputs at the micro level. This feature of substitutability at the aggregate level is a consequence of the entry and exit from the industry.

The direct impact of research-induced technical change was specified to take the form of a Hicks-neutral shift in the production function for each individual cow.[7] Indirect effects of technical change which improves the productivity of the cows themselves are reflected in producers' choices of each of the other inputs and consequently in the farm-level supply schedule. Their responses at the micro-level lead to the presence of the research variable in each of the aggregate input demand equations a well as in the aggregate supply function.

[7] The estimated elasticites are taken from Sullivan et al. (1989). The weighted average total demand for milk for the time period, 1955–1988, is –0.418.

Table 9.1. Estimated milk supply and input demand functions, 1955–1988

Independent variables	Dependent variables S	G	H	L	K	C
Pb	0.94722	1.94720	1.947200	1.94720	1.94720	1.94720
E	0.25521	0.25521	0.255210	0.25521	0.25521	0.25521
PG	–0.85502	–1.85500	–1.855000	–1.85500	–1.85500	–1.85500
PH	0.54771	0.54771	–0.452290	0.54771	0.54771	0.54771
PL	0.70151	0.70151	0.701510	–0.29849	0.70151	0.70151
PK	–0.36867	–0.36867	–0.368670	–0.36867	–1.36867	–0.36867
PC	0.02725	0.02725	0.027250	0.02725	0.02725	0.97275
PAST	–0.02470	–0.00636	–0.014244	–0.00630	0.00589	–0.00555
Trend	0.00102	0.00922	–0.012590	–0.05700	0.06087	0.00428
Interest rate	0.02512	–0.00523	0.021000	–0.00939	–0.05995	0.02172
Constant	1.54430	1.13510	4.068700	4.04910	9.29800	9.66840
R^2	0.52	0.52	0.50	0.94	0.52	0.64
Durbin Watson	0.85	0.86	0.96	0.52	0.86	1.17

Notes: S = ln (milk production in the United States) G = ln (grain concentrates fed to dairy cows) H = ln (hay fed to dairy cows) L = ln (labor allocated to milk production) K = ln (shipments of dairying equipment, $1,000 units) C = ln (milking cows) Pb = ln (price received for milk) E = ln (published research expenditures)[a] PG = ln (price of grain concentrates) PH = ln (price of hay) PL = ln (price of hired agricultural labor) PK = ln (index price of farm machinery) PC = ln (nonfeed cost of owning one cow 1 year PAST = measure of pasture conditions as a percent of normal (100% indicates normal conditions) Trend = time period (1955 = 1, ..., 1988 = 34.

All prices and dollar denominated variables are deflated to constant 1988 US dollars.

[a] This variable is constructed as a weighted average of the total public expenditures on dairy-related research each year for the 10 years prior to the current year. The weights employed are the inverse of the degree of lag of the year in which the expenditures were made. This corresponds roughly to the declining yet lingering effect of research dollars over time which other studies have adopted.

Source: Nielson et al. (1990).

The structure of the technology of milk production at the farm level is represented as follows:

$$\text{MilkProduced}_t = (\text{MilkperCow})_t * \alpha * \text{Capital}_t$$
$$\text{MilkProduced}_t = (\text{MilkperCow})_t * \beta * \text{Capital}_T$$
$$\text{MilkProduced}_t = (\text{MilkperCow})_t * \gamma * \text{CowsperHerd}_t$$
$$\text{MilkerCos}_t = A * \exp^{f(\text{RESEARCHt})} * \text{Grain}^a{}_t * \text{Roughage}^b{}_t$$

where alpha, beta, and gamma are farm-specific efficiency coefficients and {A, a, and b} are parameters of the Cobb-Douglas form of the milk-per-cow production function. Each farm's combination of values for alpha, beta, and gamma determines the farm's level of efficiency in milk production.

The aggregation procedure employed involves integrating micro-level production and input demand schedules over the frequency distributions of each of the efficiency characteristics of the individual producers. These frequency distributions were specified to be of the Pareto distribution form. From this aggregation procedure, a long-run expression for the industry supply function and for each of the aggregate input-demand schedules is obtained. These aggregate expressions are log linear in the input and output prices. As a direct implication of the theoretical model, restrictions on the parameters of these aggregate expressions exist both within and across equations.

Thus, at the aggregate level, substitution between each of the inputs is possible in response to changes in the economic environment. This is in contast to the fixed cofficient technology which prevails between some of the inputs at the micro-level. This feature of substitutability at the aggregate level is a consequence of the entry to and exit from the industry of firms with differing efficiency coefficients (and correpondingly differing levels of profitability).

The empirical results are reported in Table 9.1. The long-run elasticity of milk supply with respect to public research is estimated to be 0.25. This estimate is similar to that found in a recent study on the effects of research on milk supply in Canada by Fox et al. (1990). Their study reported long-run elasticities of Canadian milk supply to be 0.258, 0.57, and 0.707 with respect to provincial, federal, and US research expenditures, respectively. The estimate of 0.25 is somewhat below the estimate of 0.538 for the United States which was reported by Bredahl and Peterson (1976) in one of the earliest studies to report an elasticity of milk supply with respect to research expenditures. Although the elasticity of research found in this study is somewhat lower than those reported in those benchmark studies, the level of estimated responsiveness of milk production to research expenditures is still quite sufficient to allow research expenditures to have an important impact upon the industry.

The long-run own-price elasticity of milk supply is estimated to be 0.94. This estimate is lower than is reported in most of the previous studies, many of which estimate supply to be significantly own-price elastic (perhaps as high as 5) in the long run (for example, see Chavas and Klemme 1986). Most estimates of the own-price elasticity of the supply of milk have not incorporated the impact of research. The positive influence of research on productivity has been confounded with the price term in past studies. As a result, higher supply elasticities have been generated than would have occurred had the two effects been explicitly separately incorporated as in this study.

9.3.2 The Joint Determination of P_{II} and E in the US Dairy Sector

Are milk consumers better off when the government implements price supports to compensate dairy farmers for the negative effects of research on profits? To answer this question, model simulations are presented that embrace all of the theoretical features and empirical supply estimates of the US dairy sector derived above. The supply model with the expressions described in Table 9.1 are incorporated for milk supply and the input demands (grain, hay, capital, labor, and cow numbers). Constant elasticity demand functions are specified for Class I and Class II milk consumption with elasticities of –0.25 and –0.55, respectively.[8] All parameters are evaluated at their long-run values in the simulations. Specific details of US dairy policy such as the producer assessment levied in the early 1980s are also included in the analysis.[9]

From the baseline simulation, we generated the actual values of the relative political weights, γ, defined in each of the first-order conditions, (5) and (6). Note that each first-order condition is determined from the maximization of the government objective function with respect to each policy instrument, P_{II} and E. The value of γ in each case is determined independently of each other; let us denote γ_p as the relative political weight determined from Eq. (5) and γ_e as the relative political weight determined from Eq. (6). These weights can be viewed as the revealed preferences of the government with respect to the choice of each policy instrument.

The values for γ_p and γ_e are presented in Table 9.2. The estimated values of γ_p are very stable. Furthermore, it is surprising how close the values of γ_p are to those of γ_e for many of the time periods under consideration. The values of each weight are derived from very different estimated parameters as to how each instrument affects the dairy market. Nevertheless, the estimated weights are in the same range of values for the entire time period. This is a very important result and lends strong support to our hypothesis that governments rationally choose price supports and research expenditures jointly, and that the interaction effects between the two policies are explicitly recognized by politicians.

The baseline values for the key parameters are given in Table 9.3. The actual blend price received by farmers, the predicted supply, actual research expenditures, and initial (status quo) producer welfare are presented. Because the absolute value of consumer welfare is undefined in the constant elasticity case, only

---

[8] The assumed elasticities are taken from Sullivan et al. (1989). The weighted average total demand for milk for the time period 1955–1988 is –0.418. Tauer and Kaiser (1990) estimate a total demand function with an elasticity of –0.39.

[9] A complete documentation of the model simulation input and output with data are available from the authors upon request.

Table 9.2. Estimated values of γ_p and γ_E, 1955–1988

				γ_p					
1955	2.190	1962	2.374	1969	2.194	1976	2.170	1983	2.214
1956	2.198	1963	2.530	1970	2.188	1977	2.144	1984	2.196
1957	2.311	1964	2.320	1971	2.151	1978	2.119	1985	2.224
1958	2.350	1965	2.305	1972	2.080	1979	2.192	1986	2.239
1959	2.273	1966	2.186	1973	2.243	1980	2.189	1987	2.318
1960	2.271	1967	2.157	1974	2.518	1981	2.173	1988	2.275
1961	2.383	1968	2.227	1975	2.293	1982	2.112		
				γ_E					
1955	2.422	1962	5.339	1969	4.133	1976	3.563	1983	3.678
1956	2.325	1963	30.377	1970	3.802	1977	3.169	1984	3.281
1957	2.842	1964	4.269	1971	3.578	1978	2.695	1985	3.185
1958	3.261	1965	4.422	1972	2.970	1979	2.878	1986	3.201
1959	3.259	1966	3.182	1973	4.071	1980	2.510	1987	2.898
1960	3.130	1967	3.407	1974	8.327	1981	3.032	1988	2.570
1961	5.652	1968	4.505	1975	4.140	1982	2.951		

Table 9.3. Baseline values of key variables, 1969–1988

	Blend price	Predicted milk supply	Research expenditures	Initial producer welfare
	dollars per hundred weight	billion pounds	million dollars	$10 million
1969	5.490	122.301	0.50103	1,054.976
1970	5.710	122.761	94.39174	1,039.875
1971	5.870	130.145	97.34285	1,086.583
1972	6.070	151.464	100.38180	1,265.831
1973	7.140	111.208	107.14950	1,029.180
1974	8.338	6.768	119.99560	844.208
1975	8.750	104.950	130.79950	982.870
1976	9.660	130.308	137.19100	1,273.619
1977	9.720	132.899	145.41880	1,227.809
1978	10.600	140.471	158.44170	1,314.672
1979	12.000	124.828	179.54070	1,188.736
1980	13.000	125.111	202.52840	1,137.001
1981	13.800	124.358	230.73990	1,086.943
1982	13.550	143.015	244.56350	1,156.407
1983	13.570	126.941	254.16550	960.712
1984	13.460	134.356	260.21870	966.869
1985	12.750	128.014	274.60030	866.214
1986	12.510	131.880	280.66290	844.156
1987	12.540	118.829	295.53950	741.682
1988	12.240	123.663	312.73870	736.304

changes in consumer, and hence in net social welfare, are reported in the following policy simulation runs.

To determine the importance of how governments choose price supports and research expenditures jointly, a special policy simulation was undertaken whereby research expenditures were set to zero and price supports remained endogenous; i.e., determined by the political process. This scenario evaluates the implication of fixing research (in this particular example, it is set at zero) and allowing commodity policy to be determined by governments. The results are given in Table 9.4. The blend price increases substantially but not enough to offset the supply decreasing effects of zero research expenditures. Hence, output declines compared to the baseline solution. Farmers benefit in this scenario while consumers lose substantially. Taxpayers, on the whole, benefit because supplies have decreased and costs of surplus disposal declines. Farmers gain in this scenario due to the inelasticity of the demand curve. Limiting supply by reducing E to zero allows them to capture more profits than was possible in the base scenario. However, the final column in Table 9.4 indicates that society as a whole is worse off under this set of policies than it is under the base scenario.

The results in Table 9.4 confirm the theoretical model's prediction that, under conditions of an inelastic demand curve, a relatively more elastic supply curve and a significant effect of research on costs, farmers can be hurt as a result of research expenditures. Because dairy farmers do lose from research, price supports are used to partially compensate them. If price supports were not available to governments but research was determined endogenously, farmers would lose so much from the elimination of price supports that research expenditures would also be eliminated. Results in Table 9.5 confirm that, when price supports are eliminated (i.e., prices are determined where supply equals demand and no price discrimination occurs), research expenditures, although determined endogenously through the political process, are driven to zero. Farmers benefit relative to the status quo but consumer and net social welfare declines sharply. Hence, it is in society's interest to have price supports in order to compensate farmers for the profit reducing effects of public research expenditures.

Finally, the "social optimal" levels of research expenditures as prescribed by welfare economists who typically ignore the political process (i.e., set γ_e and γ_p equal to one) are given in Table 9.6. As expected, price supports are not employed in this scenario; and prices are determined solely by market supply and demand forces. The results indicate that optimal research expenditures with farmers and consumers having equal political weights are slightly over four times the actual research expenditures (see Table 9.3 for values of the latter). This result is consistent with Griliches (1964, p. 969) who estimated that a fourfold increase in research expenditures for agriculture would yield positive net social benefits for the US economy. More recently, Fox (1985) has found that optimal expenditure levels for public research were on the order of four times recent actual expenditure in US agriculture. Hence, our results are entirely consistent with recent studies. As expected, farmers lose with research expenditures inducing an increase in supply

Table 9.4. Effects of eliminating research expenditures with endogenous price supports, 1969–1988

	Milk supply	Blend price	Δ in consumer welfare	Δ in producer welfare	Δ in taxpayer welfare	Δ in net social welfare
	billion pounds	dollars per hundred weight	$10 milion			
1969	92.420	9.155	–1,127.977	274.455	123.547	–729.9760
1970	92.842	9.530	–1,108.503	272.716	114.599	–721.1880
1971	98.707	9.827	–1,107.096	292.999	77.477	–736.6200
1972	115.653	10.234	–1,175.139	363.779	–14.776	–826.1370
1973	89.161	12.674	–1,443.006	435.566	–2.376	–1,009.8160
1974	81.400	17.454	–2,113.333	815.270	–325.827	–1,623.8890
1975	87.310	16.150	–1,586.360	526.331	–90.136	–1,150.1650
1976	99.288	16.250	–1,416.303	358.876	107.094	–950.3330
1977	101.509	16.393	–1,324.148	353.870	72.390	–897.8880
1978	107.699	17.949	–1,385.764	392.095	33.129	–960.5400
1979	96.692	20.541	–1,474.183	387.422	83.275	–1,003.4860
1980	95.875	22.001	–1,347.728	337.604	107.866	–902.2580
1981	95.846	23.487	–1,329.522	339.466	93.506	–896.5808
1982	109.176	22.757	–1,212.466	317.418	70.944	–824.1046
1983	98.288	22.872	–1,191.486	312.209	82.838	–796.4399
1984	102.175	22.253	–1,118.500	267.385	124.608	–726.5077
1985	101.989	22.383	–1,200.217	350.600	8.265	–841.3538
1986	104.561	21.682	–1,145.781	329.920	22.298	–793.5637
1987	102.698	23.865	–1,379.697	489.942	–147.340	–1,037.0958
1988	103.770	22.798	–1,234.718	414.531	87.058	–907.2444

Table 9.5. Effects of eliminating price supports with endogenous research expenditures

	Milk supply	Blend price	Δ in consumer welfare	Δ in producer welfare	Δ in taxpayer welfare	Δ in net social welfare
	billion pounds	dollars per hundred weight		$10 million		
1969	90.059	8.908	–1,044.075	205.581	163.244	–675.250
1970	89.950	9.217	–1,009.017	190.044	172.191	–646.782
1971	91.781	9.100	–886.389	101.357	264.499	–520.533
1972	98.494	8.638	–695.176	–94.451	509.885	–279.745
1973	88.390	12.559	–1,412.135	409.638	–2.376	–1,004.873
1974	80.468	17.243	–2,065.589	776.461	–325.827	–1,614.955
1975	86.510	15.994	–1,552.447	498.017	–90.136	–1,144.566
1976	96.079	15.696	–1,295.264	252.255	202.716	–840.293
1977	95.508	15.372	–1,118.629	167.625	267.502	–683.502
1978	98.479	16.331	–1,081.782	105.269	358.974	–617.538
1979	96.162	20.422	–1,453.950	369.720	83.275	–1,000.955
1980	95.253	21.851	–1,325.015	317.988	107.866	–899.161
1981	95.532	23.416	–1,319.507	329.872	93.506	–896.129
1982	100.422	20.911	–971.380	96.678	323.536	–551.165
1983	98.475	22.442	–1,140.169	316.998	82.838	–740.332
1984	102.352	21.797	–1,064.527	271.913	128.442	–664.172
1985	101.701	22.187	–1,177.348	343.635	8.265	–825.448
1986	104.519	21.324	–1,104.208	329.033	22.298	–752.876
1987	102.529	23.574	–1,347.618	485.846	–147.340	–1,009.112
1988	103.228	22.673	–1,220.729	402.223	–87.058	–905.564

compared to the status quo. As a result, market prices fall substantially. However, net welfare for society as a whole improves.

9.4 Implications and Concluding Remarks

In the evaluation of agricultural policy, it should be recognized that interactions between research and commodity policy may result in these two types of policies being selected jointly. Research increases social welfare; but, under the conditions which characterize the dairy industry, producers lose while consumers gain. Because dairy farmers have significant political influence, it appears that commodity policy allows governments to partially compensate producers and increase research expenditures from what they otherwise would be so that a Pareto improvement can occur. This is the case, despite the presence of the deadweight losses which accompany price intervention policies. The model results also indicate the way in which the joint determination of the two policies can lead to underinvestment in research in US agriculture.

Strong support for the hypothesis that governments choose research and price supports jointly was given by the almost identical values of the political weights generated from each of the two policy-decision rules. Empirical results support the hypothesis of the paper that price supports and public research expenditures are used as complementing instruments. Consistent with intuition, the model predicts that governments will intervene with price supports more heavily in sectors with a more inelastic demand, elastic supply, and highly productive research. These same industries are expected to have a greater level of underinvestment.

A current empirical example of the interaction between the two types of policies in the US dairy industry has arisen in the public debate over the introduction of the bovine growth hormone. This growth hormone is a protein which occurs naturally in cows. When the natural level of the hormone is supplemented by injections, the amount of milk which each cow can produce is increased, even if all other inputs are held constant. The research efforts which have made this biotechnology available were, to a significant degree, publicly funded. The adoption of the bovine growth hormone will expand the supply schedule for milk and will increase the amount of milk which will be produced for any fixed level of price support.

It is widely recognized that such a development will increase the cost of maintaining any given level of the price support. It is also widely feared among farmers that this will lead to a lowering of the level of the price support. These farmers feel that the price of milk will be lowered to the point that, despite their now lower-per-unit costs of production, they will generate less profit in the new environment than they are currently able to generate. Accordingly, they expect the price policy to react to the presence of the new technology. They realize that the research policy has results which interact with the price policy selection.

Dairy industry participants also recognize that the research policy which has made the introduction of the growth hormone possible is endogenous to the mar-

Table 9.6. Effects of "social optimal" levels of research (prices determined where supply equals demand.)

	Milk supply	Blend price	Δ in consumer welfare	Δ in producer welfare	Δ in net social welfare
	billion pounds	dollars per hundred weight	million dollars	$10 million	
1969	125.472	3.850	371.300	–294.797	406.797
1970	125.739	3.970	397.796	–298.455	411.445
1971	128.407	3.930	407.883	–368.113	499.635
1972	139.030	3.700	448.631	–556.726	733.208
1973	124.561	5.330	491.357	–167.149	265.829
1974	113.294	7.270	564.999	118.087	168.933
1975	122.324	6.750	623.843	–98.848	170.773
1976	137.255	6.600	687.760	–355.760	534.104
1977	136.228	6.470	722.444	–389.031	560.100
1978	142.013	6.770	872.355	–465.570	660.795
1979	138.984	8.420	1,016.078	–259.397	417.132
1980	138.146	8.920	1,204.774	–275.158	417.940
1981	136.518	9.992	1,102.576	–223.021	363.145
1982	142.261	9.099	1,070.620	–380.163	543.214
1983	138.906	9.947	1,002.231	–161.846	366.567
1984	145.280	9.566	1,090.792	–195.143	414.416
1985	144.022	9.741	1,139.142	–113.996	220.385
1986	147.980	9.386	1,152.238	–113.015	256.893
1987	146.360	10.160	1,355.084	13.533	58.850
1988	147.061	9.768	1,424.773	–37.555	71.149

ket for milk. Of course, the characteristics of this market are heavily influenced by the level of the support price. The endogeneity of the research policy to such characteristics is evident in the public response to the anticipated widespread adoption of the bovine growth hormone. Many have called for a halt to the further development of the growth hormone. Wisconsin and Minnesota, both important dairying states, have passed temporary legislation to ban the use of the bovine growth hormone. While these developments have revealed some concern about the effect of the hormone on the quality of the milk, more important concerns have focused on the anticipated economic consequences of the widespread adoption of the hormone.

This episode, as it unfolds in the dairy industry, is indicative of an awareness among industry participants that interactions between the two types of policies do exist. Furthermore, participants and observers of the dairy industry believe that, to at least some extent, the two types of policies are jointly determined. They are actively engaged in attempting to influence the portfolio of policies which will be implemented. In the US dairy industry, further research-induced technical advancement is being blocked (or, at the least, slowed down) while a price-support policy continues to channel income from consumers and taxpayers to producers. This outcome is one in which research and research-induced advances may well appear to be underutilized or underfunded from a purely social welfare perspective. However, the existence of a price support indicates that the welfare of the two groups is not weighted equally by policymakers. Moreover, since interactions between the two policies necessitate that the two instruments be jointly determined, it follows that the two instruments are complements. What would otherwise appear to be a combination of "underinvestment" in the advancement of technology and "overinvestment" in a price-support intervention can be understood within the framework advocated here.

References

Alston J, Edwards G, Freebairn J (1988) Market distortions and benefits from research. Am J Agric Econ 70; 2:281–288

Bredahl M, Peterson W (1976) The productivity and allocation of research: US agricultural experiment station. Am J Agric Econ 58; 4:684–692

Chavas JP, Klemme RM (1986) Aggregate milk supply response and investment behavior on US dairy farms. Am J Agric Econ 68; 1:55–66

Cochrane WW (1958) Farm prices, myth and reality. University of Minnesota, Minneapolis

Evensen RE, Kislev Y (1975) Agricultural research and productivity. Yale University Press, New Haven

Evenson RE, Waggoner PE, Ruttan VW (1979) Economic benefits from research: an example from agriculture. Science 205:1101–1107

Fox G (1985) Optimal public investment in US agricultural research: a study of the management of technical change. PhD Diss, University of Minnesota, Minneapolis

Fox G, Roberts B, Brinkman G (1990) Canadian dairy policy and the returns to federal dairy cattle research. Department of Agricultural Economics and Business, University of Guelph, Ontario

Griliches Z (1958) Research costs and social returns hybrid corn and related innovations, J Polit Econ 66:419–431
Griliches Z (1964) Research expenditures, education, and the aggregate agricultural production function. Agric Econ Rev 54:961–974
Ippolito RA, Masson RT (1978) The social cost of government regulation of milk. J Law Econ 21:33–65
Nielson DJ (1989) Modeling aggregate milk production utilizing farm level data. PhD Diss, University of Chicago, Chicago
Nielson DJ, Rausser GC, de Gorter H (1990) A supply model for studying change in the US dairy industry. Working Paper No 551. Department of Agricultural and Resource Economics, University of California, Berkeley
Rausser GC, Foster WE (1990) Political preference functions and public policy reform. Am J Agric Econ 72, No 3:642–652
Rausser GC, Ruttan VW (1982) Agricultural research policy. University of Minnesota Press, Minneapolis
Ruttan VW (1982) Agricultural research policy. University of Minnesota Press, Minneapolis
Schlesinger A Jr (1984) The political Galbraith. J Post-Kensian Econ 7:8
Schultz TW (1945) Agriculture in an unstable economy. McGraw Hill, New York
Schultz W (1953). The economic organization of agriculture. McGraw-Hill, New York
Tauer LW, Kaiser HM (1990) Optimal agricultural policy with biotechnology: bovine somatotropin and the dairy sector. AE Staff Paper 90–9, Department of Agricultural Economics, Cornell University, Ithaca
Thurow LC (1971) The zero-sum society. Penguin Books. New York

Chapter 10

Modeling Phased Reduction of Distortionary Policies in the US Wheat Market Under Alternative Macroeconomic Environments

B. Stephen Labson and Gordon C. Rausser

10.1 Introduction

Throughout much of the developed world, macroeconomic policies in the two decades following World War II afforded a unique period of macroeconomic stability. As a result, concern regarding the macroeconomic linkages with food and agricultural systems largely disappeared. In the early 1970s, with the major changes in monetary policies and central bank behavior, macroeconomic linkages were once again recognized as prime factors complicating the performance of the agriculture and food systems.

The roller coaster ride that agriculture experienced through the 1970s and 1980s has been significantly influenced by macroeconomic and international linkages (Rausser et al. 1986). Agriculture's prosperous condition in the 1970s was followed by a recession in the early 1980s. This more recent history stands in sharp contrast to the basic stability of the 1950s and 1960s. It is important to note that the roller coaster experience of the 1970s and 1980s is not unprecedented. For example, the period from 1900 through 1950 is surprisingly similar to the 1970s; and the late 1920s through the 1930s have some basic characteristics of the 1980s.

The effects of policy adjustments (or trade liberalization) on output and input markets, especially land markets, cannot be isolated from a country's domestic macroeconomic conditions or from world economic conditions that arise during the process of phased reductions in coupled subsidies. To illustrate the importance of domestic and international economic conditions, consider US agriculture in the early 1980s. Suppose that a GATT code had been established with phase reductions in coupled subsidies of 20% in 1980, 20% in 1981, 20% in 1982, and thereafter 10% per year for the next 5 years. This reform would have been difficult indeed to implement in the face of high real rates of interest, an appreciating dollar on world markets, slow growth in worldwide income, and a domestic recession. During this period, market overshooting (Rausser et al. 1986) would have driven agricultural prices to unbelievably low levels, creating a policy disequilibrium. Interest groups could have been expected to exert pressure to change any orderly plan to phase out coupled subsidies. This would have added to the inherent instability of the market and detracted from credibility of governments in implementing reform.

Unfortunately, few studies have examined the dynamic path resulting from phased subsidy reduction,[1] and even fewer studies recognize that such paths are conditioned on the macroeconomic environment. If the dynamic path is very sensitive to such forces, the credibility of governments in implementing reform can be enhanced by the design of "flexible" agricultural policy reforms. Flexible policy rules can explicitly incorporate macroeconomic and international linkages into automatic adjustment roles. The design of such policies should make clear what the adjustments will be so that the changes can be anticipated by producers, processors, distributors, consumers, and others involved in the sector. Such flexibility could be expected to result in smoother, more orderly changes in prices.

Accordingly, the main premise of this chapter is that macroeconomic linkages are relevant to agriculture and that quantitative results of trade liberalization will diverge under various realizations of the underlying macroeconomy. Since much of the distortionary impact of US agricultural policy is conditioned on realized agricultural prices, any structural change in the form of exogenous shocks to supply or demand will alter the level of distortion for a given setting of the relevant policy instruments. More specifically, macroeconomic variables such as exchange rates, income, and interest rates may significantly alter the short-run path of prices and production in US agriculture, given removal of distortionary agricultural policies. No quantitative experiments have been designed that fully examine the short-run macroeconomic effects on agriculture under trade liberalization.

The purpose of this study is to examine the effects of phased reduction of US agricultural policy instruments under alternative macroeconomic environments. To what degree are the effects of policy reform determined by the exchange rate, income, and interest rates? Are predictions about the effect of policy reform robust to various realizations of the macroeconomic environment, and can the policy analyst simply shift the results of previous studies by a "macroeconomic constant"?

After assessing the major limitations of studies that have analyzed trade liberalization for the US agricultural sector (sect. 10.2), we turn to the policy instrument set in sect. 10.3. Based on a constructed econometric model reported in Appendix A, a policy simulation model of the US wheat market is presented in sect. 10.4. After examining the properties of the policy simulation model in sect. 10.5, simulation experiments on phased reductions in US wheat subsidies are reported in sect. 10.6.

10.2 US Agricultural Policy and Ad Valorem Equivalents

Studies on trade liberalization (such as Roningen et al. 1987; Hertel et al. 1989; Robinson et al. 1989; Anderson and Tyers 1990) aggregate distortionary agricultural policies into a scalar measure which is then normalized into an ad valorem equivalent.

---

[1] For a major exception, see Just, Rausser, and Zilberman (this Vol.).

Tyers and Anderson (1986) use the domestic to border price ratio as a measure of intervention. The methodology of Tyers and Anderson is inadequate for evaluating policy reform in "large" countries, such as the United States, that affect world price. More specifically, US acreage controls will not be measured as distortionary to the degree that the resulting domestic price effect equals the border price effect. Also, the distortionary impact of stock policy via the loan rate and Commodity Credit Corporation activity will not be captured by border measures.

Roningen et al. (1987) measure support to agricultural production as the ratio of total value of policy transfers to total producer revenue. They estimate this measure for 1984 and implicitly assume that it is an adequate proxy for the producer incentive; however, production controls that are coupled to direct payments are neglected. To simulate unilateral trade liberalization by the United States, they remove the ad valorem equivalent of the estimated producer subsidy and recalculate domestic supply, demand, trade, and market-clearing world price.

Robinson et al. (1989) and Hertel et al. (1989) estimate a measure of the distortionary aspects of US agricultural policy, rather than the total transfer which may include transfers which are relatively neutral with respect to the production decision. In computing the net producer incentive equivalent (PIE), they explicitly account for the implicit tax of the set-aside requirement, and the price-enhancing effect of acreage reduction. Robinson et al. compute this net measure of distortion for 1986, and Hertel et al. use 1984 as the base year. Both studies then solve for equilibrium values for a given base period with the distortionary policies in place, and then recompute the long-run equilibrium for an undistorted system.

Notice that the short-run effects of policy reform and the partial reduction of program instruments cannot be evaluated without accounting for endogenous participation. The main problem is that the *level* of acreage set-aside is endogenous even when the diversion rate is exogenous.

The preceding measures of distortion to production are not robust proxies for evaluating the distortionary impact of the US farm program. Whalley and Wigle (1990) have shown that ad valorem equivalent modeling may lead to inaccurate results when program participation is endogenous and conditioned on production controls.[2]

---

[2] These studies do not account for the fact that program payments are based on a producer's base acreage, which is currently defined as a 5-year moving average of acreage planted or considered planted. If the target price is to draw acreage into production which would not have been planted otherwise, the producer must take a short-run (5-year) loss (plant such that marginal cost is greater then price received) in order to gain future program rents. Following the intuition behind the formal treatment of dynamic optimization and the US farm program as shown by de Gorter and Fisher (1989), farmers would increase production only to the extent that the cost of entry to the program is outweighed by the stream of discounted program profits. As such, for a positive discount rate, the first 5 years of the "entry tax" may significantly reduce the distortionary effect of the target price even in the long run.

10.3 The Policy Set

The current set of policy instruments in the US wheat program is essentially as established by the Agricultural and Consumer Protection Act of 1973. The major instruments used in the wheat program are target price, acreage controls, and the loan rate.

While the major policy instruments have remained intact since 1973, the settings of the program instruments are subject to discretionary change by the US Congress and the Secretary of Agriculture. Participation in the program is voluntary and is also subject to change on an annual basis. A brief summary of the characteristics of the major policy instruments follows. For a more detailed explanation of the policy set, see Gardner (1987, 1989) and Wright (1984).

10.3.1 Deficiency Payments

Producers that participate in the wheat program are eligible for deficiency payments. The value of the deficiency payment is determined by the difference between the legislated "target" price and the higher of the market price or loan rate. Producers receive this deficiency payment for a pre-determined program yield times acreage planted. Acreage planted by participating producers is often (but not always) constrained to not exceed the producers base acreage minus a uniform rate of acreage to be set aside (the diversion rate). Base acreage is determined by a 5-year moving average of acreage planted and considered planted. Acreage considered planted consists of acreage planted, set aside, and diverted. The program yield has been fixed since 1987. Prior to 1987, program yield was based on a rolling average of past yields. Over the last decade, deficiency payments ranged from less than 5% of market price to over 50% of market price, even though the target price has varied by no more than 6% in any given year.

10.3.2 Acreage Controls

In order to be eligible for program benefits, one must first establish "base" acreage. Subject to legislative discretion, yearly diversion rates are established that limit the amount of one's base acreage eligible for payments through the acreage reduction program (ARP). A uniform rate is set which is the percentage of a participating producers' base acreage that must be set aside for the crop year. During the 1980s, the diversion requirement was as high as 30% of base acreage.

10.3.3 The Loan Rate

The Commodity Credit Corporation (CCC) is required by law to accept wheat as collateral from program participants in exchange for a loan. The amount of the loan is equal to the number of bushels placed as collateral times the legislated "loan rate." The terms of the loan are nonrecourse in that, if the producer wishes to default on the loan, the grain held as collateral serves as payment in full. In general, the loan rate serves as a price floor in that the CCC stands willing to purchase any amount necessary to support price at the loan rate. Within the loan period of 9 months, producers can pay back the loan plus interest and sell at the market price if they wish.

10.3.4 The Farmer-Owned Reserve

The farmer-owned reserve (FOR) has played a controversial role in US agricultural policy since its inception in 1977 (Salathe et al. 1984; Wright 1984). In general, the FOR is a loan program much like the CCC loan program. The primary difference is that grain entered into the FOR is inaccessible for 3 or more years unless a predetermined release price is hit. It is important to note that while the 3-year constraint is legislated by Congress, it has been subject to change. During the 1980s, significant quantities of grain were, in fact, released prior to the 3-year constraint within the guidelines of payment-in-kind (PIK) programs.

10.3.5 Additional Voluntary Division Programs

Several other program instruments have been used to divert acreage from production over the minimum required amount. These programs have been available to producers that comply with the basic (minimum) diversion rate. For example, added voluntary diversion has taken place under PIK programs based on an approximately one-to-one basis of diverted production. This program instrument has been used sparingly; however, for the several years that it has been used, the effect on the market has been significant. In 1983, over 18 million acres were set aside under the PIK program.

Since 1986, producers have been offered the additional flexibility of diverting additional acreage. Under the ARP, producers receive deficiency payments on 92% of diverted acreage in excess of the minimum required. When the program was instituted in 1986, the maximum amount of diversion eligible for payments under this program was 50% of the farmers' base acreage. Currently, farmers may divert 100% of their base acreage and remain eligible for payment on the acreage diverted in excess of the ARP constraint. This paid diversion program is thus called the *0-92 Program* since producers can plant zero percent of their base acreage while receiving deficiency payments on 92% of eligible acreage. Typically, less than 5% of wheat base has been entered into the 0–92 Program.

The Food Security Act of 1985 authorized the conservation reserve program (CRP) with the intent of removing 40–45 million acres of fragile cropland from production (Young and Jagger 1989). The CRP removes acreage from production for 10-year periods. Payment for this is based on individual bids by the producer which are essentially rental payments by the government. Over 10 million acres of wheat base have been entered into this long-term diversion program since its inception in 1986. Wheat program statistics for 1985–1990 are shown in Table 10.1.

The 1990 Farm Bill recognizes the need for flexible programs that are responsive to a dynamic international marketplace. Furthermore, the intent of the 1990

Table 10.1. 1985–1990 Wheat program statistics

Year	Target price	Deficiency	Loan rate	Commercial Credit Corporation	Farmer-owned reserve
	dollars per bushel			million bushels	
1985	4.38	1.08	3.30	602	596
1986	4.38	1.98	2.40	830	632
1987	4.38	1.81	2.28	283	467
1988	4.23	0.69	2.21	190	287
1989	4.10	0.32	2.06	125	150
1990	4.00	N.A.[a]	1.95	N.A.	N.A.

Year	Base acres	Set-aside[b]	Conservation reserve program[c]	Participation	Acreage reduction program
	million acres			percent	
1985	94.0	18.8	0	73	30.0[d]
1986	91.6	21.0	0.6	85	22.5
1987	87.6	23.9	4.2	88	27.5
1988	84.8	22.5	7.1	86	27.5
1989	82.3	9.8	10.3	78	10.0
1990	80.5	6.6	N.A.	80	5.0[e]

[a] Data not available.
[b] Includes acreage reduction program, payment-in-kind, and 0–92 programs.
[c] Cumulative.
[d] Includes 10% cash land diversion (CLD) program.
[e] Farmers had option to plant 105% of base in 1990.

Source: USDA ERS Wheat Outlook and Situation Yearbook (various issues, 1985 through 1991).

Farm Bill is to decouple income support from the production decision by allowing additional flexibility in the croping decision of participating producers.

10.4 A Policy Simulation Model of the US Wheat Market

An econometric-based policy-simulation model of the US wheat sector was specified and estimated. The model is an annual trade model which explains program participation, acreage planted by program participants and nonparticipants, yield, US food demand and feed demand, market storage, government storage, and export demand.

The critique of econometric policy evaluation by Lucas (1976) cuts deepest into a study of this nature and cannot be neglected or assumed away. Since the intent of this study is to evaluate large shocks to the system, explicit attention must be given to identification of the reduced form relations as specified in order to ascertain whether the parameters as measured are invariant to the policy shocks that we wish to evaluate. As Lucas stated, "... given that the structure of an econometric model consists of optimal decision rules of economic agents, and that the optimal decision rules vary systematically with changes in the structure of series relevant to the decision maker, it follows that any change in policy will systematically alter the structure of econometric models."

Following the conceptual argument of Sargent (1987) and the application to agriculture by Just et al. (1992), this econometric policy evaluation will attempt to specify the partially reduced form equations such that the parameters are sufficiently stable with regard to the macroeconomic and policy regimes to be considered. Specifically, in order to evaluate the short-run effects of policy reform or partial phased reduction of specific program instruments, certain facts pertaining to the US wheat program must be explicitly addressed. This study will focus its attention on the following aspects: acreage eligible for program payments is predetermined, participation in the program is voluntary, in order to receive program payments one must comply with acreage controls, and the supply response of producers outside the program can be significant.

10.4.1 Production

Production is separated into acreage response and yield response. Given that participation in the farm program is voluntary, acreage response must be further disaggregated into participant acreage response and nonparticipant acreage response. Participation, itself, is determined within the system. This framework follows that of the current generation of econometric policy simulation models – e.g., the US Department of Agriculture's Food and Agricultural Policy Simulator, FAPSIM (Gadson et al. 1982); the Center for Agricultural and Rural Development world wheat trade model (Devadoss et al. 1990); and the model presented in Just et al. (this Vol.).

When program participation is voluntary, based on a predetermined level of base acreage, and based on effective acreage controls, the aggregate acreage relationship can be defined as

$$AP \equiv (1-\alpha)Q^m + \alpha\bar{Q}, \qquad (1)$$

where aggregate acreage planted AP is the sum of acreage which would have been planted in a free market, Q^m, weighted by the level of acreage not entered into the program $(1-\alpha)$; and the production constrained level of program acreage, $\bar{Q}$, weighted by the level of program participation, α, since $\bar{Q}$ is defined as base acreage times one minus the diversion rate.

First, notice that, given the level of base acreage and the diversion requirement, acreage planted by program participants can be identified with knowledge of the equilibrium value of program participation. As Just et al. note, US agricultural policy regimes change often; and, given the limited observations, identification is problematic. For any given setting of the primary program instruments (target price, loan rate, diversion requirement, and additional voluntary diversion payments) the level of distortion and attractiveness of program participation is dependent on the current state of the system. In order to lessen the identification problem, we assume that farmers compare anticipated profit per acre under program compliance to that under noncompliance. As such, we presume that program participation can be described by the level of anticipated profit above that which would be anticipated from planting outside of the program provisions. Specifically, when market price is greater than or equal to the loan rate, excess profit per acre from program participation π^p is defined as the value of anticipated deficiency payments minus foregone returns due to the diversion requirement, plus payments on any additional diversion programs when in effect.

$$\pi^p \equiv [\{max(p^T, P^m) - P^m\}(1-DVR)]\,\bar{Y} - [(P^m - vc)DVR]Y + ADP, \qquad (2)$$

where P^T is the target price, P^m is the expected market price, vc is a measure of variable costs per bushel, DVR is the proportion of acreage a farmer must divert in order to participate in the program, Y is planned yield per acre, $\bar{Y}$ is program yield, and ADP is additional diversion payments.

Notice that in order to solve for total acreage planted we must identify the amount of acreage that is planted by nonparticipating producers, $(1-\alpha)Q^m$. Since we observe the outcome of a distorted market, we cannot directly observe the level of acreage which would have been planted in a free market, Q^m; however, acreage planted by nonparticipants (APNP) is approximately determined by the observable expected market price, variable costs of production, the relative price of substitute crops, and the amount of acreage entered into the program.[3]

[3] Given the fairly limited observations on additional voluntary diversion programs, profits from additional voluntary diversion are only included when anticipated profits from voluntary diversion are greater than anticipated market returns from planting. This simplification is not entirely correct; however, it should be a very good approximation. In examining the data, one finds that the most effective voluntary programs over the 1974–1988 period were the PIK programs of 1983 and 1984 which are accounted for in this representation. The

The CRP has been an important program since its inception in 1986, with a total wheat enrollment as of August, 1990, of 10.3 million acres. For this study, acreage enrolled in the CRP is operationally considered exogenous. Since only a small amount of acreage had been entered into the CRP during the sample period used for estimation (6 acres in 1986 and 4.8 acres in 1987), this should have a very limited effect on the parameters as estimated. Furthermore, by the first period of the simulation experiments to be run for this study (1991), the stated goal of the CRP should be achieved, thus CRP acreage can be held fixed at estimated 1991 levels. The effect of the CRP on planted acreage must be accounted for as an important structural shift in the reduced-form acreage decision. Operationally, we have treated CRP acreage as an additive shift in the reduced-form acreage equation.

Total production is simply total acreage planted times yield per acre. Yield per acre is presumed to be determined by expected market price, variable costs of production, acreage set-aside which enters to capture the effect of "slippage," and a time trend to capture exogenous technical change.

All values are in real, rather than nominal terms. Quoted futures prices are used as proxies for the producer incentive price in the estimation process.[4] For forward simulation, farmers' production decisions are based on that price which clears the market. The advantage of this procedure, as compared to using lagged prices in the simulation process, is that we can evaluate policies in a system where farmers are able to react to their current environment.

10.4.2 Demand

The proposition that the macroeconomy has a pervasive effect on the agricultural sector is well developed. Andrews and Rausser (1986) point out that macroeconomic disturbances have been central in agricultural policy developments at least since the late 1800s. Schultz (1945); Schuh (1974); Johnson et al. (1977); Chambers (1979); Chambers and Just (1981); and Rausser et al. (1986) demonstrate that the macroeconomic environment significantly effects agricultural prices and incomes.

Following the general theme of the literature cited, the exchange rate, real interest rate, and income are specified as exogenous shifters of export demand, food demand, feed demand, and market held storage.

Food demand is determined by own price, income, and lagged food demand. Relative price of other food grains is not entered since we found little substitutability among wheat and other similar grains in food demand.

Feed demand is determined by own price, the relative price of wheat to corn, income, and lagged feed demand. Given the nature of this project, we let income

---

0-92 program is becoming marginaly important, but too few observations exist in which to estimate the relevant effects.

[4] The survey work done by Eales et al. (1990) offers empirical evidence that farmer price expectations are not significantly different than the observed futures price.

serve as the reduced-form parameter for the derived demand for feed. Lagged feed demand captures the dynamic adjustment in the livestock sector.

For this study we have specified a single-equation, reduced-form representation of export demand. Since we are evaluating unilateral policy reform, the reduced-form coefficient on price should capture much of the response in the world wheat market with respect to US policy shocks. Large trade models such as the models described by Baily (1989) and Devadoss et al. (1990) typically disaggregate the world market into regional submodels. Operationally, these models basically identify various intercepts; however, if we have little a priori knowledge of how these intercepts will change given policy reform, the added identification is superfluous. Excess demand is specified as being a function of US price, the exchange rate, world income, and lagged excess demand. Price and the exchange rate have been entered as separate regressors for two reasons. Following Chambers and Just (1979), we will allow for separate effects of exchange rates and prices. As a practical matter, deflating price by the exchange rate would add undue complexity to system estimation.

The export enhancement program (EEP) is not considered to have a direct effect on export demand. As Gardner (1989) explains, so long as any transactions by countries receiving EEP subsidies are occurring at the market price, it is the market price which is the relevant price at the margin, and thus the market price is indeed the proper proxy. The EEP will have an important indirect effect, however. The transfer of wheat from CCC-held storage to the market represents additional availability and thus lower equilibrium price. When private storage co-exists with public storage measurement of this effect is difficult since the market may have anticipated this outflow, and adjusted privately held stocks accordingly. This price-depressing effect of EEP storage transfers and adjustment of private storage is modelled explicitly in the storage equations.

Storage has typically been an important means of government intervention in the wheat market since the early 1950s. Inflow to government storage is solved for by first solving for equilibrium price. If price is less, then the predetermined loan rate stocks flow into government control until equilibrium price is equal to the loan rate. Government release is assumed to be exogenous. This study is meant to evaluate policy, not forecast it; thus, a behavioral release equation is unnecessary.

When private rational speculative storage is allowed to exist, private storers will hold stocks such that current price equals the discounted expected value of the commodity in the next period unless there is a stockout. Since the government does not generally destroy its reserves of grains, any government acquisition will eventually be returned to the market. The market (private storers) must form an expectation on the means and timing of government storage disposal. Market storage is determined by the endogenous current price of wheat and expected future price, along with the exogenous interest rate. Future net government release serves as an instrument for future expectations and ties together the government/market storage relationship in an explicit form.

10.5 Properties of Simulation Model

10.5.1 Model Validation

Conventional validation techniques have been employed to analyze the dynamic and parametric stability of the simulation model. Given the nonlinearity of the system, analytical representation of the dynamic properties of the model is difficult. As an alternative, the dynamic properties were simulated by imposing a once-and-for-all shock to the model and then running the model forward to determine whether the values of the endogenous variables converge. The stability checks showed strong convergence in less than ten periods, implying that the model is dynamically stable.

In order to assess the parametric stability of the model, in-sample simulations were run on both the individual structural equations and the simulation model as a whole. As estimated, the structural equations track the observed relationship well as measured by goodness-of-fit and anticipation of observed turning points. The model as a whole also did well in explaining movement in variables endogenous to the system. For a complete description of the validation techniques and results, see Labson (1990).

10.5.2 Multiplier Analysis

In order to understand the properties of the estimated model, reduced-form impact multipliers were computed. The multipliers were computed by first solving the system for equilibrium values with the exogenous and predetermined variables at their mean values over the 1974–1987 sample period. The system was then shocked by increasing a given exogenous variable by 10%. The proportional difference in the first period is the impact multiplier. The results are reported in Table 10.2.

10.5.3 Macroeconomic Shocks

The exchange rate is shown to have an important effect on exports and price. A 10% appreciation of the dollar leads to a 7.7% decrease in the real price of wheat. The 7.7% change in price has a limited effect on US food and feed demand for wheat (0.1% and 2.8%). Even though the 10% change in the exchange rate has a less then unit effect on exports (8.2%), the observed variation in the exchange rate over the last 20 years has been of a magnitude to make the realized effects substantial. Notice that the 7.7% decrease in price makes participation in the program more attractive since the opportunity cost of diversion decreases with market price, thus program participation increases. The 2.1% increase in participation increases acreage set-aside by 1.5%, thus the decrease in price has two important effects on total acreage planted. The standard own-price effect leads nonparticipating producers to curtail production; furthermore, since participation is endogenous, the increase in acreage set-aside leads to further reduction in total acreage planted.

Table 10.2. Impact response to a 10% shock to exogenous variables evaluated at their mean values

Endogenous variables	Exogenous variables				
	Exchange Rate[a]	Gross national product: United States	Gross national product: Rest-of-world	Interest rate	Diversion rate
	percent				
Real price of wheat	−7.7	2.1	0.6	−0.5	0.8
Food demand	0.1	1.8	0.0	0.0	0.0
Feed demand	2.8	7.0	−0.2	0.2	−0.3
Export demand	−8.2	−0.4	0.7	0.1	−0.2
Market storage	0.4	−1.4	−0.4	−1.3	−0.6
Participation rate	2.1	−0.6	−0.2	0.2	−1.5
Nonparticipant acreage	−5.9	1.6	0.5	−0.4	2.1
Total acreage	−2.2	0.9	0.2	−0.2	−0.4
Production	−3.3	0.9	0.3	−0.2	−0.3
Acreage set-aside	1.5	−0.4	0.4	0.1	6.1
Crop receipts	−10.7	3.0	0.9	−0.8	0.5
Government payments	0.0	0.0	0.0	0.0	0.0

[a] Rest-of-world currency per dollar.
Notes: 10.0 = unit elasticity. All values are rounded to nearest tenth of a percent.

A 10% increase in the real interest rate decreases market storage by 0.6%, thus price by 0.5%. The impact response to real interest rates is relatively small; however, the observed volatility of observed ex post real rates over the last 20 years suggest that interest rate shocks can and do have a measurable effect on storage, price, production, and crop receipts.[5] A 100% increase in the ex post real interest rate evaluated at its sample mean value of 1.64% would decrease mean price by 5%, or 21 cents per bushel.

10.5.4 Policy Shocks

In the initial period, as the diversion rate is increased, program participation decreases (1.5%) which, in turn, dampens the positive effect on diverted acreage and results in a less-than-unit response in set-aside with respect to a change in the diversion rate (6.1%).[6] Since set-aside acreage is typically less then 25% of total acreage planted in the United States, a 6.3% change in set-aside acreage translates to only a 0.4% change in acreage planted and a 0.8% change in price. This result does not imply that the diversion requirement is unimportant. Given that we have observed changes in the diversion requirement of more then 100%, the realized effect on the sector can be significant. At mean values, a 100% increase in the diversion requirement would decrease acreage by 3.2 million acres in the first year and increase market price by 34 cents per bushel.

When the exogenous variables are set at their mean sample values, equilibrium price is 53 cents above the average target price.[7] Given the deterministic nature of the model, a 10% change in the target price has no effect on the system as specified. The combination of a fixed target price and random market price implies that a positive option value of participation exists even when expected price is greater than the target price. The option value of the fixed target price is likely to be very small when the expected price is so much greater than the target price, however. A casual inspection of quoted option values on wheat futures contracts leads us to believe that a 10% change in the mean target price represents less than a 5 cent per bushel change in the option value of program participation, or less then a 1% change in the participation rate.

10.5.5 Structural Change and Impact Response Multipliers

The reduced form relations exhibit significant nonlinearities which stem primarily from endogenous participation, the loan rate, and the target price. Given the nonlinearity of the reduced form relations, the effect of an exogenous shock on the

---

[5] Note that, as specified, the real interest rate enters only through the effect on market-held stocks. If interest rate shocks have an effect on farm production costs as Tweeten (1980) suggests, the full effect of interest rates may be tempered by the change in endogenous production.

[6] Since the diversion rate is defined as a ratio, interpretation of the reported impact response must be made with care. Note that a 10% change in the diversion rate at its mean value of 13.4%. We are not evaluating a change to 23.4% required diversion.

[7] This does not imply that the target price was below the market price on average over the sample period.

system is conditioned on the state of the system. This is precisely the point made in the introduction of this chapter that the effects of policy reform on the sector are conditioned on the macroeconomy. Furthermore, the effects of macroeconomic shocks on the sector are conditioned on the policy environment. Table 10.3 reports the computed impact multipliers evaluated at observed 1985 values for the exogenous and predetermined variables. This setting represents an adverse macroeconomic environment with low equilibrium price.

Comparison of the effect of an exogenous change in the exchange rate on the sector evaluated at 1985 values clearly demonstrates that multipliers in distorted markets are conditioned on the state of the system. Table 10.3 shows that a 10% appreciation of the dollar results in a 2.5% reduction in price. This is only about one third of the effect when equilibrium price is well above the loan rate. When equilibrium price is near the loan rate, the fall in price is constrained due to CCC acquisition. Quantities exported fall and government payments rise significantly as compared to the more favorable environment as represented in Table 10.2.

A 10% increase in the diversion requirement has a greater effect on total acreage planted as compared to the more favorable equilibrium. Given the relatively low market price, the opportunity cost of diversion is of little consequence, thus farmers are less likely to leave the program and plant at market price. Notice that even though the percentage change in participation is greater, given the 1985 values, the absolute change in participation is smaller.

Given the 1985 settings, the target price is effective. Since deficiency payments are based on fixed base acreage and program yields, as the target price is increased, production decreases as program participation becomes more attractive, which, in turn, serves to divert acreage which would otherwise have been planted. Notice that this result is in stark contrast to the overly stylized model of per unit subsidy often used in trade liberalization studies which suggests that lowering the US target price unambiguously decreases total US production.[8]

10.6 Simulation Experiments

As specified and estimated, the model can simulate the effects of various US agricultural policies and macroeconomic environments on the US wheat sector. The simulation exercises that follow are not meant to be forecasts of future equilibrium values, rather to study the system given contrasting settings of the exogenous and predetermined variables.

[8] A dynamic model of intertemporal profit maximization following the model of de Gorter and Fisher (1989) may admit a positive relationship between the target price and production.

Table 10.3 Impact response to a 10% shock to exogenous variable evaluated at observed 1985 values

	Exogenous variables					
	Gross national product					
Endogenous variables	Exchange rate[a]	United States	Rest-of-world	Interest rate	Diversion rate	Target price
	percent					
Real price of wheat	−2.5	1.8	0.5	−1.1	1.4	3.3
Food demand	0.0	1.9	0.0	0.0	0.0	0.0
Feed demand	0.5	5.4	−0.1	0.2	−0.3	−0.6
Export demand	−15.7	−0.4	1.0	0.2	−0.3	−0.7
Market storage	1.3	−0.9	−0.3	−3.2	−0.7	−1.6
Participation rate	3.5	−2.5	−0.8	1.5	−5.2	7.4
Nonparticipant acreage	−6.3	4.6	1.4	−2.7	8.5	−10.2
Total acreage	−1.3	1.0	0.3	−0.6	−0.6	−1.2
Production	−1.6	1.2	0.3	−0.7	−0.4	−0.9
Acreage set-aside	3.5	−2.5	−0.8	1.5	4.3	7.4
Crop receipts	−4.0	3.0	0.9	−1.8	1.1	2.4
Government payments	44.5	−8.8	−2.7	5.4	−13.9	42.6

[a] Rest-of-world currency per dollar.
Notes: 10.0 = unit elasticity. All values are rounded to nearest tenth of a percent.

In order to examine and identify the conditional effects that the macroeconomy, sectoral policy, and initial conditions have on the US wheat sector, the exogenous variables of concern have been grouped into distinct regimes, or sets. These sets are then used in simulation experiments in order to obtain the time path of the endogenous variables.

The time span of the experiments is 5 years. For the phase-out experiments, the target price, loan rate, and diversion requirement are reduced in equal increments such that they would be eliminated within 10 years. As such, the phase-out experiments covered in this study represent the first half of complete phase-out of the distortionary programs. This research strategy fits the intent of the project since we wish to quantify the comparative short-run effects (1 to 5 years) of various macro/policy regimes, rather then the comparative long-run steady state values of the system.

Since the thrust of this project is to understand the combined effects of the macroeconomy and policy on agriculture, rather then to explain the macroeconomy itself, the macroeconomic sets used for the experiments correspond to past observed macroeconomic environments. Two distinct regimes will be used which roughly correspond to the "tax subsidy" application of Rausser et al. (1986). The dichotomous relationships between the macroeconomic environments of the late 1970s and the early 1980s serve well to punctuate the role of the macroeconomy on agriculture. A third macroeconomic set will be used which is defined by a fairly static macroeconomic environment. This "base" set will be helpful in comparing our results to other studies which commonly evaluate policy shocks over static macroeconomic environments. Rather than list the values for the exogenous variables for each set, a summary is given which describes the important areas of interest. In particular,

- Base Macroeconomic Set: This set is defined by a static macroeconomic environment. The base macroeconomic set describes a macroeconomic environment similar to that observed in 1989. US gross national product (GNP) is fixed at a constant 1.2% growth rate. Rest-of-world (ROW) GNP is assumed to closely follow that of the United States and is also set at 1.2%. The exchange rate is fixed at its 1989 level, and real interest rates are set at the 1989 ex post rate of 3.52%.
- 1981–1985 Macroeconomic Set: This set corresponds to the observed macroeconomic environment for 1981–1985. US and ROW GNP increases at the average rate of increase observed over the 1981–1985 period at 2.56% and 2.14% per year, respectively. The exchange rate (dollar) appreciates at the average rate observed over the 1981–1985 period of 10.5% per year. Real interest rates trend from an initial value of 3.52% to the average ex post rate observed from 1981–1985 of 4.85%.
- 1975–1979 Macroeconomic Set: This set corresponds to the observed macroeconomic environment for 1975–1979. US and ROW income increase at their observed averages of 3.22% and 2.87% per year, respectively. The exchange rate (dollar) depreciates at 2.6% per year, and the real interest rate trends

downward at 1.07% per year to the average ex post rate observed from 1975–1979 of –1.07%.

The policy sets evaluated are:

- Base Policy Set: This is a static policy set which can be used to look at the macroeconomic experiments abstracting from policy shocks. The target price and loan rate have been fixed at 1990 values. The diversion rate is set at 12.5%, and CCC stock outflow is fixed at 25 million bushels per year. Voluntary set-aside is set at 4 million acres per year for all simulations. For all policy sets, acreage entered into the conservation reserve program is fixed at an anticipated level of 12.1 million acres for 1991.
- All Instruments: This policy set reduces the major policy instruments (loan rate, target price, and diversion requirement) in equal increments.
- Payment Acreage: This set is defined as a 1960s-style payment acreage which is proportional to program base acreage. Program base acreage, the target price, loan rate, and diversion requirement are frozen at 1991 levels, as payment acreage is reduced in equal increments over the simulation experiments.

10.6.1 BASE Macroeconomic Set Simulations

Two simulations were run with the exogenous macroeconomic variables of interest set as defined by the base macroeconomic set. First, equilibrium values for the 1991–1995 period were computed given the base policy set. Then, under the same base macroeconomic set, the target price, loan rate, and diversion requirement were reduced as defined by the "all instruments" policy set. Simulation results for these two experiments are reported in Tables 10.B1 and 10.B2 of Appendix B.

The results depicted in Fig. 10.1 and Appendix Tables 10.B1 and 10.B2 demonstrate the effect of phased reduction of the major policy instruments in a macroeconomic environment similar to that observed in 1989. When the program instruments are held at their 1991 base setting, the real price of wheat increases by 21%, primarily due to the relatively low value of the dollar and 1.2% rate of growth in income, which increase export and domestic demand. Given sustained strength in prices and crop receipts, total acreage planted increases steadily throughout, starting at 79 million acres in 1991 and increasing to 94 million acres by 1995. No government payments are incurred.[9]

Given the settings of the base macroeconomic set, the effect of policy reform on the sector is overshadowed by the price effects of the favorable macroeconomic environment. When the program instruments are held constant, the real price of wheat increases by 21% through 1995. When phased reduction of the program instruments is combined with a favorable macroeconomic environment, the price of wheat increases, but by slightly less as compared to the base policy scenario. As shown in Fig. 10.1, by the fifth year of phased reduction, equilibrium price is only 6 cents per bushel lower due to policy reform.

[9] For this study, government payments are defined as simply the sum of CCC acquisitions and deficiency payments.

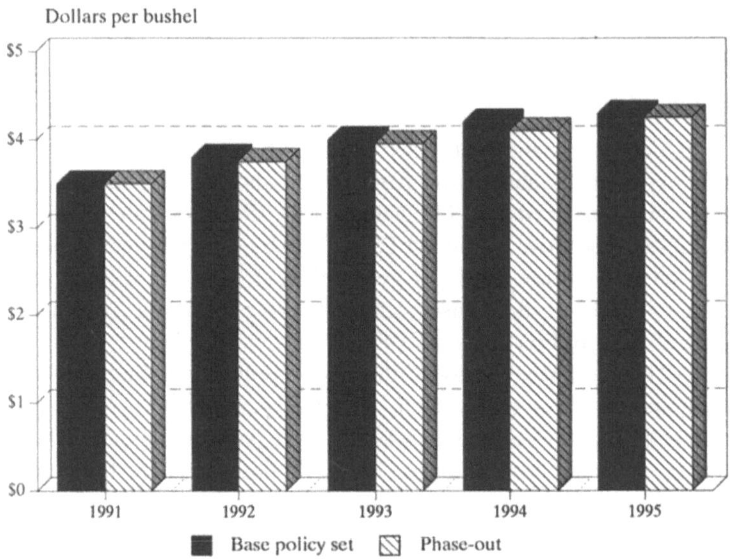

Fig. 10.1. Base macro set – real price of wheat

The cause of this apparent "noneffect" can be directly attributed to the combination of a low-valued dollar and the corresponding high market price. A strong market price represents a high opportunity cost of diversion and, thus, a low rate of program participation. Given the relatively low distortionary impact of the program in the base policy scenario, it should come as no surprise that the effect of policy reform will be minimal. It is important to remember that the level of distortion is directly related to the observed macroeconomic environment. The next set of experiments demonstrates that policy reform can have a significant impact on a weak agricultural sector.

10.6.2 The 1981–1985 Macroeconomic Set

The early 1980s exhibited a sharp decline in agricultural prices and farm income. Rausser et al. (1986) attribute much of this decline to an appreciating dollar, low income growth, and high real interest rates. Given the 1981–1985 macroeconomic set and base policy set, equilibrium prices and crop receipts drop dramatically (Appendix B, Table 10.B3). The appreciating dollar contributes to a 53% decrease in export demand. The high real interest rates help to further depress price since market stocks increase only slightly. Over the 5-year simulation period, price falls by 44% and yearly crop receipts by more than 50%. As price falls below the target price and the costs of diversion decrease, acreage enters the program and over 9 million acres are diverted from production. Government payments rise to almost $3 billion per year due to deficiency payments alone.

Now compare the effects of phase-out, given the adverse 1981–1985 macroeconomic set as shown in Appendix B, Table 10.B4. Again, as the dollar appreciates, export demand falls and thus price, production, and crop receipts. Program participation rises over the simulation period due to weak prices; however, this is tempered by a decreasing target price. The declining diversion rate increases participation, but not as much as in the case of the base policy set. The combination of lower participation and a smaller diversion rate as compared to the base policy scenario mean that almost 5 million acres that would have been set aside come into production depressing price by 53 cents per bushel in 1995. Gross annual income of wheat producers, as defined by crop receipts and government payments, by 1995 fall by almost $2.7 billion as a result of policy reform.

Contrary to the implications of the base macroeconomic scenario, in Fig. 10.2 it is clearly demonstrated that phaseout of the major program provisions can have a very severe impact on the sector and that the degree of impact is conditional on the macroeconomic environment. Since we have imposed an adverse macroeconomic set, prices, production, and farm income are expected to fall, all else are held constant. The important point, however, is that, given the 1981–1985 macroeconomic set, prices fall by an extra 31% from what they would have been if the wheat program were left at its 1991 setting. Contrast this to the base scenario in which phaseout has only a negligible effect on prices, production, and farm income. The reason for this conditional effect is that program participation, acreage set-aside, and price are jointly determined. Given this joint dependence, the distortionary impact of a given policy set is partly determined by the realized macroeconomic environment. Since the costs and benefits of program participation are not constant across various economic sets, neither is the effect of reducing the program settings.

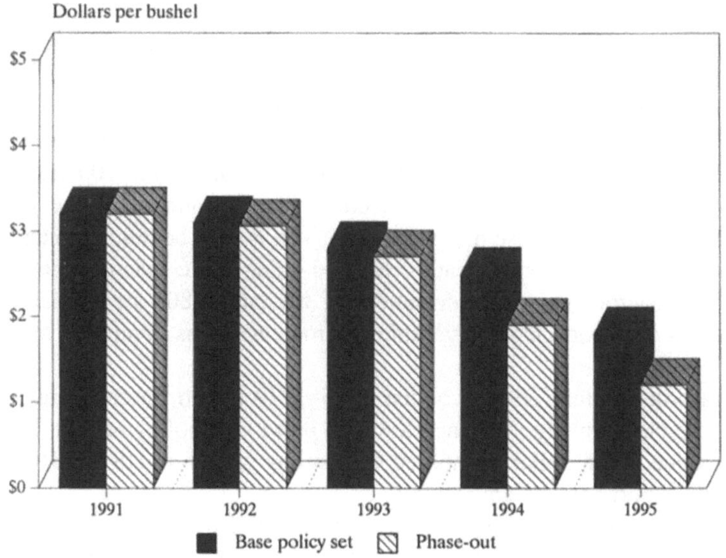

Fig. 10.2. 1981–1985 macro set – real price of wheat

In order to better understand the mechanism driving these results, we can compare participation rates and prices under the base macroeconomic set and the 1981–1985 macroeconomic set during phase-out. Under the favorable macroeconomic conditions as described by the base macroeconomic set, phased reduction of the policy instruments leads to increased participation. This seemingly perverse relationship is primarily due to the high opportunity cost of diversion and the lack of anticipated deficiency payments due to the high market price. As the diversion requirement is lowered, participation increases. The additional acreage entered into the program and the attendant required set-aside help to offset the acreage coming into production due to the exogenous decline in the diversion rate. This effect tempers the price depressing effect of phase-out over the simulation period. The converse holds under an adverse macroeconomic environment. As the price of wheat declines due to the adverse macroeconomic set, the opportunity cost of diversion is less important, and the reduced target price overshadows the smaller opportunity costs of diversion as the program instruments are reduced. In this case, participation does decrease as compared to the base policy set. Given the smaller diversion rate under phase-out, less acreage is enrolled in the program so that total acreage set-aside is almost 5 billion acres less than it would have been without phased reduction of the policy instruments, which leads to a decrease in price of 57 cents per bushel by 1995.

10.6.3 The 1975–1979 Macroeconomic Set

The macroeconomy of the late 1970s provided an extremely beneficial environment for agriculture (e.g., Rausser et al. 1986). The exchange rate (dollar) starts at a relatively low level and declines by about 1.5% per year, which increases export demand, price, production, and crop receipts. The low (negative) ex post real interest rate induces storage in the face of high prices, thus contributing to strong prices.

The 1975–1979 macroeconomic set is much like the base macroeconomic set. Given the similarity, prices, production, and crop receipts are much like those reported in the base simulation experiment shown in Appendix Table 10.B.1. An important point to notice is that the macroeconomic environment of the late 1980s, as approximated by the base macroeconomic set, is very beneficial to agriculture relative to the last several decades. Studies using static macroeconomic sets based on the macroeconomic environment of the late 1980s are implicitly basing their results on a particulary favorable macroeconomy as compared to its past performance.

Given the macroeconomic settings of the 1975–1979 period, prices, production, and crop receipts show vigorous increase with or without policy reform (Fig. 10.3). The low value of the dollar, combined with strong growth in US and ROW income, drives up export demand appreciably.

When the program instruments are held constant in the face of increasing price, program participation and total acreage set-aside decrease due to the opportunity cost of the diversion requirement. However, given the large opportunity cost of diversion, participation increases as the program is phased out. This increase in

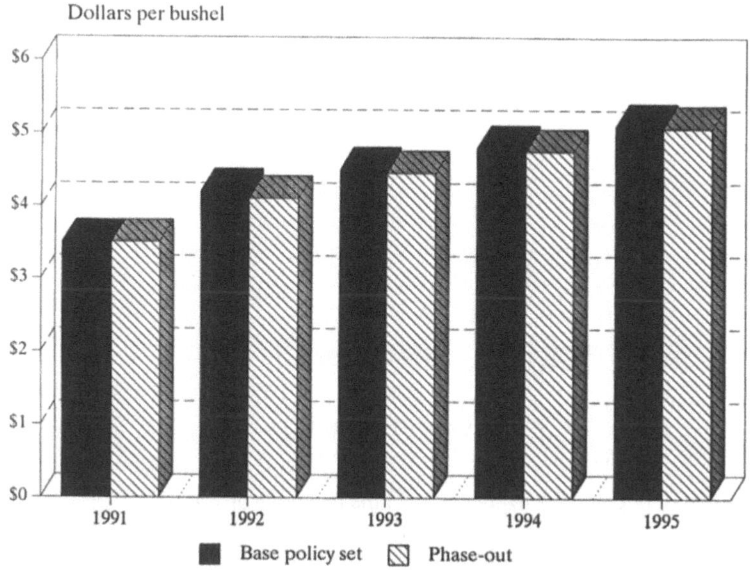

Fig. 10.3. 1975–1979 macro set – real price of wheat

participation helps to keep acreage out of production due to the declining but positive diversion rate, and to hold prices to that which would have been obtained under the experiment holding the wheat program in place.

10.6.4 The Conditional Effects of Phase-out

Table 10.4 and Figs. 10.4 and 10.5 summarize the conditional effects of phased reduction of the major policy instruments as reported in the preceding six experiments. The table shows comparisons between final period values (1995 values) for a given macroeconomic set with program instruments held constant, and those un-

Table 10.4. Conditional effects of phased reduction of program parameters

Macroeconomic set	Δ Real price of wheat	Δ Production	Δ Crop receipts	Δ Government payments	Δ Gross farm income
			percent		
Base	−1.4	0.4	−1.1	0	−1.1
1981–1985	−31.4	5.2	−27.7	−57.0	−40.5
1975–1979	−0.6	0.2	−0.4	0	−0.4

Note: Δ is the change in the endogenous variable attributed to phased reduction of the program parameters, evaluated in 1995.

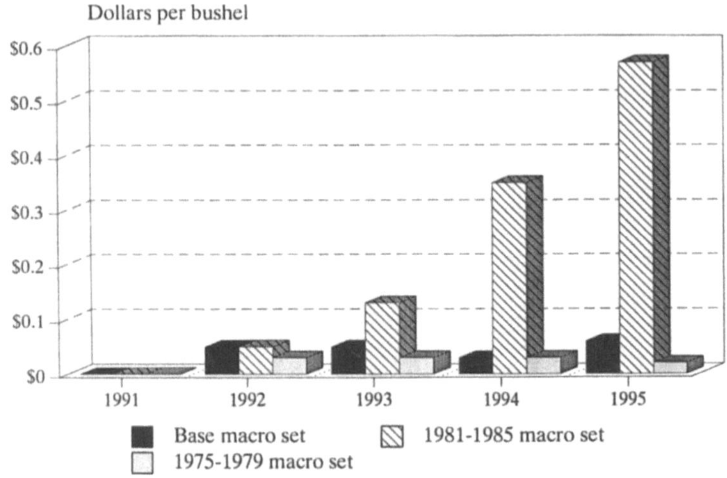

Fig. 10.4. Price effect of policy reform – decrease in the real price of wheat

der phased reduction of the program instruments. The reported values of Table 10.4 are in percentage terms and should be interpreted as the relative effect of phase-out on the relevant variables, conditioned on the realized macroeconomic environment.

As presented in Table 10.4, the results show that in a benign macroeconomic environment (base and 1975–1979 macroeconomic sets) current levels of the policy instruments contribute only a small measure of distortionary impact on the sector. Since the level of distortion is limited, phased reduction has only a minor impact on short-run prices, production, crop receipts, and government payments.

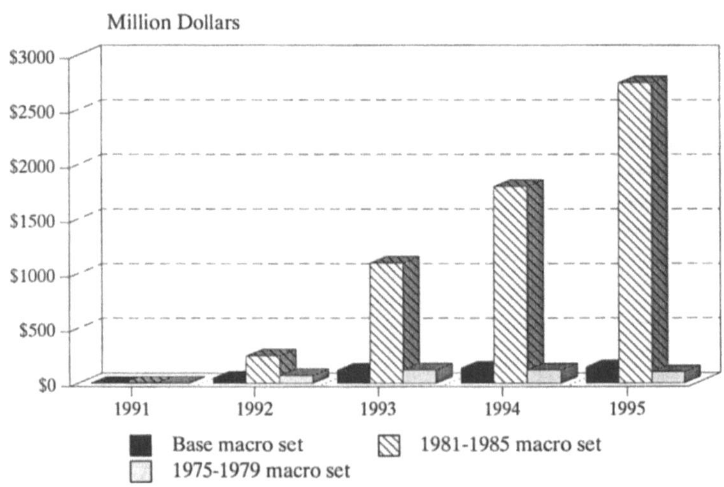

Fig. 10.5. Price effect of policy reform – decrease in the real price of wheat

Added to the low level of distortion is the perverse result that phased reduction of all policy instruments actually increases program participation in the short run as the decreases in the opportunity costs of diversion outweigh the effect of a lower target price and loan rate. This rise in participation helps to offset the added acreage coming into production as the diversion rate is lowered and helps to hold up prices.

The converse holds for the case of an adverse macroeconomic environment (1981–1985 macroeconomic set). As prices deteriorate due to the adverse macroeconomic set, the current policy setting becomes very distortionary. We can separate the effects of the macroeconomy on the sector into direct and indirect effects. The direct effect is the more obvious one in which the adverse macroeconomic environment depresses prices, production, and crop receipts and increases government payments. The indirect effect is due to the change in participation and, thus, set-aside acreage and price. Since the opportunity cost of diversion (participation) is much less under the adverse macroeconomic set, the perverse result of increased participation due to phased reduction is not obtained. The perverse result helped to further dampen the already limited effects of phase-out. However, in an adverse macroeconomic environment program, participation and acreage set-aside are less than under a fixed policy scenario; thus, added pressure on prices occurs as added production in generated, not only by the smaller diversion requirement but also by the endogenous decrease in participation.

10.6.5 Alternative Design in Phased Reduction of Policy Parameters

Various schemes for reducing the distortionary aspects of US agricultural programs exist. The specific design used in the preceding simulation experiments reduced target price, loan rate, and diversion requirement in equal increments. This particular design is tractable and consistent with the GATT framework.

In order to assess the relative importance of alternative designs in phased reduction of distortionary policies, a payment acreage scheme was simulated. For these simulation experiments, target price, loan rate, and diversion requirement are held constant; however, the amount of acreage on which a producer is entitled to receive deficiency payments is a declining proportion of his fixed base acreage.

Phased reduction of payment acreage was simulated under the various macroeconomic sets. When the favorable macroeconomic sets (base and 1975–1979) are considered, we find that this specific form of policy adjustment has no effect on the sector as measured by our model. Operationally, this is due to the deterministic nature of the model. Since producers act as though they have perfect foresight, they attribute no option value to target prices and loan rates when they are below the market price. The base and 1975–1979 macroeconomic sets lead to a market price which is greater than the target price; hence, deficiency payments are zero[10]. A model that admits stochastic optimization would offer different results; how-

[10] The primary reason the model obtains a 1991 market price greater than the target price is the presupposed low carry-in from the 1990 crop. The realized growing conditions of 1990 greatly diminished the realized market price.

ever, Gardner (1989) notes that such option value implied by the program is probably relatively small.

Given the adverse 1981–1985 macroeconomic set, a payment arcreage scheme helps to support price over the simulation period as compared to the phased reduction of the all-instruments policy scenario. As the opportunity cost of diversion falls because of the weak equilibrium price, participation and total acreage set-aside as shown in Appendix Table 10.B8 increase by a larger margin as compared to the all-instruments scheme (Appendix Table 10.B4). The main force driving the differential effects is that the diversion requirement is fixed in the payment-acreage experiment, where it is gradually reduced in the all-instruments experiments. Given the parameters of the system, the endogenously determined set-aside acreage increases market price and crop receipts. That is to say, price and crop receipts fall by less, given phased reduction of payment acreage, than under phased reduction of all major program instruments during the period of phase-out.

It is to be stressed that the goal of this research is to examine the short-run effects of phased reduction of the major program instruments. Given that the short run in this particular case is at least 10 years long due to the extended nature of phased reduction, the short run becomes empirically important.

The major results of the experiments dealing with the differential effects of policy design conditioned on the realized macroeconomy are summarized in Table 10.5. Final period (1995) values are compared on the basis of the effect of phase-out design relative to a fixed policy scenario under the base macroeconomic set, and the 1981–1985 macroeconomic set.

The relative effects of two policy designs conditioned on specified macroeconomic environments are shown in Table 10.5. The experiments show that, given phased reduction of all major policy instruments, the preceding results are robust

Table 10.5. Alternative design in phased reduction of policy parameters

Macroeconomic set	Δ Real price of wheat	Δ Production	Δ Crop receipts	Δ Government payments	Δ Gross farm income
	percent				
Base (All)	−1.4	0.4	−1.1	0	−1.1
Base (Pay)	0	0	0	0	0
1981–1985 (All)	−31.4	5.2	−27.7	−57.0	−40.5
1975–1979 (Pay)	−8.7	1.3	−7.0	−39.7	−21.3

[a] Gross farm income is defined as crop receipts plus deficiency payments.

Note: Δ is the change in the endogenous variable attributed to phased reduction of the program parameters, evaluated in 1995. All = "all instruments" phased out. Pay = "paying acreage" phased out.

in a qualitative sense at least over the two phase-out designs considered and the the quantitative results are conditioned on the realized macroeconomy. That is to say that, for both policy designs, the macroeconomy matters; and the magnitude of the short-run effect of phased reduction is partly determined by the policy design.

Given the base macroencomic set, the relative effects of program design are neglibible. The deterministic nature of this model leads to the result in which reduction of payment acreage has no effect on the sector when market price is above the target price and the diversion rate is held constant. If the option value of program participation was accounted for, we suspect that the two policy designs under the base macroeconomic set would still lead to similar outcomes because the reduced option value of program participation is probably small relative to the opportunity cost of diverting acreage.

For the 1981–1985 macroeconomic set, prices and crop receipts fall as payment acreage is reduced, but not nearly as much as under the reduction of all instruments. Government payments over the simulation period are very different over the two phase-out schemes as might be expected since the settings of the program instruments are different within each period. The interesting point is that, in order to understand the short-run effects of phased reduction of distortionary policies on equilibrium price, production, and income, one must be explicit about the manner in which those policies are to be phased out.

10.7 Conclusion

This study has offered a framework for policy analysis in which large shocks to the system can be evaluated. Sufficient attention has been given to the underlying structure of the system such that the parameters of the model are invariant to the class of policy evaluations which we have studied. This framework has been applied to the evaluation of phased reduction of the US wheat program and has led to the following results.

10.7.1 The Macroeconomy Has a Large Influence on US Agriculture

Following the results of many previous studies cited throughout this chapter, we have shown that the macroeconomy can have a severe impact on agriculture prices, production, and income. Our simulations showed that, given a macroeconomic environment similar to that as observed during the early 1980s, the real price of wheat could fall by as much as 42% – even with the wheat program held in place.

10.7.2 The Effect of Policy Reform is Conditioned on the Macroeconomy

Given the current macroeconomic environment, policy reform will have little effect on agricultural prices, production and income. This result is not robust to various realizations of the macroeconomy. Given the adverse environment of the early 1980s, phased reduction of the major program instruments will further decrease real price by 31% and total farm revenue by 40% (including government payments) as compared to holding the program in place.

10.7.3 The US Farm Program Cannot be Modeled as an Ad Valorem Equivalent

Since program participation is voluntary, even when the diversion rate is held constant, total acres set aside are endogenous. Ad valorem equivalents will not be useful in simulating phased reduction or partial policy reform unless further attention is given to the participation decision. Furthermore, ad valorem equivalents as commonly measured are not appropriate for evaluating policy reform in the long run. It is unreasonable to assume that the target price is the rational producer incentive price in the face of pre-announced policy reform. Evaluations based on commonly used measures of ad valorem equivalents overestimate the production incentives of US farm program and thus underestimate the net effect of production controls. Contrary to many previous studies on unilateral policy reform, we find that phased reduction of the major policy instruments will lead to decreased market price and crop receipts as land is brought back into production in a market facing inelastic demand.

The normative implications of the above empirical results are simply that subsidy reductions be conditioned upon the macroeconomic environment. In particular, flexible adjustments should be introduced so as to admit the possibility of some variance in the implementation of reform, depending upon the external economic conditions. Total liberalization by a fixed time, without some conditionality or variance, is a very risky proposition indeed. It threatens government credibility directly and endangers the implementation of reform strategies. To be sure, an inability to implement reform in a few countries can even undermine an externally binding GATT code. Government or political failure in implementing reform in some nations could lead to revisions in a GATT code that would make one-binding constraints totally ineffective.

References

Anderson K, Tyers R (1990) How developing countries could gain from agricultural trade liberalization in the Uruguay round. In: Goldin I, Knudsen O (eds) Agricultural trade liberalization: implications for developing countries. OECD, Paris, pp 41–75

Andrews MS, Rausser GC (1986) Some political aspects of macroeconomic linkages with agriculture. Am J Agric Econ 68:413–417

Baily KW (1989) A structural econometric model of the world wheat market. Tech Bull 1763. US Department of Agriculture, Economic Research Service, Washington, DC

Chambers RG (1979) An econometric investigation of the effect of exchange rate and monetary fluctuations on US agriculture. Dissertation, University of California, Berkeley

Chambers RG, Just RE (1979) A critique of exchange rate treatment in agricultural trade models. Am J Agric Econ 63:32–46
Chambers RG, Just RE (1981) Effects of exchange rate changes on US agriculture: a dynamic analysis. Am J Agric Econ 63:32–46
de Gorter H, Fisher EO (1989) The dynamic effects of agricultural subsidy programs in the United States. Working Paper 89–9. Cornel University, New York
Devadoss S, Helmar M, Myers WH (1990) The world wheat trade model: specification, estimation, and validation. CARD Tech Rep #90-TR14. Iowa State University, Ames
Eales JS, Engel BK, Hausser RJ, Thompson SR (1990) Grain price expectation of Illinois farmers and grain merchandises. Am J Agric Econ 71(3):701–708
Gadson KE, Price JM, Salathe LE (1982) Food and agricultural policy simulator (FAPSIM): structural equations and variable definitions. US Department of Agriculture, Staff Rep AGES820506, Economic Research Service, Washington, DC
Gardner BL (1987) Protection of US agriculture: why, how, and with what consequences. Department of Agriculture and Resource Economics, University of Maryland, Ames (unpub)
Gardner BL (1989) Gains and losses from the wheat program. Department of Agriculture and Resource Economics, University of Maryland, Ames
Hertel TW, Thompson RL, Tsigas ME (1989) Economywide effects of unilateral trade and policy liberalization in US agriculture. In: Stoeckel BA, Vincent D, Cuthbertson S (eds) Macroeconomic consequences of farm support policies. Duke University Press, Durham, pp 260–292
Johnson PR, Grennes T, Thursby M (1977) Devaluation, foreign trade controls, and domestic wheat prices. Am J Agric Econ 59:619–627
Labson BS (1990) Applied agricultural policy evaluation in an uncertain macroeconomic environment. PhD Dissertation, Department of Agricultural and Resource Economics, University of California, Berkeley
Lucas R (1976) Econometric policy evaluation: a critique. The Phillips curve and labor markets. Carnegie-Rochester conference series
Rausser GC, Chalfant JA, Love HA, Stamoulis K (1986) Macroeconomic linkages, taxes and subsidies in the US agricultural sector. Am J Agric Econ 69:399–412
Robinson S, Adelman I, Kilkenny M (1989) The effect of trade liberalization in agriculture on the US economy. In: Stoeckel AB, Vincent D, Cutherbertson S (eds) Macroeconomic consequences of farm support policies. Duke University Press, Durham, pp 222–259
Roningen V, Sullivan J, Waino J (1987) The impact of the removal of support to agriculture in developed countries. Paper presented at the American Agricultural Economics Association summer meetings, East Lansing, Michigan, August
Salathe L, Price JM, Banker DE (1984) An analysis of the farmer-owned reserve program, (1977–82). Am J Agric Econ 66:1–11
Sargent TJ (1987) Dynamic macroeconomic theory. Harvard University Press, Cambridge
Schuh EG (1974) The exchange rule and US agriculture. Am J Agric Econ 56:1–13
Schultz TW (1945) Agriculture in an unstable economy. McGraw-Hill, New York
Sims CA (1974) Distributed labs. In: Intrilligator MD, Kendrick DA (eds) Frontiers of quantitative economics, vol 2, 2nd edn. North Holland, Amsterdam, pp 289–332
Tyers R, Anderson K (1986) Distortions in world food markets: a quantitative assessment. World Development Report background paper. World Bank, Washington, DC
Tweeten LG (1980) An economic investigation into inflation passthrough to the farm sector. West J Agric Econ 5:89–106

US Department of Agriculture (various issues) Wheat: situation and outlook report. Economic Research Service, Washington, DC

Whalley J, Wigle R (1990) Terms of trade effects, agricultural trade liberalization and developing countries. In: Golding I, Knudsen O (eds) Agricultural trade liberalization: implications for developing countries. OECD, Paris

Wright BD (1984) An assessment of the United States farmer-owned reserve and Commodity Credit Corporation storage as market stabilization policies. Draft, American Interprise Institute, Washington, DC

Young E, Jagger C (1989) Effects of the CRP: focus on wheat production. Wheat situation and outlook report. US Department of Agriculture, Economic Research Service, Washington, DC

Appendix A: Parameter Estimation

Production

The production block has been estimated over the sample period of 1974–1987 using the three-stage, least-squares estimator. Given the nonlinearity of real net profit from participation, the variable had been specified as exogenous for estimation purposes. Given this procedure, the three-stage, least-squares estimator is operationally equivalent to the seemingly unrelated regression estimator since the production system as specified for estimation contains no endogenous right-hand-side variables.[1] Note that for simulation purposes real net profit from participation is endogenous. All estimated coefficients are of the assumed sign, and generally are significant as judged by the asymptotic t-ratio.

Following Sims (1974), the Durbin-Watson statistic presents a low power test for autocorrelation. The low Durbin-Watson statistic for the acreage planted by nonparticipants suggest autocorrelation and biased parameter estimates. The production equations were re-estimated using a standard autocorrelation correction technique. The re-estimated parameters did not vary much from those reported above and indicate that the autocorrelation bias is minimal.

Program Participation Rate

PRTR = 0.625 + 0.00824*RNPFP + 0.375*DDVR

(9.659) (17.622)

[0.04]

T-ratio ( ) elasticity [ ]

$R^2 = 0.974$ D.W. = 2.05

PRTR = Program participation rate.
RNPFP = Real net profit from participation
DDVR = Indicator variable for zero mandatory diversion

Yield per Planted Acre

YLD = -426.9 + 1.051*RFPW - 0.0163*RVC + 0.065*STASD + 10.506*t - 0.061*tq

(2.236) (0.161) (1.110) (2.785) (2.855)

[0.18] [-0.03] [0.02]

T-ratio ( ) elasticity [ ]

$R^2 = 0.86$ D.W. = 2.80

---

[1] In the production block, the futures price enters exogenously. Simultaneity bias should be minimal due to the large variance in excess demand as compared to US supply. Note that, in the demand block, current price is endogenous and instruments are formed for the reduced form relations.

YLP = Yield per planted acre
RFPW = Real futures price of wheat (October quote for September delivery)
RVC = Real variable costs
STASD = Acreage set-aside (mandatory and voluntary programs)
t = Time
tq = Time squared

Acreage Planted by Nonparticipants

APNP = 120 + 0.26*LAPNP	+ 4.579*RFPW	- 1.66*RVC	- 1.039*PACP	+ 26.339*SUB
(1.811)	(1.619)	(2.985)	(8.802)	(1.871)
[0.28]	[0.97]	[-4.11]	[-2.72]	[1.48]

T-ratio ( ) elasticity [ ]

$R^2 = 0.913$ D.W. = 1.122

APNP = Acreage planted by nonparticipants
LAPNP = Lagged acreage planted by nonparticipants
RFPW = Real future price of wheat
RVC = Real variable costs
PACP = Program acreage considered planted
SUB = Relative price (future price of wheat divided by future price of corn.

Demand

The demand block has been estimated over the sample period of 1963–1987 using the three-stage, least-square estimator. The estimated coefficients are of the assumed sign and generally are significant. The distributed lag specification appears to capture much of the underlying dynamics of the demand system. The Durbin-Watson statistics are all close to two, and the estimated parameters proved to be vary robust to alternative autocorrelation specifications.

Food

FOOD = 74.728 + 0.720*LFOOD	- 2.044*RPW	+ 0.0362*USGNP
(3.881)	(0.900)	(1.948)
[0.71]	[-0.02]	[0.18]

T-ratio ( ) elasticity [ ]

$R^2 = 0.962$ D.W. = 1.968

FOOD = US wheat used for food
LFOOD = Lagged FOOD
RPW = Real price of wheat (price received by farmers, annual average weighted by marketings)
USGNP = Real US gross national product

Feed

FEED = 308.7 + 0.364*LFEED - 26.798*RPW + 0.0723*USGNP - 190.45*SUBF

(2.381)	(2.776)	(2.555)	(2.302)
[0.35]	[-0.47]	[0.81]	[-0.9]

T-ratio ( ) elasticity [ ]

$R^2 = 0.80$ D.W. = 2.32

FEED = US wheat used for feed and seed
LFEED = Lagged FEED
RPW = Real price of wheat
USGNP = US gross national product
SUBF = July price of wheat divided by July price of corn

Export Demand

ED = 1954 + 0.679*LED - 69.845*RPW - 12.155*EX + 0.013*WGNP

(3.427)	(1.597)	(3.396)	(0.264)
[0.65]	[-0.29]	[-1.27]	[0.08]

T-ratio ( ) elasticity [ ]

$R^2 = 0.728$ D.W. = 2.11

ED = Export demand
LED = Lagged export demand
RPW = Real price of wheat (US currency)
EX = Rest of world currency per dollar (Federal Reserver Board trade-weighted exchange rate)
WGNP = World gross national profit. (O.E.C.D. gross national profit serves as a proxy)

Market Stocks

KMARK = 1017 - 95.387*RPW - 56.453*RRATE + 0.533*DKGOV

(3.374)	(3.655)	(5.670)
[-0.82]	[-0.14]	[-0.01]

T-ratio ( ) elasticity [ ]

$R^2 = 0.549$ D.W. = 1.74

KMARK= Market held of crop year carry-over
RPW = Real price of wheat
RRATE = Real rate of return on 6-month T-Bill (ex post)
DKGOV = Net change in government held stocks in period t + 1

Appendix B

Table 10.B1. Base macroeconomic set and base policy set

Year	Real price of wheat	Food demand	Feed demand	Export demand	Market storage	Production
	dollars/bushel	million bushels				
1991	3.50	772	422	1,494	471	2,450
1992	3.88	778	433	1,635	435	2,764
1993	4.02	784	437	1,722	421	2,883
1994	4.14	790	438	1,774	409	2,945
1995	4.24	796	440	1,805	400	2,986

Year	Participation	Nonparticipant acreage	Total acreage	Acreage set-aside	Crop receipts	Government payments
		million acres			million dollars	
1991	0.55	42	79	5.8	8,569	0
1992	0.54	52	88	5.7	10,713	0
1993	0.53	56	91	5.6	11,598	0
1994	0.53	58	93	5.6	12,207	0
1995	0.52	59	94	5.5	12,662	0

Calculated.
Note: Government payments are the sum of deficiency payments and the cost of Commercial Credit Corporation acquisition.

Table 10.B2. Base macroeconomic set and all instruments policy set

Year	Real price of wheat	Food demand	Feed demand	Export demand	Market storage	Production
	dollars/bushel	million bushels				
1991	3.50	772	422	1,494	471	2,450
1992	3.84	778	434	1,637	438	2,770
1993	3.97	784	438	1,727	426	2,891
1994	4.09	790	441	1,782	415	2,956
1995	4.18	796	443	1,814	407	2,999

Year	Participation	Nonparticipant acreage	Total acreage	Acreage set-aside	Crop receipts	Government payments
		million acres			million dollars	
1991	0.55	42	79	5.8	8,569	0
1992	0.55	51	88	5.2	10,648	0
1993	0.55	54	91	4.5	11,490	0
1994	0.56	54	93	3.9	12,080	0
1995	0.57	54	94	5.3	12,522	0

Calculated.
Note: Government payments are the sum of deficiency payments and the cost of Commercial Credit Corporation acquisition.

Table 10.B3. 1982–1985 macroeconomic set and base policy set

Year	Real price of wheat	Food demand	Feed demand	Export demand	Market storage	Production
	dollars/bushel	million bushels				
1991	3.20	775	434	1,391	487	2,377
1992	3.12	786	466	1,356	482	2,557
1993	2.74	798	496	1,208	508	2,482
1994	2.33	813	526	968	534	2,288
1995	1.85	828	559	655	567	2,030

Year	Participation	Nonparticipant acreage	Total acreage	Acreage set-aside	Crop receipts	Government payments
		million acres			million dollars	
1991	0.56	40	77	5.6	7,598	40
1992	0.57	46	83	6.1	7,971	146
1993	0.65	38	82	6.9	6,794	755
1994	0.76	25	77	8.1	5,331	1,613
1995	0.88	8	69	9.4	3,747	2,905

Calculated.
Note: Government payments are the sum of deficiency payments and the cost of Commercial Credit Corporation acquisition.

Table 10.B4. 1981–1985 macroeconomic set and all instruments policy set

Year	Real price of wheat	Food demand	Feed demand	Export demand	Market storage	Production
	dollars/bushel	million bushels				
1991	3.20	775	434	1,391	487	2,377
1992	3.08	786	467	1,359	486	2,564
1993	2.60	799	500	1,220	519	2,506
1994	1.98	814	537	1,001	567	2,353
1995	1.27	830	578	718	622	2,135

Year	Participation	Nonparticipant acreage	Total acreage	Acreage set-aside	Crop receipts	Government payments
		million acres			million dollars	
1991	0.56	40	77	6.0	7,598	40
1992	0.57	45	84	5.5	7,898	0
1993	0.59	42	83	5.0	6,510	0
1994	0.63	35	80	4.7	4,658	439
1995	0.75	20	74	4.8	2,708	1,250

Calculated.
Note: Government payments are the sum of deficiency payments and the cost of Commercial Credit Corporation acquisition.

Table 10.B5. 1975–1979 macroeconomic set and base policy set

Year	Real price of wheat	Food demand	Feed demand	Export demand	Market storage	Production
	dollars/bushel	million bushels				
1991	3.72	775	422	1,513	502	2,502
1992	4.23	785	436	1,689	505	2,867
1993	4.53	798	443	1,822	528	3,041
1994	4.82	812	449	1,925	552	3,163
1995	5.09	826	455	2,008	578	3,269

Year	Participation	Nonparticipant acreage	Total acreage	Acreage set-aside	Crop receipts	Government payments
		million acres			million dollars	
1991	0.54	44	80	5.7	9,298	0
1992	0.52	55	90	5.5	12,127	0
1993	0.51	61	94	5.4	13,769	0
1994	0.50	64	97	5.3	15,242	0
1995	0.49	68	100	5.2	16,649	0

Calculated.
Note: Government payments are the sum of deficiency payments and the cost of Commercial Credit Corporation acquisition.

Table 10.B6. 1975–1979 macroeconomic set and all instruments policy set

Year	Real price of wheat	Food demand	Feed demand	Export demand	Market storage	Production
	dollars/bushel	million bushels				
1991	3.72	775	422	1,513	502	2,502
1992	4.20	785	437	1,692	508	2,873
1993	4.49	798	445	1,826	532	3,048
1994	4.78	812	451	1,931	556	3,170
1995	5.06	827	456	2,014	581	3,267

Year	Participation	Nonparticipant acreage	Total acreage	Acreage set-aside	Crop receipts	Government payments
		million acres			million dollars	
1991	0.54	44	80	5.7	9,298	0
1992	0.54	54	90	5.2	12,065	0
1993	0.54	58	95	4.6	13,675	0
1994	0.54	60	98	4.0	15,151	0
1995	0.55	62	100	3.5	16,576	0

Calculated.
Note: Government payments are the sum of deficiency payments and the cost of Commercial Credit Corporation acquisition.

Table 10.B7. Base macroeconomic set and payment acreage policy set

Year	Real price of wheat	Food demand	Feed demand	Export demand	Market storage	Production
	dollars/bushel	million bushels				
1991	3.50	772	422	1,494	471	2,450
1992	3.88	778	433	1,635	435	2,764
1993	4.02	784	437	1,722	421	2,883
1994	4.14	790	438	1,774	409	2,945
1995	4.24	796	440	1,805	400	2,986

Year	Participation	Nonparticipant acreage	Total acreage	Acreage set-aside	Crop receipts	Government payments
		million acres			million dollars	
1991	0.55	42	79	5.8	8,569	0
1992	0.54	52	88	5.7	10,713	0
1993	0.53	56	91	5.6	11,598	0
1994	0.53	58	93	5.6	12,207	0
1995	0.52	59	94	5.5	12,662	0

Calculated.
Note: Government payments are the sum of deficiency payments and the cost of Commercial Credit Corporation acquisition.

Table 10.B8. 1981–1985 macroeconomic set and payment acreage policy set

Year	Real price of wheat	Food demand	Feed demand	Export demand	Market storage	Production
	dollars/bushel	million bushels				
1991	3.20	775	434	1,391	487	2,377
1992	3.12	786	466	1,356	482	2,557
1993	2.73	798	496	1,209	508	2,484
1994	2.27	813	528	973	539	2,298
1995	1.69	829	563	669	581	2,507

Year	Participation	Nonparticipant acreage	Total acreage	Acreage set-aside	Crop receipts	Government payments
		million acres			million dollars	
1991	0.56	40	77	6.0	7,598	40
1992	0.57	46	83	6.1	7,971	131
1993	0.64	39	82	6.8	6,771	607
1994	0.72	28	77	7.7	5,225	1,140
1995	0.80	15	70	8.5	3,484	1,751

Calculated.
Note: Government payments are the sum of deficiency payments and the cost of Commercial Credit Corporation acquisition.

Chapter 11

Alternative Subsidy Reduction Paths: The Role of Fiscal and Monetary Policy Linkages

Pier Giorgio Ardeni and Gordon C. Rausser

11.1 Introduction

There is much debate about the potential effects of phased reductions in governmental intervention in US agriculture. Unfortunately, there is little evidence that this debate has taken into account the linkages of the agricultural sector with the balance of the US and international economies. The purpose of the analysis presented in this chapter is to give a structural interpretation to the macroeconomic linkages – both forward and backward – between the agricultural sector and the aggregate economy; to review and criticize the structural exploration of macroeconomic time series concerning the agricultural sector and the aggregate economy, and to draw out the implications of alternative macroeconomic shocks on the phased reduction of governmental intervention in agriculture.

The substantial variation in exchange rates, inflation rates, relative farm prices, and agricultural incomes since the early 1970s has induced a new stream of research on the relationships between macroeconomic policy and the agricultural sector (Schuh 1974, 1976; Tweeten 1980; Chambers 1981, 1984; Gardner 1981; Chambers and Just 1982; Barnett et al. 1983; Bessler 1984; Rausser 1985; Rausser et al. 1986). In all of these studies, the exchange rate has been recognized as an important determinant of real farm prices through its effects on the trade balance (Schuh 1974). A series of theoretical and empirical studies on the effect of exchange rates has shown, for instance, the importance of an overvalued currency on US agriculture production and exports (Vellianitis-Fidal 1976; Chambers and Just 1979, 1982; Devadoss et al. 1986; Orden 1986).

Studies on relative prices and aggregate inflation have supported the hypothesis that the variability in real farm income and prices increases with the general price level variability (Vining and Elwertowski 1976; Parks 1978; Cukierman 1979; Cukierman and Wachtel 1979, 1982; Hercowitz 1981; Fischer 1982; Mussa 1982; Rotemberg 1982; Stockton 1988). At the core of this research is the idea that, if an unanticipated exogenous shock (e.g., monetary expansion) occurs, all the price and interest rate adjustments will happen in some sectors earlier than in others. Assuming prices adjust more quickly in competitive markets than in imperfectly competitive markets, farm prices can be expected to rise faster than nonfarm prices, provided of course that agricultural markets are indeed more competitive.

Various explanations for these relative price movements have included differences in the supply and demand elasticities of specific products (Cairnes 1874) and, more recently, the effects of contract length on the speed of adjustment (Bordo 1980). According to Bordo, a change in money supply causes a faster response for farm commodity prices than industrial prices and a faster response for nondurable than durable prices.

The existence of nominal influences on real variables in agricultural markets has been tested in a numerous studies (e.g., Tweeten 1980; Gardner 1981). In a more general setting, Fischer (1982) has studied three sets of hypotheses linking aggregate price changes to relative price variability: the adjustment cost hypothesis, the rational expectation unanticipated disturbance hypothesis, and the asymmetric price response hypothesis. The first two hypotheses imply that relative price variability is affected by macroeconomic disturbances; the third hypothesis implies that autonomous relative disturbances have macroeconomic effects. Under the first two hypotheses, both price level changes and relative price variability are caused by the same aggregate supply and demand interactions.

The third hypothesis is based on the assumption that prices respond asymmetrically to disturbance, for instance, they may be downward inflexible. Under this hypothesis, as Fischer notes (1982, p. 180), "If the disturbances that move relative prices were primarily supply side, resources should be moving out of the industries where prices have risen towards the industries where prices have yet to fall. If the disturbance were demand side, resources should be moving towards the higher prices sectors." Hence, differential responses in prices in this case are due to more than price stickiness alone. In Fischer's empirical work, the available evidence is not totally supportive of the first two hypotheses and the third hypothesis could not be rejected.

A number of other studies address the broader macroeconomic scenario. In some cases, this broader perspective includes an examination of the linkages with agricultural commodity prices (Rausser 1985; Frankel 1986; Rausser et al. 1986). Stemming from Dornbusch's (1976) overshooting models of exchange rate determination, these studies attempt to capture the linkages among exchange rates, money, interest rates, and commodity prices. This work begins with the fixed/flex price specification (following Hicks 1974; Okun 1975), modeling the farm sector as a set of auction markets while the nonfarm sector is characterized by gradual adjustment of prices. In this framework, agricultural market dynamics is studied, taking into account not only the real demand and supply forces directly related to the farm sector but also the effects of monetary and fiscal policies. The results show that monetary and fiscal policies can have substantial effects on prices and income in the agricultural sector over the short run, whereas sector-specific policies appear to have more significant influences in the long run. Regardless, both sets of policies can have dramatic effects on the dynamic path of the agricultural sector.

Unfortunately, the "state of the art" in examining macroeconomic linkages and the role of monetary, fiscal, and commodity-specific policies on the performance of the US agricultural sector is still unsatisfactory. This is, in large part, because not all linkages have been either conceptually or empirically investigated. In particular, the fix/flex specification neglects the structure of commodity-specific policies which limit the downward movement in many agricultural prices. Moreover, the major emphasis in this work has been on what can be referred to as the *forward linkages*, i.e., those effects that run from the aggregate economy to the agricultural sector. The *backward linkages* have been almost completely neglected.

As noted above, one of the purposes of our analysis is to identify and analyze the backward linkages (from the agricultural sector to money and foreign exchange markets) in conjunction with the forward linkages. Although Gardner (1981, p. 876) might have been correct when he stated, "A fully specified model is not necessary to identify macroeconomic effects upon agriculture; because agriculture is a small part of the general economy (when measured in terms of total value at the farm gate)," there are many other reasons why agriculture could have significant feedback effects on the monetary side of the economy. In particular, a sufficiently large subsidization program for some commodities can have a significant effect on governmental budgets and, thus, fiscal policy. A priori, the fiscal policy effect can in turn influence monetary policy, especially if the monetary authorities' reaction function is not completely exogenous.

With the above motivation as background, we first turn to a theoretical framework that formally incorporates the major features of agricultural policy in the dynamics of commodity, exchange rate, interest rate, money, and manufacturing good markets. This provides the basis for the specification of a vector error correction model with exogenous variables which is empirically estimated in section 11.4. Based on tests of specific hypotheses regarding identifying restriction as well as forward and backward linkage relationships, a policy simulation model is constructed. This policy simulation model is used to investigate different rates of reduction in governmental subsidization of commodity markets in the face of alternative macroeconomic shocks. From these policy simulation results, a number of concluding remarks and insights are offered in section 11.7.

11.2 The Theoretical Framework

The theoretical structural model developed here is a two-sector model in which a number of interactions between the money and foreign exchange markets and the goods market are potentially present. These interactions are both direct and indirect and operate through several different channels. In the analysis, the entire set of interactions are admitted – both direct and indirect and among money, exchange rate, and prices – in a fully comprehensive model incorporating all the relevant exogenous variables. The major theoretical features of the model can be summarized as follows. Manufacturing output is demand determined, while farm output is partly demand determined and partly supply determined where the supply conditions depend on the degree of intervention of the government in agriculture. Prices adjust slowly to changes in money. A balance-of-payment equation determines the rate of accumulation of reserves as a fraction of the total money stock. Since capital mobility is imperfect, either the capital account or the current account balances can be nonzero in the short run.

The monetary authority intervenes on the foreign exchange market in order to keep the rate of depreciation of the exchange rate in line with the domestic trend of monetary growth. Total money supply growth is given by the rate of credit creation (controlled by the monetary authority) and the rate of accumulation of reserves (controlled through the foreign exchange intervention rule). Price inflation in the two sectors depends on excess demand pressures and on the money growth rate. In the long run, price inflation is equal in the two sectors (as well as output growth) and is equal to the rate of monetary growth. This is equal to the target rate of credit creation, as well as to the rate of exchange depreciation.

Price *inflation* and output *growth* in the two sectors, money *growth*, and the exchange *depreciation rate* are the endogenous variables. The money *stock*, the price *levels* in the two sectors, and the exchange rate *level*, as well as interest rates, foreign prices (we assume the home country is small), total farm stocks, and government expenditure in agriculture, are exogenously given.

Changes in the exchange rate have a direct effect on prices since they imply changes in relative prices. They also have indirect effects, through the foreign exchange intervention rule, since the latter implies a change in domestic supply; a consequent change in income; and, thus, a pressure on prices. Changes in money also have an effect on prices (although not a direct effect as prices are sticky in the short run), since they induce changes in domestic demand. Thus, money is nonneutral in the short run. Changes in money supply have an indirect effect also through the change in interest rates, the change in the capital account balance, the consequent pressures on the exchange rate, and therefore on relative prices. Finally, changes in money also have depreciating effects on the exchange rate through the nonsterilized foreign exchange intervention.

Autonomous changes in prices have an effect on both money and exchange rates. An exogenous supply shock to the entire economy (for example, an oil shock) which has stag-flation effects, induces changes in the terms of trade (the real exchange rate and sectorial relative prices) and results in changes in the trade balance; in domestic output; and, hence, in money demand. An accommodating monetary policy and a "leaning against the wind" foreign exchange policy will let the changes in prices be fully reflected in changes in money and exchange rates. Sectoral changes in prices, due to autonomous supply shifts, also have effects on money and exchange rates through the trade balance and domestic demand.

Within the two-sector model, we incorporate the effects of government farm support programs on the dynamics of agricultural prices in response to changes in monetary and exchange rate policy. The effect of the target price is such that, if the government fully "protects" agriculture, then all downward changes in relative prices are paid back to domestic producers. Thus, changes in market prices are dampened; and the supply reduction measure helps producers adjust to exogenous falls in demand and to alleviate excessive stock accumulation. Reducing excess supply thus has dampening effects on inflation variability. In the limit, if the agricultural output is kept at the market-clearing level, price inflation in the farm sector is equal to general trend inflation. The two policy variables can be proxied by two variables whose actual effect turns out to be even more composite – total (private and public) farm stocks and government expenditure in agriculture.

With no intervention policy in agriculture, following an exogenous reduction in the foreign price of agricultural products we would have a shift of internal demand from domestic to foreign goods, a trade balance deficit, and thus depreciating pressure on the exchange rate. The monetary authority would then intervene on the foreign exchange market by contracting the supply of domestic money in the world market according to the intervention rule.

The devaluing pressure on the exchange rate would create an interest differential in favor of the foreign countries and a capital outflow which would increase the depreciating pressure on the exchange rate. Money supply growth would thus decrease proportionally and so would price inflation, at least initially. Real depreciation would then lead to an increase in demand (and output). The increasing output demand would raise inflation, and thus real money balances would start to fall. Nominal interest rates would then increase, thereby restoring the capital account. Ultimately, real depreciation of the exchange rate and falling real money balances would bring the system back to equilibrium.

With government intervention in agriculture, the government is able to "neutralize" any effect of foreign disturbances on domestic prices and demand. That is, the government can fix the target price at the existing domestic price level, allowing domestic producers to sell on the domestic (and world) market at the world price, with the latter below the former, and paying the difference. This kind of intervention amounts to a complete "sterilization" of foreign disturbances on the trade balance, the exchange rate, and on monetary growth and domestic inflation.

In the long run, this strategy would lead to unsustainable cumulative budget deficits; if the foreign price decrease is permanent, an increase in domestic taxes is then necessary in order to finance the deficit. This, however, would lead to a decrease in disposable income with the consequent deflationary effects. Therefore, we define an intervention mechanism that allows zero-sum deficits in the long run. We will assume all nonfarm government expenses are exactly balanced by tax revenues so that all government spending in agriculture amounts to a deficit in its budget. The resulting budget deficit must be either monetized or debt financed.

Both operations have obvious effects on the money market, the capital account, and the trade balance. Increases in government expenditures which are debt financed directly increase domestic absorption and income but do not have any effect on the monetary base. Increases in government expenditure financed by money creation directly increase domestic absorption and income and obviously affect money supply. In the former case, the overall long-run effect depends on the degree of capital mobility and be either contractionary or expansionary. In the latter case there are no real long-run effects. Without ruling out any of the two possibilities, the budget deficit is specified to be debt financed and partly monetized.

The introduction of government expenditure in agriculture has several implications. First, the real effects of changes in money and in the exchange rate appear to be dampened. This accounts for the sluggishness of movements in prices, output, and the trade balance that occur because of "institutional" factors. In particular, by acting on the way agricultural output reacts to changes in the monetary variables, the standard model is modified and all price deflationary or inflationary effects are lessened. Under this new framework, money is still neutral in the long run; but it has nonneutral effects in the short run which are smaller than in the standard model, and the overshooting in the exchange rate is smaller as well.

Second, the entire dynamics of farm prices is altered. In the standard model, the differences in the dynamics of the two prices are ultimately due to their different degrees of stickiness and to the overall GNP share of the two sectors. These differences can, for instance, put the farm sector in a "cost-price" squeeze if manufacturing prices increase more than farm prices in the short run. With government intervention, farm prices are more protected; and if the degree of intervention is high, the overall real effect on farm prices of monetary contractions or of exchange rate appreciations can be nil, and thus the differential speed effect can turn in favor of farm prices.

Third, several feedback effects other than the ones already present in the standard model can occur in the new formulation. Since the main scope of government intervention is to counter unfavorable movements in relative prices and to dampen the negative effects of monetary shocks on the farm sector, the impact on the money market and on the entire economy resulting from the financing of the budget deficit now represents one more channel of feedback from prices to money and the exchange rate.

11.2.1 Government Intervention

In order to introduce explicitly government intervention in agriculture, we assume that agricultural producers have a notional supply function of the type defined by Barro and Grossman (1976). This supply function depends mainly on two types of forces: a combination of policy indicator variables and excess demand pressures. The policy indicator variables represented here are target price, q, and a land reduction premium (proxy for set-aside acreage programs), v. Regardless of the market prices, agents who participate in government programs are assured a certain price of q. The second policy indicator serves to lessen the financial burden of accumulating stocks resulting from target prices well above market prices, e.g., under current legislation, an acreage reduction program is employed to assist in reducing producers' supply.[1]

From the point of view of total aggregate domestic supply of sector A, the two government instruments work in opposite ways. The higher q, the higher producers' supply. The higher v, the lower the actual supply. For sector A, the policy forces are represented by G, leaving the form of the function unspecified as $G(q, v)$. Hence, aggregate supply of sector A, Y_A, will depend positively on the

[1] In reality, set-aside acreage programs have been carried on a voluntary basis.

policy function $G(q, v)$ on relative prices $(p_A - e - p^*_A)$, and on demand pressures $(\ln D_A)$, i.e.,

$$y_A = (1 - \omega)\left[\psi G(q,v) + \zeta\left(p_A - e - p_A^*\right)\right] + \omega\left[\ln D_A\right]. \tag{1}$$

We restrict the model to satisfy the following assumptions:

$$G_q > 0;\ G_v < 0. \tag{1a}$$

In the long run, goods markets clear; and, hence, supply is entirely determined by policy variables and relative prices. In the short run, however, supply depends also on excess demand pressures.

The coefficients $(1 - \omega)$ and ω measure the weight that producers attach to the two arguments in the supply function. If ω is close to one, agricultural output is essentially demand determined. Alternatively, if ω is close to zero, supply would be essentially determined by policy variables and relative prices.

In a competitive world (we have a small country that faces given world prices), if foreign prices (in domestic currency) fall below a certain level, then domestic producers will be paid the price given by q fixed at that level (excluding operating trade barriers). The difference between the target price, q, and the world price, $e + \rho^*_A$, is paid to domestic producers by the government. Target prices set above market prices create excess supply and large government expenditures to finance the implied level of subsidy.

In order to reduce wide excess supply accumulations from target price incentives, the government affects producers' decisions through the acreage reduction program. The higher stock accumulation, the stronger will be the action of the government to reduce output supply. Although inventories are exogenous in our framework, it is clear that it is the interaction between the inventories and production costs (and revenues) on one side and between the inventories and interest rates on the other that plays a major role in determining the amount of acreage reduction intervention in agriculture.

The workings of commodity program interventions can be embedded in a macroeconomic model by first examining the case of $G(q, v) \approx 0$, and the equilibrium conditions, $\bar{\rho}_A = \bar{e} + \rho^*_A$, $y_A = \bar{y}_A$, $\dot{\rho}_A = \dot{\rho} = \dot{m} = \dot{e} = \dot{c}$, $B = T = 0$. Given no trade barriers, an exogenous reduction in the foreign price of agricultural products would result in a shift of internal demand from domestic to foreign goods; a trade balance; and, thus, depreciating pressure in the exchange rate. With no intervention policy in agriculture, the monetary authority would then intervene by contracting the supply of domestic money on the world market.

The pressure on the exchange rate would create an interest differential in favor of the foreign countries and a capital outflow which would increase the pressure on the exchange rate. Money supply growth would thus decrease proportionately, as would price inflation (at least initially). Real depreciation would lead to an increase in demand (and output). The increasing output demand would raise inflation, and thus real money balances would start to fall. Nominal interest rate would then increase, thus restoring the capital account. Ultimately, real depreciation of the exchange rate and falling real money balances would bring the system back to equilibrium.

With government intervention in agriculture, we can have a quite different scenario. Suppose that, in the extreme case, the government wants to neutralize any effect of foreign disturbances on domestic prices and demand. That is, suppose the government fixes the target price, q, at the existing level, p_A, and allows domestic producers to sell on the domestic (and world) market at the world price, $e + p^*_A$, where $e + p^*_A < p_A$ and pays them the difference (per unit of product). This kind of intervention amounts to a complete sterilization of foreign disturbances on the trade balance, the exchange rate, and on monetary growth and domestic inflation.

However, this could not last forever since, in the long run, this would lead to unsustainable cumulative budget deficits. If the foreign price decrease is permanent, an increase in domestic taxes would then be necessary in order to finance the deficit, but would lead to a decrease in disposable income with the consequent deflationary effects. Therefore, we need to define an intervention mechanism that would allow zero-sum deficits in the long run.

In the following, it is assumed that $G(q, v) = G(q) + G(v)$:

$$G(q) = g_1(\dot{p}_A - \dot{e})\; g_1 \geq 0, \tag{2}$$

$$G(v) = g_2(y_A - \bar{y}_A)\; g_2 \geq 0, \tag{3}$$

and that world price inflation is zero. Hence, whenever the increase in domestic agricultural prices is greater than the increase in foreign prices (due only to the exchange rate component in this case), the target price will increase at the rate given by difference of the two rates and proportionately to the intervention coefficient, g_1. This formulation ensures that, in the long run,

$$G(q) = 0 \tag{2a}$$

since $\dot{p}_A = \dot{e}$. Similarly, whenever we have an excess supply of farm products, the acreage reduction subsidization will increase at the rate given by the difference between money growth and domestic farm price inflation.[2] Even in this case, in the long run,

$$G(v) = 0. \tag{3a}$$

Notice also that the two rules are not symmetric. Only if $\dot{p}_A > e$, will $G(q) > 0$, which means that favorable conditions for the home country imply the lack of any need for government intervention. Similarly, if $\dot{p}_A > m$, $G(v) = 0$; this means that, with excess demand pressures having inflationary and output-expanding effects, there is no need for supply reduction incentive programs.

These equalities highlight the relationships among the farm policy instruments. The two types of intervention can be interpreted as follows: The rate of increase in government-financed price-target programs is guided by a rule that relates to the excess of the domestic price increase over the increase of the exchange rate. The

[2] Recall that the higher the excess supply in agriculture, the lower $\dot{p}_A$ is with respect to m.

higher the "support" (intervention) coefficient, the higher the response in the rate of growth of government expenditure in price-target programs. Hence, in the limit, as g_1 tends to infinity, the rate of increase in domestic agricultural prices is kept equal to the rate of depreciation.

In the above case, the rate of increase in relative prices would be basically zero. What such an intervention rule shows is that the higher government support is, the lower will be the gains (or, alternatively, the losses) in competitiveness due to movement in the exchange rate and/or in foreign prices. However, although temporary shifts in competitiveness are minimized, the negative effects of the increase in the government expenditure for agriculture will have impacts on the government budget deficit and, ultimately, on the money market.

On the other hand, the rate of increase in government-financed supply-reduction programs is guided by a rule that relates it to the excess of money supply growth over farm price inflation. The higher the intervention coefficient, g_2, the higher the response of government expenditure in supply-reduction programs. In the limit, all excess supply is "absorbed" by government intervention so that, ultimately, any excess supply is eliminated and the price growth rate is kept equal to the money supply growth rate. This amounts to assuming, in the limit, that agriculture supply is kept at the market-clearing level. It is clear, however, that even in this case all the budget effects of the supply-reduction programs will have an impact on the money market, depending on the magnitude of the intervention coefficients.

11.2.2 Trade Linkages

In order to incorporate all the potential linkages, the relationship between the government budget deficit and other variables in the model must be specified. Assume first that domestic taxes exactly match government spending except for the portion of the government expenditure that is allocated to agriculture. In other words, assume that the whole budget *deficit* is spent in farm-support programs. The rate of increase in the budget deficit is actually the cumulative deficit (since in equilibrium it is zero).

Secondly, we know from simple national account identities that the current account balance (CAB) can be associated with the gap between disposable income and private domestic absorption and the gap between taxes and government spending

$$CAB \equiv (Y - TX - A) - (G - TX), \tag{4}$$

where $Y - TX$ is disposable income, A is private domestic absorption $(Y - E + M)$, and $G - TX$ is the government budget deficit. Hence, the government budget deficit is equal to that part of the trade deficit (assuming unilateral transfers are zero) that exceeds total private domestic excess demand.

An increase in government spending naturally affects domestic aggregate demand. It affects only investment and savings if the increased spending is financed by the sale of government bonds. The sale of bonds does not affect the domestic money supply since the funds obtained by the government from the bond sale returns to the public as the government spends it. Thus, the LM curve is not directly affected by the changes in government debt-financed spending. Of course, if the increase in government expenditure is financed by the issuance of money, then both aggregate demand and demand for money will be affected.

Increases in government expenditure which are debt financed directly increase domestic absorption and income but do not have effects on the monetary base. However, the increase in income heightens the demand for money, driving interest rates up. At the same time, all else constant, the increase has a negative impact on the trade balance, raising the demand for imports. The rise in the interest rates attracts capitals from abroad, restoring the capital account and counterbalancing the current account. Whether the balance of payments will be in surplus or in deficit will depend ultimately on the degree of capital mobility, the magnitude of the income multiplier, the willingness to save, and the propensity to import.

If capital mobility is low, an increase in government expenditure has a negative effect on the balance of payments. Money supply decreases over time, due to the decumulation of international reserves, inducing income to decrease and partially offsetting the initial expansionary effect. If capital mobility is high, a balance of payment surplus will arise and money supply will increase, generating an additional income expansion over time.

Increases in government expenditure financed by money creation directly increase domestic absorption and income and also affect money supply. The income effect generates a trade balance deficit through the increased demand for imports. The excess supply of money will be spent on foreign goods and also on foreign assets, thereby generating a capital account deficit. Complete adjustment will occur over time by means of net purchases of foreign goods. This slow adjustment will be reflected by trade deficits. In essence, the increase in the money supply through the printing of money will induce both the balance of trade and the balance of payments to worsen over the short run.

If no other disturbances occur, however, these payments and trade deficits will be eliminated over time without any major real consequence. The initial adjustment of the balance of payments represents mostly a portfolio-composition move, operating through the capital account, while the subsequent adjustment represents mostly a move to reduce the portfolio size, operating through the trade balance.

11.2.3 Debt Financing and Money Creation

In the framework presented here, the government budget deficit is presumed to be partly debt financed and partly monetized through money creation. Since the trade balance is a linear function of the log of real income and the log of relative prices, the budget deficit as an explicit determinant of the trade balance, in addition to real income and relative prices, can be represented as

$$T = -(\alpha_A - \beta_A)(e - p_A + p^*_A) - (\alpha_B - \beta_b)(e - p_B + p^*_B) - \tau y - \tau_g G(g, v). \quad (5)$$

For a 1% increase in the budgeted deficit, the trade deficit is assumed to increase proportionately (not considering the income effect). Money growth is now given by

$$\dot{m} = \dot{c} + \frac{\dot{R}}{M} + sG(q,v), \tag{6}$$

where $sG(q,v)$ is the fraction of the cumulative budget deficit which is monetized, c is the rate of domestic credit creation, $\dot{R}$ is the rate of change of reserves, and M is the money stock. Since $G(q,v) = G(q) + G(v)$ from the *intervention rules in agriculture* defined in Eqs. (2) and (3), we have

$$G(q,v) = g_1(\dot{p}_A - \dot{e}) + g_2\left(\frac{\pi_B}{\pi\pi_A}\right)(\dot{m} - \dot{p}_A), \tag{7}$$

where π_B represents the degree of stickiness of prices in the sector, π_A is the degree of stickiness of prices in the agricultural sector, and π is the total degree of stickiness in prices.

If the degree of intervention (g_i) is high, then the government response to any change in the rate of depreciation and/or in the rate of money supply growth will be such that farm price inflation will be brought to the same new rate. That is, the higher g_i, the more $d\dot{p}_A \to d\dot{e}$ and $d\dot{p}_A \to d\dot{m}$.

The effect on the overall economy of such rules can be sketched as follows: If the degree of intervention is high and $d\dot{p}_A/d\dot{e} \approx 1$, then

$$\frac{d\dot{p}}{d\dot{e}} = \rho\frac{d\dot{p}_A}{d\dot{e}} + (1 - \rho)\frac{d\dot{p}_B}{d\dot{e}} = \rho + (1 - \rho)\frac{d\dot{p}_B}{d\dot{e}} \tag{8}$$

where ρ_B represents the log of the domestic prices of the manufacturing sector and ρ represents the weight of the agricultural sector on gross national product. Similarly, since $d\dot{p}/d\dot{m} \approx 1$,

$$\frac{d\dot{p}}{d\dot{m}} = \rho\frac{d\dot{p}_A}{d\dot{m}} + (1 - \rho)\frac{d\dot{p}_B}{d\dot{m}} = \rho + (1 - \rho)\frac{d\dot{p}_B}{d\dot{m}}. \tag{9}$$

Now, money growth depends on domestic credit growth and on exchange rate growth (through accumulation of reserves and foreign exchange intervention). But, as shown in Eq. (6), if the government monetizes part of its budget deficit through money creation it also depends on budget growth which, in turn, depends on government intervention rules in agriculture. Therefore,

$$\dot{m} = \dot{c} - \theta(\dot{e} - \dot{c}) + sg_1(\dot{p}_A - \dot{c}) + sg_2\left(\frac{\pi_B}{\pi\pi_A}\right)(\dot{m} - \dot{p}_A). \tag{6a}$$

If both g_1 and g_2 are high (agriculture is fully protected), then all short-run discrepancies between $\dot{p}_A$ and $\dot{e}$ and $\dot{p}_A$ and m will be financed by government. Thus, the actual market price, $\dot{p}_A$, will increase at the same rate as $\dot{e}$ and $\dot{m}$, and the difference between *actual* and *potential* price will be fully paid by government. Money growth is given by

$$\dot{m} = (1+\theta)\dot{c} - \theta\dot{e} + s(g_1 - g_2)\dot{p}_A - sg_1\dot{e} + sg_2\left(\frac{\pi_B}{\pi\pi_A}\right)\left[(1+\theta)\dot{c} - \theta\dot{e}\right]$$

$$= \left[1 + sg_2\left(\frac{\pi_B}{\pi\pi_A}\right)\right]\left[(1+\theta)\dot{c} - \theta\dot{e}\right] - sg_1\dot{e} + s\left[g_1 - g_2\left(\frac{\pi_B}{\pi\pi_A}\right)\right]\dot{p}A; \tag{6b}$$

where θ represents how active government intervention is in the foreign exchange market. Note that, if the budget deficit is fully monetized,

$$\dot{m} = \left[1 + g_2\left(\frac{\pi_B}{\pi\pi_A}\right)\right]\left[(1+\theta)\dot{c} - \theta\dot{e}\right] - g_1\dot{e} + \left[g_1 - g_2\left(\frac{\pi_B}{\pi\pi_A}\right)\right]\dot{p}_A. \tag{6c}$$

11.2.4 Domestic Market Linkages

For agriculture, the equilibrium condition is

$$\ln D_A = \delta\left(\frac{\delta_A}{\delta_B}\right)\left(e - p_A + p_A^*\right) + \gamma y - \sigma r, \tag{10}$$

where δ, δA, and δ_B represent the elasticity of total demand, and the elasticity of demand for agricultural and manufactured products, respectively. If Eq. (10) is combined with Eq. (1), we have

$$y_A = (1-\omega)\left[\psi G(q,v) - (1-\omega)\zeta\left(p_A - e - p_A^*\right)\right] + \omega\delta\left(\frac{\delta_A}{\delta_B}\right)\left(e - p_A + p_A^*\right) + \omega\gamma y - \omega\sigma r. \tag{10a}$$

Given a similar expression for y_B, we have

$$y = \rho y_A + (1-\rho)y_B \tag{11}$$

and

$$y = \mu\rho\left\{(1-\omega)\psi G(q,v) - [1-\omega]\left[\delta\frac{\delta_A}{\delta_B} + \xi\right]\left[p_A - e - p_A^*\right] - (1-\omega)\gamma y + (1-\omega)\sigma r\right\} + \mu\delta\left(e - p + p^*\right) - \mu\sigma r \tag{11a}$$

where $\mu = 1(1-\gamma)$, as before. If agricultural producers do not place any weight on supply conditions ($\omega = 1$), then (11a) reduces to

$$y = \mu\delta(e - p + p^*) - \mu\sigma r, \tag{11b}$$

where total aggregate output is a function of relative prices and real interest rate. Therefore, if no weight is put on supply conditions, we are back to the case we have analyzed above. Recall that we are now dealing with a different framework: business sector output is still demand determined (and market B clears), whereas agricultural output is now both supply and demand determined.

Specifying the real interest, r, as the difference between the nominal interest rate, i, and the expected rate of inflation, $\dot{\rho}^e$, which, under perfect foresight, is equivalent to $\phi y - \lambda(m-p) - \dot{\rho}$, in long-run-deviation terms, Eq. (11b) becomes

$$\begin{aligned} y - \bar{y} = {} & \mu\rho(1-\omega)\left\{\psi G(q,v) - \left[\delta\frac{\delta_A}{\delta_B} + \xi\right][x_A - \bar{x}_A]\right. \\ & - \mu[\sigma\phi - \rho(1-\omega)(\sigma\phi - \lambda)](y - \bar{y}) + \mu[\sigma\lambda - \rho(1-\omega)\sigma\lambda](h - \bar{h}) \\ & + \mu[\sigma - \rho(1-\omega)\sigma]\dot{p} - \dot{c}) + \mu\sigma(x - \bar{x}), \end{aligned} \tag{12}$$

where $\bar{h}$ represents the steady-state real money balances, i.e., $h = m - p$, and $\bar{x}$ represents the steady-state real exchange rate, i.e., $x = e - p$. Now, if agricultural producers do not place any weight on supply conditions ($\omega = 1$), then (12) reduces to

$$y - \bar{y} = \mu\delta(x - \bar{x}) - \mu\sigma\phi(y - \bar{y}) + \mu\sigma\lambda(h - \bar{h}) + \mu\delta(\dot{p} - \dot{c}). \tag{12a}$$

Conversely, if farmers do not place any weight on demand conditions ($\omega = 0$), that is, agricultural output is completely supply determined, then (12) reduces to

$$\begin{aligned} y - \bar{y} = {} & \mu\rho\left\{\psi G(q,v) - \left[\delta\frac{\delta_A}{\delta_B} + \xi\right](x_A - \bar{x}_A)\right\} - (1-\rho)\mu\sigma\phi(y - \bar{y}) \\ & - \mu\rho\gamma(y - \bar{y}) + (1-\rho)\mu\sigma\lambda(h - \bar{h}) + (1-\rho)\mu\sigma(\dot{p} - \dot{c}) + \mu\sigma(x - \bar{x}) \end{aligned} \tag{12b}$$

since $\delta(x - \bar{x}) = \rho\delta(\delta_A/\delta_B)(x_A - \bar{x}_A) + (1-\rho)\delta(\delta_B/\delta_A)(x_B - \bar{x}_B)$. Rearranging

$$y - \bar{y} = \rho[y - \bar{y}_A] + (1-\rho)[y_B - \bar{y}_B]. \tag{12c}$$

Eq. (12c) makes it clear that, even in this case ($\omega = 0$) total output is equal to the weighted average of the sectoral outputs. However, the significant difference here is that agricultural output is completely supply determined [as can be seen by looking at the first term in Eq. (12c)] while nonfarm output is completely demand determined [second bracketed term in Eq. (12c)]. Going back to Eq. (12) and substituting for p, we have

$$\begin{aligned} y - \bar{y} = \frac{1}{S}\Bigg\{ & \mu\rho(1-\omega)\left\{\psi G(q,v) - \left[\delta\left(\frac{\delta_A}{\delta_B}\right) + \xi\right](x_a - \bar{x}_A)\right\} + \mu\delta(x - \bar{x}) \\ & + \mu[\sigma\lambda - \rho(1-\omega)\sigma\lambda](h - \bar{h}) + \mu[\sigma - \rho(1-\omega)\sigma](\dot{m} - \dot{c})\Bigg\}, \end{aligned} \tag{13}$$

where $S \equiv 1 + \mu\sigma(\phi - \pi) - \mu\rho(1 - \omega)[\sigma(\phi - \pi) - \gamma]$. Substituting the monetary intervention equation, $\dot{m} - \dot{c} = -\theta(\dot{e} - \dot{c})$, we get

$$y - \bar{y} = \frac{1}{D}[\mu\rho(1 - \omega)]\Bigg\{ - \psi G(q,v) - \xi(x_A - \bar{x}_A)$$

$$-\left[\delta\frac{\delta_A}{\delta_B}(x_A - \bar{x}_A)\right] - \gamma(y - \bar{y}) - \sigma(\phi - \pi)(y - \bar{y}) \tag{14}$$

$$-\sigma\lambda(h - \bar{h}) + \sigma\theta(\dot{e} - \dot{c})\Bigg\} + \frac{1}{D}(y - \bar{y})_D,$$

where $D \equiv 1 + \mu\sigma(\phi - \pi)$ and $(y - \bar{y})_D = \mu[\delta(x - \bar{x}) + \sigma\lambda(h - \bar{h}) + \sigma\theta(\dot{e} - \dot{c})]$. Here $(y - \bar{y})_D$ is just the long-run deviation aggregate output when both outputs are demand determined. Also, as

$$r - \bar{r} = (\phi - \pi)(y - \bar{y}) - \lambda(h - \bar{h}) + \theta(\dot{e} - \dot{c}),$$

we have

$$y - \bar{y} = \frac{1}{D}[\mu\rho(1 - \omega)]\{\psi G(q,v) - \xi(x_A - \bar{x}_A) - (y_A^d - \bar{y}_A^d) + \frac{1}{D}(y - \bar{y})_D;$$

since

$$y_A^d - \bar{y}_A^d = \delta\left(\frac{\delta_A}{\delta_B}\right)(x_A - \bar{x}_A) - \gamma(y - \bar{y}) - \sigma(r - \bar{r}),$$

where $y^d{}_A$ indicates demand for good A. Hence, (14) reduces to

$$y - \bar{y} = \left[\frac{1}{D}(\mu\rho)\right]\left[(y_A^s - \bar{y}_A^s) - (y_A^d - \bar{y}_A^d)\right] + \frac{1}{D}\left[(y - \bar{y})_D\right]. \tag{14a}$$

But, since $\bar{y}^d{}_A = \bar{y}^s{}_A$,

$$y - = \left[\frac{1}{D}(\mu\rho)\right]\left[y_A^s - y_A^d\right] + \frac{1}{D}(y - \bar{y})_D. \tag{14b}$$

Now, we can write (14a) as

$$y - \bar{y} = \frac{\mu}{D}\left[\rho(y_A^s - \bar{y}_A^s) + (1 - \rho)(y_B^d - \bar{y}_B^d)\right]. \tag{14c}$$

This is the reduced-form equation for output, given by a weighted average of long-run deviations of agricultural output (supply determined) and long-run deviations of business sector output (still demand determined). In the long run, supply matches demand in both sectors. In the short run, deviations from the long-run equilibrium values are due to producers' decisions in the agricultural sector and in the nonfarm sector they are due to demand pressures.

Appreciation of the exchange rate by increasing money growth (through the balance of payments and the intervention rule) raises inflationary expectations, reduces the real rate of interest, and therefore expands aggregate demand in both sectors. Eq. (14a) shows that a real expansion in the money stock raises output in the nonfarm sector (which is demand determined) and raises demand in the farm sector. According to ω, output will rise proportionally in the farm sector also. However, real depreciation has the opposite effect on agricultural supply, whose elasticity is given by $(1-\omega)\xi$.

Hence, when the economy is away from the stationary equilibrium, a change in relative prices (or in the real exchange rate) has an ambiguous effect on the supply of agricultural output. While an improvement in the domestic relative agricultural price will directly induce higher output, it could also indirectly lead to its reduction through the effect on demand. The total effect of a change in relative prices on the supply of agricultural output will depend on the relative size of these effects.

11.2.5 Model Solution

Turning to the solution of the model, the balance-of-payments equation, together with the foreign-exchange intervention rule, is now given by

$$-\theta(\dot{e} - \dot{c}) = \eta(i - \dot{e} - i*) - (\alpha - \beta)(e - p + p*) - \tau y - \tau_g G(q,v), \tag{15}$$

where η is the degree if capital mobility; α, β are trade balance elasticities; τ is the trade balance elasticity of income; and τ_g is the proportional change in the trade balance from changes in the governmental budget deficit. From this we get

$$\dot{e} = \left[\frac{1}{\theta - \eta}\right]\left[\dot{c} + \eta\lambda h + \eta i * - (\alpha - \beta)(e - p + p*) + (\tau - \eta\phi)y - \tau_g G(q,v)\right]. \tag{15a}$$

In the long run, $\dot{e} = \dot{c}$ and $G(q,v) = 0$. Hence,

$$\dot{e} - \dot{c} = \left[\frac{1}{\theta - \eta}\right]\left[(\tau - \eta\phi)(y - \bar{y}) + \eta\lambda(h - \bar{h}) + (\alpha - \beta)(x - \bar{x}) + \tau_g G(q,v)\right]. \tag{15b}$$

Long-run deviations of output are given by Eq. (14), rewritten as

$$y - \bar{y} = a_1(x - \bar{x}) + a_2(h - \bar{h}) + a_3(\dot{e} - \dot{c}) + a_4 G(q,v) + a_5(x_A - \bar{x}_A) \tag{16}$$

where now

$$a_1 = \frac{\mu\delta}{S} > 0; \quad a_2 = \frac{\mu\delta\lambda R}{S} > 0; \quad a_3 = -\frac{\mu\sigma R\theta}{S} < 0;$$

$$a_4 = \frac{\mu\psi(1 - R)}{S} > 0; \quad a_5 = -\frac{\left[\mu\left(\delta\frac{\delta_A}{\delta_B} + \xi\right)(1 - R)\right]}{S} < 0; \tag{16a}$$

$$R = 1 - \rho(1 - \omega); \quad S = 1 + \mu\sigma(\phi - \eta)R + \mu\gamma\,(1 - R).$$

Similarly, rewrite (15a) as

$$\dot{e} - \dot{c} = b_1(y - \bar{y}) + b_2(h - \bar{h}) + b_3(x - \bar{x}) + b_4 G(q,v), \tag{17}$$

where

$$b_1 = \frac{\tau - \eta\phi}{\theta - \eta} > 0; \quad b_2 = \frac{\eta\lambda}{\theta - \eta} > 0;$$

(17a)

$$b_3 = \frac{\alpha - \beta}{\theta - \eta} < 0; \quad b_4 = \frac{\tau_g}{\theta - \eta} > 0.$$

Thus, substituting for $\dot{e} - \dot{c}$ from Eq. (17) into Eq. (16) yields

$$y - \bar{y} = \left[\frac{1}{1 - a_3 b_1}\right]\left[(a_1 a_3 b_3)(x - \bar{x}) + (a_2 a_3 b_2)(h - \bar{h})\right] \tag{18}$$

$$+ (a_4 a_3 b_4) G(q,v) + a_5(x_4 - \bar{x}_A).$$

Two of the four coefficients are unambiguously signed. The first is positive, and the fourth is negative. The second coefficient (the coefficient of real money balances) will depend, again, on the sign of the term, $\theta - \eta - \eta\theta$. Thus, if $\theta - \eta - \eta\theta > 0$, then $\theta/(1 + \theta) > \eta$; that is, the intervention coefficient is greater than the coefficient that measures the degree of capital mobility. Hence, $(a_2 + a_3 b_2)/(1 - a_3 b_1)$ will be positive.

Similarly, the coefficient of the budget deficit term will depend on the sign of $(a_4 + a_3 b_4)$ and, therefore, on various parameters of the model. If $R \to 0$, then $(a_4 + a_3 b_4) > 0$; but, if $R \to 1$, the whole coefficient will be negative. A value of R close to 1 is either due to a small value of the GNP share of the agricultural sector, ρ, or to a high value of ω, regardless of ρ.

Thus, if agricultural producers attach a greater weight to the supply conditions in their supply function ($\omega \to 0$) but the agricultural sector has little importance in the national economy, then the budget deficit will have a negative impact on the domestic output. The positive effect on the agriculture is, in fact, more than offset by negative effect on the rate of depreciation.

Eq. (17) now becomes

$$\dot{e} - \dot{c} = \frac{a_1 b_1 + b_3}{1 - a_3 b_1}(x - \bar{x}) + \frac{a_2 b_1 + b_2}{1 - a_3 b_1}(h - \bar{h}) + \frac{a_4 b_1 + b_4}{1 - a_3 b_1} G(q,v) + \frac{a_5 b_1}{1 - a_3 b_1}(x_A - \bar{x}_A). \tag{19}$$

Again, three of the coefficients are ambiguously signed. The second and third are positive, and the fourth is negative (as is the first). This implies that a depreciation in the real exchange rate (either in one sector or in the aggregate) reduces the rate of depreciation as compared to the trend growth in domestic credit.

The novelty of the above results with respect to the existing literature is that output deviations now depend directly on policy intervention measures in agriculture (and therefore on the budget deficit) and on relative farm prices. Deviations in the rate of depreciation are positively associated with increasing intervention in agriculture and negatively associated with relative prices in the agricultural sector.

Also, since $\dot{h} \equiv \dot{m} - \dot{\rho} = -\pi(y - \bar{y})$, we have

$$\begin{aligned} \dot{m} - \dot{p} = & -\pi \frac{a_1 + a_3 b_3}{1 - a_3 b_1}(x - \bar{x}) - \pi \frac{a_2 + a_3 b_2}{1 - a_3 b_1}(h - \bar{h}) \\ & -\pi \frac{a_4 + a_3 b_4}{1 - a_3 b_1} G(q,v) - \frac{\pi a_5}{1 - a_3 b_1}(x_A - \bar{x}_A); \end{aligned} \tag{20}$$

and, noting that $\dot{m} = (1 + \theta)\dot{c} - \theta\dot{e} + sG(q,v)$

$$\dot{m} - \dot{p} = (1 + \theta)\dot{c} - \theta\dot{e} + sG(q,v) - \dot{p}.$$

Substituting from (19a) for $\dot{c}$ and equating to yields

$$\begin{aligned} \dot{x} \equiv \dot{e} - \dot{p} = & \frac{(1 + \theta)(a_1 b_1 + b_3) - \pi(a_1 + a_3 b_3)}{1 - a_3 b_1}(x - \bar{x}) \\ & + \frac{(1 + \theta)(a_2 b_1 + b_2) - \pi(a_2 + a_3 b_2)}{1 - a_3 b_1}(h - \bar{h}) \\ & + \frac{(1 + \theta)(a_4 b_1 + b_4) - \pi(a_4 + a_3 b_4) + s}{1 - a_3 b_1} G(q,v) \\ & + \frac{(1 + \theta)a_5 b_1 - \pi a_5}{1 - a_3 b_1}(x_A - \bar{x}_A). \end{aligned} \tag{21a}$$

The signs of the coefficients are as follows: In Eq. (20), the first is negative, the fourth is positive, and the third can be either positive or negative. Also, if capital mobility is low and foreign exchange intervention is high, then $a_2 + a_3 b_2 > 0$; so the second coefficient is negative. In Eq. (21a), the first coefficient is certainly negative (it is the sum of two negative terms). The sign of the second coefficient is ambiguous. If $(a_2 b_1 + b_2)$ is positive, then real money balances will have a positive effect on real exchange rate growth if

$$(1 + \theta)(a_2 b_1 + b_2) > \pi(a_2 + a_3 b_2).$$

We can also rewrite this inequality as

$$a_2 \, [(1 + \theta)b_1 - \pi] + b_2 \, [(1 + \theta) - \pi a_3] > 0.$$

Since the second term is positive, the inequality will hold depending on the magnitude and the sign of the first term. For the fourth coefficient in Eq. (21a) –

assuming that the effects of $(x_A - \bar{x}_A)$ and of $(x - \bar{x})$ on the rate of real depreciation go in the same direction – as a_5 is negative, it will be negative only if $[(1+\theta)b_1 - \pi] > 0$. But then the coefficients of $(h - \bar{h})$ and $G(q,v)$ in Eq. (21a) will both be positive.

Notice that the relative price of agricultural products $(x_A - \bar{x}_A)$ has a negative effect on total domestic output in Eq. (18a) and a positive effect on real balance growth in Eq. (20), as opposed to the real exchange rate, whose effects are opposite in sign in the two equations. The reason for this is, since agricultural output is supply determined, the effect of relative prices are reversed. Conversely, both $(x_A - \bar{x}_A)$ and $(x_A - \bar{x})$ have a negative effect on the rate of real depreciation in Eq. (21a).

The effect of the budget deficit on output and real money balances can be either positive or negative. If the agricultural sector GNP share is small, the budget deficit effect is likely to be negative in either case. Alternatively, its effect on the rate of real depreciation is positive. This result is consistent with the characterization given of the relationships between budget deficit and balance of payments, whose changes are reflected primarily by movements in the exchange rate.

The characterization of the steady state is given by $\dot{\rho}_A = \dot{\rho}_B = \dot{\rho} = \dot{e} = \dot{m} = \dot{c}$. This implies that $y_A = \bar{y}_A$, $y_B = \bar{y}_B$, $y = \bar{y}$, $B = 0$, $T = 0$, $x = \bar{x}$, $h = \bar{h}$, $\dot{x} = 0$, $\dot{h} = 0$, *and* $G(q,v) = 0$. The long-run values of x and h are still given by

$$\bar{h} = \left(\frac{\phi}{\lambda}\right)\bar{y} - \left(\frac{1}{\lambda}\right)(i* + \dot{c}) \tag{22}$$

and

$$\bar{x} - \left[\frac{1}{\beta - \alpha}\right]\left[\eta\dot{c} + (\tau - \eta\phi)\bar{y} + \eta\lambda\bar{h} + \eta i^{*}\right] + p^{*}. \tag{23}$$

The long-run value of y can be recovered from (12), viz.,

$$\bar{y} = \left[\frac{\mu}{1 + \mu\sigma\phi R + \gamma - \gamma R}\right] \left[-(1 - R)\left(\delta\left(\frac{\delta_A}{\delta_B}\right) + \xi\right)\left(\bar{x}_A + p_A^{*}\right) + \delta(\bar{x} + p*) + \mu\sigma\lambda R\bar{h} + \mu\sigma R\dot{c}\right], \tag{24}$$

where $R = 1 - \rho(1 - \omega)$ as before. Therefore, $\bar{x}$, $\bar{h}$, and $\bar{y}$ are functions of $\bar{x}_A$, $p^*{}_A$, i^*, p^*, and $\dot{c}$, all of which are exogenously given. The three endogenous variables, $\dot{m}$, $\dot{\rho}$ and $\dot{e}$, can be derived in terms of $\bar{x}$, $\bar{h}$, and $\bar{y}$. The long-run steady-state representation of $\bar{y}_A$ is given by

$$\bar{y}_A = \left[\omega\delta\left(\frac{\delta_A}{\delta_B}\right) - (1 - \omega)\xi\right]\left(\bar{x}_A + p_A^{*}\right) - \omega\left[(\sigma\phi - \gamma)\bar{y} - \sigma\lambda\bar{h} - \sigma\dot{c}\right]. \tag{25}$$

11.3 Empirical Estimation

The structural model representing all of the above interactions and feedback effects can be written as

$$A_0 Y_t = A(L)Y_{t-1} + B(L)Z_t + \varepsilon_t$$

$$\varepsilon_t = C(L)\,\varepsilon_{t-1} + h_t \tag{26}$$

and its reduced form as

$$A_0 Y_t = \{C(L)A_0 + [1 - C(L)]A(L)\}Y_{t-1} + [1 - C(L)]B(L)Z_t + \eta_t \tag{27}$$

or, equivalently,

$$Y_t = D(L)Y_{t-1} + E(L)Z_t + \xi_t, \tag{28}$$

where

$$D(L) = A^{-1}{}_0\{C(L)A_0 + [1 - C(L)]A(L)\} \tag{28a}$$

$$E(L) = A^{-1}{}_0[1 - C(L)]B(L)], \tag{28b}$$

$$\xi_t = A^{-1}{}_0\eta_t, \tag{28c}$$

and A_0 is assumed to be nonsingular. The stacked vector of exogenous variables incorporates the interest rate, foreign agricultural prices, farm stocks, and government expenditure in agriculture. Our ultimate concern is not only about the own and cross dynamics of prices, exchange rates, and money – that is, about A_0, *A(L)*, and the covariance matrix of η – but also about *B(L)*. Also, if *C(L)* is diagonal, all structural disturbances depend only on their respective lagged values and not on the lagged values of the other disturbances.

A standard vector auto regressive (VAR) approach to (27) would not include the *Z* variables and, instead, would focus only on the (unconditional) four-variable representation of *Y*. To be sure, the unconditional multivariate distribution of *Y* is likely to be different from the conditional distribution of *Y* given *Z*, implying a dependence of the data-generating process of *Y* on the set, *Z*. The set, *Z*, is a set of weakly exogenous variables which we assume to be *jointly* exogenous.[3] In other words, we look at the set, *Z*, in its interactions with *Y* without considering all the possible interactions within the set.[4] Thus, if the *Z*'s are indeed relevant, there would be no way to recover the *E(L)* matrix polynomial from the unconditional VAR. In contrast, by estimating a four-variable VAR representation of *Y* conditional on *Z*, we can recover the *E(L)* matrix polynomial.

The issue of exogeneity is a delicate one. To formally test, some hypothesis on the joint distribution of the four endogenous variables and the set, *Z*, is required.

---

[3] We assume that the Z_t variables are weakly exogenous in the sense of Engle et al. (1983). The nonweak exogeneity of the Z_t variables would, of course, pose problems of consistency of the estimates.

[4] There can be interactions between the variables either due to their joint distributions or due to some feedback from the Y. We do not take the latter feedback into account here.

As we know, the implications of exogeneity vary, depending on the actual context in which the variables are jointly specified. In the present context in which the money supply variable is endogenously defined, the interest rate, which has already been determined in the model and has been equalized to the foreign rate (small country assumption), can be taken as exogenous. Foreign farm prices are the counterpart of domestic farm prices in the trade balance equation and are also assumed to be exogenous. Due to the endogeneity of prices and output, farm stocks also are taken as exogenous. Finally, government expenditure in agriculture is defined as exogenous, in that it is operated as a policy instrument to achieve some policy target.

A VAR representation of Y_t conditional on Z_t, cannot be directly implemented since the issues of nonstationarity and cointegration must be examined. Since the four endogenous variables have been shown to be nonstationary and cointegrated (Ardeni and Rausser 1990), the correct specification is then that of a *conditional vector error correcting model*, i.e., a VEC model in the four endogenous variables, Y_t conditional on a set of exogenous variables Z_t, as represented in

$$\Delta Y_t = D^*(L)\Delta Y_{t-1} + \gamma U'Y_{t-1} + E(L)Z_t\,\xi_t. \tag{29}$$

This representation is the *full* model, i.e., the more general formulation of the conditional VEC model. Our ultimate interest is not only the effect of those exogenous variables on the system but also how they affect its dynamic behavior. Thus, we will look at the MA representations of the model conditional on a set of exogenous variables and compare them to the unconditional MA representations of the basic model.

The interval period chosen for the estimates will be of quarterly data from 1972:1 to 1988:3. The endogenous variables will be LM2 (log of money supply as proxied by M2), LER (log of exchange rate), LNP (log of nonfarm price), and LFP (log of farm price). The exogenous variables will be IR (90-day treasury bill), LAIP (log of foreign agricultural prices), LAS (log of farm inventories), and LGEA (log of government expenditure in agriculture).

As far as the lag structure of the model is concerned, the evidence from Table 11.1 shows that IR and LGEA should enter the model with current values, whereas LAIP and LAS should not. This result is sensible because both the money supply and the exchange rate are expected to react quickly to changes in the interest rate brought about by changes in the domestic and foreign capital markets. Government expenditure, as a policy variable, is expected to counter expected movements in prices, not only to accommodate them. Finally, both foreign prices and farm stocks enter with lagged values since they tend to anticipate (predict) future values of actual market prices. Therefore, IR and LGEA will enter the model with their current and lagged values while LAIP and LAS will only enter with lagged values.

Several different models were estimated, all with a fourth-order lag structure (both in the endogenous and in the exogenous variables), with different sets of exogenous variables. The results of the estimation of every single model are not reported here. Here, we report only the MA representations of the full model. The

Table 11.1. Tests of significance of current values of the exogenous variables (fourth-order vector error correction model)

Variables[a]	$X^2(k)$	Significance value
	k = 4	
Current IR	10.522	0.032
Current LAIP	3.845	0.427
Current LAS	3.620	0.459
Current LGEA	10.198	0.037
	k = 8	
Current IR and LAIP	8.110	0.422
Current IR and LAS	5.580	0.694
Current IR and LGEA	31.266	0.000
Current LAS and LGEA	15.979	0.042
Current LAIP and LAS	6.967	0.540
Current LAIP and LGEA	21.452	0.006
	k = 12	
Current IR, LAIP, and LAS	8.772	0.722
Current IR, LAIP, and LGEA	27.718	0.004
Current IR, LAS, and LGEA	22.782	0.029
Current LAIP, LAS, and LGEA	18.084	0.113
Current LAIP, LAS, and LNP	18.078	0.113
	k = 16	
Current IR, LAIP, LAS, and LGEA	20.668	0.191

[a] IR = 90-day Treasury Bill;
LAIP = log of foreign agricultural prices;
LAS = log of farm inventories;
LER = log of exchange rate;
LFP = log of farm price;
LGEA = log of government expenditure in agriculture;
LM2 = log of money supply as proxied by M2;
LNP = log of nonfarm price.

impulse response functions of the full model are shown in Table 11.2. The model is estimated with the four exogenous variables, IR, LAIP, LAS, and LGEA. Here, we focus on only the MA representations obtained from the orthogonalized model (with a money-to-prices ordering). In most of the cases, in fact, the innovation correlation matrix remains almost unaltered and nearly diagonal.

Money innovations have a persistent decreasing effect on manufacturing prices, more than proportional to the increase in money supply in the long run.

The effect on farm prices is increasing in the short run but nil in the long run. Money supply expands in the short run but not in the long run. Thus, money does not appear to be neutral, both in the short and in the long run with respect to manufacturing prices, but is neutral with respect to farm prices. Farm prices increase faster than manufacturing prices in the short run, which means that farm prices appear to be less sticky than nonfarm prices. Finally, following a monetary unanticipated expansion, the exchange rate depreciates in the short run but appreciates in the long run.

The effects of a monetary shock in the conditional model are quite different from those obtained with the unconditional model. The depreciation of the exchange rate lasts only a few quarters, whereas its long-run appreciation is more than proportional to the initial increase in money supply. However, as the theoretical model would predict, after an increase in money supply we have an initial depreciation in the exchange rate is experienced which is followed by an appreciation in the long run. Conversely, the effect of a monetary shock on prices is ambiguous. Farm prices increase in the medium term but are unchanged in the long run; manufacturing prices steadily decrease.

There are two possible explanations for this result. The first is that the initial advantage in relative prices over the foreign sector is turned in favor of the farm sector when there is government intervention in agriculture. The second, more plausible, explanation is that expansions in money supply are subsumed by the interest rate and the fall in domestic manufacturing prices correctly "anticipates" the exchange rate appreciation that will come in the long run. In any case, this result seems to contradict the hypothesis that monetary shocks put the farm sector in a cost-price squeeze.

Exchange rate innovations have the expected effect. Manufacturing prices increase initially (by a small amount) but then decrease in the long run. It seems that nonfarm prices are stickier, although this result is more attributable to the slow pass-through of exchange rate changes to prices.[5] As in the unconditional model, farm prices respond faster but decrease less in the long run. Money supply does change in the short run. This result is consistent with the view that the effect of an exchange rate shock is reflected in the interest rate only after a while, while it contracts in the long run. Hence, as the model would predict, exchange rate appreciation has a deflationary effect which is dampened by government intervention for farm prices. Also, money supply increases in the short run but not in the long run. Finally, the trade balance indeed appears to adjust slowly to movements in the exchange rate.

The effects of manufacturing price innovations on farm prices and on the exchange rate are also as expected. However, the final money supply increase tends to be more similar to the long-run increase in manufacturing prices. This confirms the proposition of a neutral effect of price movements on money. The negative ef-

---

[5] This is, in effect, a J-curve effect. A negative 1% shock, in fact, would have the opposite effect of an initial decrease followed by an increase.

Table 11.2. MA representation – vector error correction model – exogenous variables: IR, LAIP, LAS, LGEA orthogonalized innovations (ordering: DLM2-DLER-DLNP-DLFP)

Effects of shocks to ξ_{M2} on:					Effects of shocks to ξ_{ER} on:			
LM2	LER	LNP	LFP	Quarter	LM2	LER	LNP	LFP
1.000	–0.011	–0.180	–0.122	1	0.000	1.000	0.063	–0.252
1.270	–0.022	–0.153	0.167	2	0.169	1.320	–0.053	–0.366
1.420	–0.298	–0.398	0.447	3	0.412	1.330	–0.258	–0.651
1.480	–0.212	–0.477	0.505	4	0.525	1.270	–0.285	–0.667
1.300	0.135	–0.609	0.607	5	0.372	1.140	–0.584	–0.494
1.300	0.465	–0.765	0.566	6	0.429	1.130	–0.848	–0.551
1.220	0.828	–0.716	0.467	7	0.349	1.390	–0.982	–0.632
1.080	1.020	–0.749	0.401	8	0.377	1.600	–1.090	–0.711
0.993	1.070	–0.830	0.374	9	0.375	1.760	–1.030	–0.745
0.823	1.120	–0.913	0.343	10	0.250	1.800	–0.973	–0.781
0.732	1.170	–1.070	0.276	11	0.173	1.750	–1.010	–0.787
0.705	1.260	–1.190	0.199	12	0.077	1.720	–1.050	–0.794
0.672	1.350	–1.240	0.146	13	0.030	1.740	–1.120	–0.822
0.660	1.390	–1.250	0.096	14	0.050	1.780	–1.160	–0.882
0.625	1.380	–1.240	0.076	15	0.067	1.810	–1.160	–0.904
0.575	1.350	–1.240	0.073	16	0.074	1.790	–1.140	–0.911
0.546	1.320	–1.260	0.064	17	0.068	1.750	–1.110	–0.903
0.529	1.310	–1.270	0.049	18	0.045	1.710	–1.100	–0.893
0.536	1.310	–1.270	0.035	19	0.042	1.680	–1.119	–0.887
0.550	1.300	–1.250	0.028	20	0.048	1.680	–1.110	–0.892

Table 11.2 (continued)

Effects of shocks to ξ_{NP} on:					Effects of shocks to ξ_{FP} on:			
LM2	LER	LNP	LFP	Quarter	LM2	LER	LNP	LFP
0.000	0.000	1.000	–0.427	1	0.000	0.000	0.000	1.000
–0.222	–0.254	1.510	–0.217	2	–0.483	0.173	0.104	0.782
–0.111	–0.957	1.870	–0.188	3	–0.529	0.472	–0.135	0.742
–0.135	–1.510	2.120	0.036	4	–0.742	0.914	–0.154	0.625
0.027	–1.950	2.110	–0.012	5	–0.842	1.170	–0.174	0.719
0.438	–2.170	2.120	0.149	6	–0.992	1.320	–0.156	0.469
0.657	–2.270	2.330	0.197	7	–1.070	1.350	–0.267	0.478
0.962	–2.420	2.490	0.329	8	–1.220	1.350	–0.373	0.436
1.070	–2.540	2.710	0.426	9	–1.270	1.370	–0.492	0.422
1.120	–2.680	2.820	0.621	10	–1.320	1.450	–0.571	0.296
1.150	–2.740	2.870	0.673	11	–1.290	1.500	–0.630	0.284
1.210	–2.710	2.870	0.735	12	–1.330	1.520	–0.597	0.241
1.250	–2.640	2.860	0.736	13	–1.350	1.490	–0.582	0.243
1.320	–2.580	2.870	0.762	14	–1.390	1.450	–0.572	0.222
1.320	–2.540	2.890	0.764	15	–1.400	1.420	–0.579	0.242
1.320	–2.530	2.870	0.791	16	–1.410	1.400	–0.579	0.222
1.290	–2.500	2.840	0.794	17	–1.390	1.400	–0.577	0.218
1.270	–2.460	2.790	0.795	18	–1.380	1.390	–0.557	0.210
1.260	–2.400	2.760	0.769	19	–1.360	1.360	–0.537	0.224
1.250	–2.350	2.730	0.755	20	–1.360	1.330	–0.517	0.228

For abbreviations, see Table 11.1.

fect on the exchange rate is strong and persistent, whereas the effect on farm prices is positive although small.

Farm price innovations have very unexpected effects. The resulting contractionary long-run decrease in money supply is more than proportional to the increase in farm prices so that farm prices are not neutral with respect to money. The decrease in manufacturing prices is, in any case, lower. In sum, price shocks are not neutral (in the sense that money supply does not increase proportionally in the long run), but regardless the agricultural sector does not seem to be pushed into a cost-price squeeze. Again, this is in line with the theoretical model of section 11.3 where movements in farm prices are fully reflected in movements in money and the exchange rate. However, the effects of the two prices are different, which seems to confirm that farm prices are supply driven and manufacturing prices are more demand driven.[6]

In conclusion, the analysis of the impulse response functions obtained with the conditional VEC model shows that, by explicitly accounting for the influence of relevant exogenous variables, some of the most important results already acknowledged in past studies are confirmed while some others are not. First, the neutrality proposition seems generally not to hold, particularly in the long run. Second, farm prices react more quickly to changes in money and the exchange rate; and, in the long run, their change is more than proportional to the change in manufacturing prices. Third, the backward effect from prices to money is less significant for manufacturing prices than for agricultural prices. Fourth, following a monetary shock, neither the exchange rate nor nonfarm prices overshoot their long-run value. Conversely, farm prices do overshoot their long-run value. These issues are explored more thoroughly in the next section through a set of specific hypotheses.

11.4 Tests of Specific Hypotheses

Both the *forward effect* and the *backward effect* hypotheses can be simply tested as linear restrictions on the block coefficients in the conditional VEC model. The *forward effect* hypothesis, i.e., the hypothesis that changes in money and in the exchange rates result in changes in prices, can be examined by testing the null hypothesis that lags in either money or the exchange rate do not affect prices. Similarly, the *backward effect* hypothesis, i.e., the hypothesis that changes in prices (output) result in changes in money and/or the exchange rate can be investigated by testing the null hypothesis that neither price effects money or the exchange rate. The first hypothesis encompasses a test of the *forward linkages*, whereas the second hypothesis encompasses a test of the *backward linkages*.[7]

---

[6] In the model, we specified agricultural output as supply determined and manufacturing output as demand determined.

[7] These tests are equivalent to those proposed by Sims (1980a, b); the issue of exogeneity is examined through tests of block restrictions; that is, the significance of a group of variables is tested against the null that the entire set of variables entering the equation is significant.

Table 11.3. Tests of two hypotheses of no feedback, vector error correction model – no exogenous variables

	No feedback from prices to money and exchange rate		
	Null hypothesis		Significance level
(1)	Lags in DLFP do not affect DLM2, DLER, DLNP	$\chi^2(12) = 36.13712$	0.00
(2)	Lags in DLNP do not affect DLM2, DLER, DLFP	$\chi^2(12) = 29.08430$	0.00
(3)	Lags in DLFP and DLNP do not affect DLM2 and DLER	$\chi^2(16) = 34.47610$	0.00
	No feedback from money and exchange rate to prices		
	Null hypothesis		Significance level
(4)	Lags in DLER do not affect DLM2, DLNP, DLFP	$\chi^2(12) = 11.60462$	0.47
(5)	Lags in DLM2 do not affect DLER, DLNP, DLFP	$\chi^2(12) = 11.76562$	0.46
(6)	Lags in DLER and DLM2 do not affect DLNP and DLFP	$\chi^2(16) = 17.24337$	0.37

For abbreviations, see Table 11.1.

Table 11.4. Tests of two hypotheses of no feedback, vector error correction model – exogenous variables: IR, LAS, LAIP, LGEA

	No feedback from prices to money and exchange rate		
	Null hypothesis		Significance level
(1)	Lags in DLFP do not affect DLM2, DLER, DLNP	$\chi^2(12) = 13.17504$	0.36
(2)	Lags in DLNP do not affect DLM2, DLER, DLFP	$\chi^2(12) = 20.09229$	0.07
(3)	Lags in DLFP and DLNP do not affect DLM2 and DLER	$\chi^2(16) = 22.99380$	0.11
	No feedback from money and exchange rate to prices		
	Null hypothesis		Significance level
(4)	Lags in DLER do not affect DLM2, DLNP, DLFP	$\chi^2(12) = 13.83154$	0.31
(5)	Lags in DLM2 do not affect DLER, DLNP, DLFP	$\chi^2(12) = 11.73416$	0.47
(6)	Lags in DLER and DLM2 do not affect DLNP and DLFP	$\chi^2(16) = 18.10038$	0.32

For abbreviations, see Table 11.1.

We have computed the tests of the feedback and the feedforward hypotheses for two VEC models. The first is the VEC model with no exogenous variable included. The second is the VEC model with four exogenous variables (IR, LAS, LAIP, and LGEA). The tests for the first model are reported in Table 11.3. All of the dynamic interactions are analyzed in a "closed" system, with no intervening effect from the outside. The hypothesis that prices do not affect money or the exchange rate cannot be rejected at any reasonable significance level. This implies that we are not able to reject the null hypothesis that there is no feedback from either price to money or the exchange rate. The hypothesis that money and the exchange rate do not affect prices, on the other hand, can be rejected. This implies that there is feedback from money or the exchange rate to prices.

Table 11.4 reports the results of the tests for the conditional VEC model. Here, the evidence does support both hypotheses. Both the hypothesis that lags in prices do not affect money and the exchange rate and the opposite hypothesis, that lags in money and in the exchange rate do not affect prices, can be rejected at any acceptable level.

In summary, while the evidence from these two sets of tests is mixed, it confirms that a proper test of the feedback hypotheses conducted with the fully con-

Table 11.5. The money equation

Variable	Lag	(1)	(2)
DLM2	1	0.124	0.243
DLM2	2	0.385	0.315
DLM2	3	–0.078	–0.042
DLM2	4	0.024	–0.004
DLER	0		–0.001
DLER	1	0.010	0.001
DLER	2	0.009	–0.009
DLER	3	–0.046	–0.035
DLER	4	–0.026	–0.032
DLNP	0		–0.383
DLNP	1	–0.643	–0.371
DLNP	2	0.732	0.687
DLNP	3	–0.158	–0.193
DLNP	4	0.355	0.255
DLFP	0		–0.189
DLFP	1	–0.424	–0.406
DLFP	2	0.009	–0.003
DLFP	3	–0.209	–0.167
DLFP	4	–0.038	0.066
		Sum of cofficients of	
DLM2		0.456	0.512
DLER		–0.052	–0.087
DLNP		0.273	–0.035
DLFP		–0.651	–0.700

For abbreviations, see Table 11.1.

Table 11.6. The exchange rate equation

Variable	Lag	(1)	(2)
DLM2	0		–0.315
DLM2	1	–0.068	0.500
DLM2	2	–1.990	–1.630
DLM2	3	0.768	0.728
DLM2	4	0.359	0.256
DLER	1	0.386	0.340
DLER	2	–0.007	–0.085
DLER	3	0.064	0.066
DLER	4	–0.017	0.024
DLNP	0		–0.460
DLNP	1	–1.270	–0.836
DLNP	2	–3.260	–2.970
DLNP	3	–0.389	–0.413
DLNP	4	0.245	–0.279
DLFP	0		–1.160
DLFP	1	0.831	0.624
DLFP	2	1.240	1.330
DLFP	3	0.897	1.020
DLFP	4	0.372	0.815
		Sum of cofficients of	
DLM2		–0.943	0.470
DLER		0.433	–0.346
DLNP		–4.784	–4.966
DLFP		3.439	2.632

For abbreviations, see Table 11.1.

ditional model would give different but more trustworthy results. In the unconditional case, we are expectedly led to the conclusion that money affects prices but not vice versa. In the conditional case, the conclusion is that, overall, the linkages are evident in both directions.

In moving from the estimated reduced form of the model to the structural form, one obvious issue that arises is simultaneity. If all the variables entered the right-hand side of the model in lagged form, then the form of the exogeneity tests reported above are appropriate for testing those restrictions in the reduced form needed to recover the structural form. However, because the matrix of the contemporaneous coefficients is not necessarily null, we must proceed by testing restrictions that enable us to recover the structural coefficient matrix.

In Tables 11.5 to 11.8, the four equations in our model are reported. In the first column of each table, the coefficients are given for the reduced-form equation estimated with the four endogenous variables, then each equation could be interpreted as a structural equation.

If we interpret the money equation as a structural equation, then it is not expected that the sum of the coefficients of nonfarm prices will be positive. By al-

Table 11.7. The manufacturing price equation

Variable	Lag	(1)	(2)
DLM2	0		–0.174
DLM2	1	0.108	0.226
DLM2	2	–0.281	–0.157
DLM2	3	0.097	0.084
DLM2	4	–0.049	–0.062
DLER	0		–0.006
DLER	1	–0.016	–0.022
DLER	2	–0.021	–0.031
DLER	3	0.024	0.017
DLER	4	–0.045	–0.036
DLNP	1	0.593	0.529
DLNP	2	–0.072	0.081
DLNP	3	–0.112	–0.129
DLNP	4	–0.095	–0.153
DLFP	0		–0.238
DLFP	1	0.067	–0.023
DLFP	2	–0.128	–0.880
DLFP	3	0.003	0.007
DLFP	4	0.089	0.167
	Sum of cofficients of		
DLM2		–0.125	–0.083
DLER		–0.058	–0.080
DLNP		1.074	0.328
DLFP		0.041	–0.175

For abbreviation, see Table 11.1.

lowing for simultaneity, we can examine the robustness of this result. We can then re-estimate the same equation with contemporaneous values of the other endogenous variables and investigate the sum of the coefficients. If contemporaneous values are indeed irrelevant, then the sum of coefficients at the bottom of columns 1 and 2 should be equal. Only for nonfarm prices does this not seem to be the case. Nonfarm prices appear to be simultaneously determined with money, at least to a degree. If contemporaneous values of money are significant in the nonfarm equation, then we should conclude that either money and nonfarm prices are simultaneously determined or both forward and backward linkages are present.

The analysis of Table 11.5 shows that lagged values of exchange rate and farm prices are significant in the money equation. This, by the way, confirms the findings of feedback from farm prices to money obtained above. Table 11.6 shows the results of the estimation of the exchange rate equation. The sum of the coefficients of farm prices do not have the expected sign. As it turns out, it appears to be fairly robust to the simultaneity assumption as it decreases but remains positive. On the contrary, the sum of money coefficients is very sensitive turning from negative to

Table 11.8. The farm price equation

Variable	Lag	(1)	(2)
DLM2	0		–0.299
DLM2	1	0.409	0.520
DLM2	2	0.309	0.015
DLM2	3	–0.051	0.110
DLM2	4	–0.074	–0.068
DLER	0		–0.010
DLER	1	–0.035	–0.037
DLER	2	–0.060	–0.056
DLER	3	0.005	0.011
DLER	4	0.061	0.009
DLNP	0		–0.828
DLNP	1	0.316	0.396
DLNP	2	0.073	0.281
DLNP	3	0.066	–0.120
DLFP	4	–0.506	–0.472
DLFP	1	–0.089	–0.114
DLFP	2	0.127	0.095
DLFP	3	0.165	0.122
DLFP	4	0.353	0.415
		Sum of cofficients of	
DLM2		0.600	0.276
DLER		–0.037	–0.135
DLNP		0.045	–0.743
DLFP		0.470	0.518

For abbreviations, see Table 11.1.

positive. Clearly, there is a simultaneity problem between money and the exchange rate.

Table 11.7 reports the results of the estimation of the manufacturing price equation. Here, money and exchange rate appear fairly robust to the assumption of simultaneity, while farm prices do not. Since manufacturing prices appear to have a significant impact oncontemporaneous money, but money does not, we may conclude that there is a substantial feedback from money to prices. Finally, Table 11.8 shows the estimation results for the farm price equation. Here the sum of money coefficients is positive, but robust to simultaneity. Thus, feedback from money to prices is also confirmed.

In conclusion, both the exogeneity tests and the tests of identifying restrictions confirm the importance of the statistical implications of using the conditional VEC model, as opposed to the unconditional model. Through the latter, in fact, we would have rejected the hypothesis that there is any feedback from prices to money and the exchange rate in any case. However, as we have shown above, given that the endogenous variables of the model are jointly conditional on a given set of exogenous variables, any such unconditional model appears misspecified. Since the "unconditional" is the approach taken in most of the literature on VAR models, the earlier results must be viewed suspiciously. The conditional model has a theoretical basis; without this basis, it would be difficult to justify any set of particular exogenous variables. As it turns out, in fact, the results from the conditional model confirm that much of evidence is being swept aside when relevant exogenous variables are omitted in the unconditional model. The tests of the forward and the backward hypotheses confirm that all such linkages are indeed present.

11.5 Some Policy Simulations

Another purpose of this study is the analysis of the effects of government intervention in agriculture; in particular, how the economy would react if such an intervention were reduced. Obviously, this is a matter of model "experimental" analysis involving a set of simulations of the possible effects of such changes. That is, we can simulate hypothetical values of government expenditure and examine the changes in the dynamic relationships among the endogenous variables. The scope of such an exercise is clear: tracing how the dynamic paths of the endogenous variables would have changed if government expenditure in agriculture had been different.

To implement such an exercise, we have to specify possible scenarios. Of course, it is not realistic to assume that government expenditure could have been different while the other variables remain unchanged. With no significant loss in realism, we have taken into account only the likely direct effect that a change in government expenditure would have on farm inventories. Since total farm stocks include stocks accounted for under loans to the Commodity Credit Corporation (part of total government expenditure in the farm sector), when specifying possible scenarios we have considered alternative effects of changes in government expenditure on farm inventories. Hence, the simulations we report are based on the effects that a combined action on government expenditure and farm inventories might have on the dynamic behavior of the model.

In particular, since our interest relates to how a *decrease* in government intervention would affect the farm sector, we have simulated the model under six different scenarios. We have considered the hypothetical situation in which, following the 1981 Farm Bill, the entire US government policy toward the farm sector would have changed, causing a dramatic decrease in government expenditure on agriculture. The six scenarios, all beginning in 1981:1, can be summarized as follows:

1. Government expenditure in agriculture decreases $150 million a year, starting from $1,200 million in the first quarter of 1981 (approximately the trend value at that time), decreasing every year during the second and third quarters and rising again in the fourth quarter, and ending with almost no expenditure in 1988:3. Total farm inventories are unchanged.

2. Government expenditure follows the same pattern as above. Farm inventories are 1.5% less the first year (1981), 3.0% less the second year (1982), and so forth. In 1988 they are 12% less (assuming this is how much absence of government expenditure would affect farm inventories).
3. Government expenditure is 10% less than what it actually was in 1981, 20% less in 1982, 30% less in 1983, and so forth. Farm inventories are unchanged.
4. Government expenditure follows the same pattern as above. Farm inventories are 1.5% less the first year (1981), 3.0% less the second year (1982), and so forth. In 1988 they are 12% less (assuming this is how much absence of government expenditure would affect farm inventories).
5. Government expenditure falls to zero (approximately) by the end of 1981. Farm inventories are unchanged.
6. Government expenditure follows the same pattern as above. Farm inventories are 12% less than what they actually were from 1981:1 to 1988:3.

All scenarios represent policies of decreasing intervention in agriculture, where the patterns of such "withdrawal" are different. In the first four they are gradual; in the last two, they take effect almost immediately. Under the first scenario (results shown in Table 11.9), monetary shocks have a quite different impact (for all comparisons, see Table 11.2). The effect on the exchange rate is the same in the short run, but the long-run appreciation effect is less pronounced. Farm prices increase almost by the same amount in the short run, but their long-run value is now higher. Manufacturing prices, after an initial decrease, now increase. Thus, the gradual decrease in government expenditure would have made the long-run effect of monetary shocks on prices more pronounced. Exchange rate shocks have a persistent effect on money in the long run, a now positive but small effect on manufacturing prices, and the same decreasing effect on farm prices. Even in this case, the sign of nonfarm price changes is reversed.

Shocks to nonfarm prices have the same short-run effect but a different long-run effect. Money supply now contracts, while farm prices are basically unchanged. The exchange rate, instead of steadily depreciating, reverts to its initial level; thus, reduced government expenditure in agriculture would have made unexpected price increases less effective. Shocks to farm prices have basically the same effect on money supply, in both the short and the long run. Conversely, the effect on the exchange rate is now basically nil, while it was actually much stronger. The effect on nonfarm prices is basically the same (deflationary) as is the own effect on farm prices.

In conclusion, it seems that having reduced government expenditure in agriculture beginning in 1981 would have had some significant effects. The feedbacks from money to prices would have been more evident and lasted longer, while those from the exchange rate would change only mildly. The feedbacks from prices to money, on the other hand, would have been just the same, while those from prices to the exchange rate would have been strengthened.

The second scenario is similar to the previous one, but now farm inventories are assumed to have decreased, although by a small amount, beginning in 1981 (results are shown in Table 11.10). A monetary shock now has a longer deprecia-

Table 11.9. MA representation – full model – simulation 1 (ordering: DLM2-DLER-DLNP-DLFP)

Effects of shocks to ξ_{M2} on:					Effects of shocks to ξ_{ER} on:			
LM2	LER	LNP	LFP	Quarter	LM2	LER	LNP	LFP
1.000	–0.044	–0.110	–0.143	1	0.000	1.000	0.111	–0.242
1.240	0.014	0.109	0.126	2	0.210	1.440	0.194	–0.518
1.320	–0.172	–0.018	0.323	3	0.437	1.570	0.269	–0.743
1.430	–0.184	0.012	0.474	4	0.527	1.560	0.361	–0.826
1.130	–0.073	0.169	0.457	5	0.417	1.610	0.345	–0.724
1.120	0.036	0.163	0.454	6	0.350	1.630	0.321	–0.810
1.050	0.104	0.341	0.443	7	0.363	1.720	0.297	–0.942
0.977	0.105	0.373	0.412	8	0.420	1.790	0.297	–1.030
1.030	0.101	0.318	0.405	9	0.496	1.800	0.322	–1.010
1.000	0.121	0.284	0.434	10	0.515	1.760	0.365	–1.040
1.030	0.154	0.205	0.389	11	0.537	1.730	0.349	–1.040
1.090	0.199	0.172	0.364	12	0.536	1.720	0.347	–1.030
1.080	0.216	0.176	0.354	13	0.544	1.730	0.353	–1.020
1.080	0.207	0.183	0.355	14	0.567	1.740	0.368	–1.050
1.040	0.193	0.214	0.355	15	0.589	1.740	0.379	–1.050
1.000	0.177	0.217	0.355	16	0.597	1.730	0.393	–1.030
1.000	0.172	0.217	0.352	17	0.600	1.730	0.401	–1.030
0.997	0.173	0.216	0.347	18	0.597	1.730	0.403	–1.030
1.010	0.173	0.204	0.339	19	0.602	1.740	0.400	–1.030
1.030	0.171	0.197	0.342	20	0.605	1.740	0.405	–1.030

Effects of shocks to ξ_{NP} on:					Effects of shocks to ξ_{FP} on:			
LM2	LER	LNP	LFP	Quarter	LM2	LER	LNP	LFP
0.000	0.000	1.000	–0.265	1	0.000	0.000	0.000	1.000
–0.195	0.056	1.580	–0.241	2	–0.595	–0.121	0.152	0.836
–0.034	–0.240	1.860	–0.127	3	0.677	–0.100	–0.248	0.817
0.003	–0.411	1.970	0.079	4	–0.947	0.061	–0.338	0.840
0.235	–0.381	1.780	–0.025	5	–1.070	0.197	–0.427	0.920
0.637	–0.153	1.570	0.047	6	–1.110	0.253	–0.433	0.677
0.844	0.023	1.580	–0.000	7	–1.180	0.214	–0.526	0.703
1.050	0.125	1.610	–0.010	8	–1.290	0.138	–0.593	0.742
0.996	0.151	1.770	0.008	9	–1.360	0.109	0.636	0.762
0.839	0.134	1.870	0.088	10	–1.430	0.102	–0.683	0.669
0.695	0.141	1.950	0.062	11	–1.410	0.086	–0.751	0.679
0.607	0.182	1.980	0.026	12	–1.420	0.062	–0.735	0.681
0.620	0.221	1.950	–0.024	13	–1.420	0.026	–0.744	0.681
0.693	0.251	1.910	–0.039	14	–1.420	–0.008	–0.754	0.683
0.752	0.264	1.880	–0.065	15	–1.430	–0.003	–0.761	0.715
0.808	0.268	1.840	–0.062	16	–1.440	–0.047	–0.761	0.715
0.813	0.271	1.830	–0.057	17	–1.430	–0.050	–0.760	0.717
0.798	0.274	1.840	–0.057	18	–1.430	–0.056	–0.759	0.720
0.775	0.275	1.860	–0.066	19	–1.420	–0.063	–0.757	0.732
0.750	0.270	1.880	–0.066	20	–1.430	–0.067	–0.752	0.735

For abbreviations, see Table 11.1.

ting effect on the exchange rate. Nonfarm prices increase less than farm prices in the short run but more than farm prices in the long run. Thus, under this scenario, not only is the effect of money persistent, it tends to put the farm sector in a cost-price squeeze over the long run.

Exchange rate shocks have a persistent effect on money in the long run, a positive and persistent effect on manufacturing prices, and the same decreasing effect on farm prices. Simulating different values of farm inventories does not seem to have any significant additional effects in this case, as all the dynamic responses are the same as under the first scenario. Also, the same seems to be true for nonfarm price shocks, since they have the same short-run effect but a different long-run effect.

Conversely, under this scenario the effects of farm price shocks are quantitatively quite different from what they were over the simulated history. Money supply decreases, the exchange rate appreciates (but less), and nonfarm prices decrease (but more). Hence, a reduction in government expenditure in agriculture coupled with a reduction in total farm inventories would have had two combined effects. On one hand, the protection of the farm sector would have been lessened. Monetary shocks would have had a more pronounced effect; and while farm prices would have reacted faster in the short run, in the long run, the farm sector would have been pushed in a cost-price squeeze. On the other hand, while all the feedback from farm prices to money would have been just the same as they have actually been, the feedback to the exchange rate would have been stronger, accentuating the instability in the foreign exchange market.

The third scenario does not generate any new insights. Apparently, little change would have occurred in the examined variables if the reduction in government expenditure in agriculture had been set very gradually, as the scenario hypothesizes. This is also interpreted as an indication of the significant influence that government intervention has on the dynamics of the model. For the fourth scenario, the same basic results are generated. In this case, it is the variation in the farm inventory variable that strengthens the feedback from farm prices to the exchange rate.

In the fifth scenario, government expenditures on agriculture would have been reduced to zero by the end of 1981 (results given are in Table 11.11). The effect of such a policy would have been dramatic. The effect of a monetary shock would have had almost no impact on both prices, in both the short and the long run, and the same unaltered effect on the exchange rate. Exchange rate revaluations would have had a more negative effect on farm prices in the medium term, but a stronger negative effect on manufacturing prices in the long run. Conversely, the effects of farm price innovations would have been dampened, particularly those on the exchange rate.

In conclusion, under this scenario, feedbacks from money to prices would have lessened as well as those from farm prices to the exchange rate, while feedbacks from the exchange rate to prices and those from farm prices to money would have been the same. The apparent contradiction with the results obtained under the first scenario can be explained in terms of how the public forms expectations about

Table 11.10. MA representation – full model – simulation 2 (ordering: DLM2-DLER-DLNP-DLFP)

Effects of shocks to ξ_{M2} on:					Effects of shocks to ξ_{ER} on:			
LM2	LER	LNP	LFP	Quarter	LM2	LER	LNP	LFP
1.000	–0.063	–0.057	–0.143	1	0.000	1.000	0.116	–0.263
1.190	–0.034	0.246	0.139	2	0.209	1.450	0.191	–0.560
1.270	–0.269	0.193	0.359	3	0.409	1.570	0.274	–0.780
1.370	–0.330	0.333	0.558	4	0.491	1.530	0.373	–0.856
1.080	–0.271	0.569	0.560	5	0.333	1.570	0.384	–0.781
1.110	–0.189	0.612	0.573	6	0.249	1.590	0.359	–0.882
1.090	–0.135	0.810	0.604	7	0.259	1.670	0.353	–1.020
1.060	–0.131	0.857	0.591	8	0.319	1.720	0.367	–1.120
1.170	–0.120	0.812	0.604	9	0.407	1.720	0.389	–1.120
1.170	–0.084	0.794	0.651	10	0.442	1.670	0.428	–1.140
1.220	–0.036	0.738	0.622	11	0.467	1.620	0.422	–1.130
1.260	0.024	0.730	0.605	12	0.477	1.600	0.436	–1.120
1.230	0.053	0.746	0.597	13	0.486	1.600	0.468	–1.110
1.220	0.058	0.758	0.594	14	0.510	1.600	0.506	–1.130
1.170	0.059	0.784	0.586	15	0.533	1.590	0.540	1.110
1.140	0.058	0.778	0.574	16	0.545	1.570	0.569	–1.090
1.130	0.063	0.766	0.562	17	0.557	1.560	0.588	–1.080
1.140	0.071	0.755	0.549	18	0.566	1.560	0.601	–1.070
1.160	0.074	0.735	0.537	19	0.580	1.570	0.610	–1.070
1.170	0.730	0.725	0.537	20	0.593	1.580	0.624	–1.060

Table 11.10 (continued)

Effects of shocks to ξ_{NP} on:					Effects of shocks to ξ_{FP} on:			
LM2	LER	LNP	LFP	Quarter	LM2	LER	LNP	LFP
0.000	0.000	1.000	–0.229	1	0.000	0.000	0.000	1.000
–0.231	0.025	1.640	–0.232	2	–0.540	–0.081	0.119	0.865
–0.112	–0.318	2.000	–0.123	3	–0.594	–0.170	–0.330	0.840
–0.083	–0.551	2.220	0.112	4	–0.790	0.184	–0.505	0.882
0.155	–0.594	2.190	0.053	5	–0.909	0.382	–0.676	0.924
0.558	–0.446	2.120	0.159	6	–0.929	0.515	–0.801	0.687
0.821	–0.345	2.210	0.181	7	–1.030	0.546	–0.981	0.681
1.050	–0.290	2.330	0.237	8	–1.170	0.529	–1.130	0.681
1.060	–0.286	2.530	0.306	9	–1.300	0.553	–1.240	0.643
0.952	–0.302	2.680	0.410	10	–1.430	0.582	–1.360	0.518
0.852	–0.285	2.820	0.416	11	–1.460	0.585	–1.490	0.478
0.794	–0.234	2.900	0.398	12	–1.500	0.570	–1.540	0.442
0.825	–0.183	2.910	0.368	13	–1.530	0.540	–1.610	0.406
0.903	–0.142	2.910	0.367	14	–1.540	0.502	–1.680	0.387
0.969	–0.113	2.890	0.354	15	–1.570	0.467	–1.720	0.393
1.020	–0.089	2.850	0.358	16	–1.580	0.443	–1.750	0.379
1.030	0.066	2.850	0.361	17	–1.590	0.424	–1.770	0.373
1.010	–0.047	2.870	0.355	18	–1.600	0.399	–1.780	0.376
0.994	–0.034	2.890	0.342	19	–1.600	0.375	–1.780	0.386
0.965	–0.032	2.910	0.335	20	–1.610	0.356	–1.780	0.391

For abbreviations, see Table 11.1.

Table 11.11. MA representation – full model – simulation 5 (ordering: DLM2-DLER-DLNP-DLFP)

Effects of shocks to ξ_{M2} on:					Effects of shocks to ξ_{ER} on:			
LM2	LER	LNP	LFP	Quarter	LM2	LER	LNP	LFP
1.000	–0.048	–0.174	–0.222	1	0.000	1.000	0.102	–0.242
1.260	0.073	–0.023	0.240	2	0.231	1.450	0.135	–0.562
1.260	–0.046	–0.069	0.078	3	0.478	1.630	0.112	–0.773
1.400	–0.008	0.011	0.100	4	0.561	1.570	0.108	–0.883
1.160	0.040	0.064	0.064	5	0.463	1.670	–0.042	–0.825
1.210	0.092	0.005	0.053	6	0.465	1.740	–0.255	–0.919
1.190	0.148	0.102	0.005	7	0.527	1.880	–0.445	–1.030
1.190	0.176	0.102	–0.050	8	0.598	1.940	–0.578	–1.140
1.270	0.170	0.101	–0.042	9	0.645	1.960	–0.661	–1.120
1.230	0.163	0.111	–0.019	10	0.592	1.900	–0.730	–1.090
1.260	0.167	0.081	–0.056	11	0.537	1.830	–0.849	–1.040
1.280	0.196	0.072	–0.055	12	0.453	1.790	–0.963	–1.010
1.280	0.206	0.063	–0.069	13	0.397	1.790	–1.050	–0.999
1.310	0.212	0.063	–0.078	14	0.363	1.790	–1.100	–1.000
1.300	0.209	0.068	–0.076	15	0.332	1.770	–1.130	–0.984
1.300	0.200	0.052	–0.067	16	0.288	1.710	–1.140	–0.947
1.290	0.198	0.046	–0.066	17	0.234	1.666	–1.170	–0.902
1.290	0.203	0.037	–0.071	18	0.181	1.630	–1.200	–0.877
1.290	0.207	0.030	–0.074	19	0.149	1.620	–1.220	–0.863
1.290	0.207	0.032	–0.072	20	0.129	1.610	–1.220	–0.860

Table 11.11 (continued)

Effects of shocks to ξ_{NP} on:					Effects of shocks to ξ_{FP} on:			
LM2	LER	LNP	LFP	Quarter	LM2	LER	LNP	LFP
0.000	0.000	1.000	−0.327	1	0.000	0.000	0.000	1.000
−0.235	0.025	1.670	−0.283	2	−0.582	−0.049	0.105	0.897
−0.770	−0.358	1.980	−0.006	3	−0.594	0.041	−0.124	0.769
−0.143	−0.592	2.220	0.220	4	−0.874	0.148	−0.154	0.657
−0.074	−0.510	2.310	−0.031	5	−0.937	0.247	−0.151	0.644
0.263	−0.169	2.460	−0.131	6	−0.943	0.302	−0.067	0.477
0.473	0.029	2.770	−0.292	7	−0.957	0.236	−0.134	0.589
0.771	0.082	3.040	−0.390	8	−1.000	0.128	−0.196	0.646
0.865	0.039	3.240	−0.372	9	−1.040	0.132	−0.246	0.650
0.894	0.021	3.270	−0.316	10	−1.060	0.179	−0.311	0.598
0.963	0.120	3.300	−0.411	11	−1.020	0.202	−0.340	0.619
1.070	0.299	3.320	−0.552	12	−1.040	0.184	−0.315	0.604
1.230	0.426	3.360	−0.662	13	−1.050	0.152	−0.319	0.626
1.350	0.472	3.440	−0.699	14	−1.090	0.114	−0.336	0.668
1.400	0.461	3.480	−0.715	15	−1.120	0.096	−0.362	0.687
1.430	0.455	3.450	−0.703	16	−1.130	0.107	−0.375	0.669
1.430	0.483	3.410	−0.723	17	−1.130	0.120	−0.379	0.666
1.470	0.543	3.380	−0.776	18	−1.130	0.114	−0.370	0.670
1.510	0.595	3.370	−0.821	19	−1.140	0.099	−0.358	0.680
1.550	0.607	3.370	−0.832	20	−1.150	0.082	−0.357	0.692

For abbreviations, see Table 11.1.

government policy. Under the first scenario, the reduction in the expenditure by the government is gradual and, therefore, its effects are anticipated. Under this scenario, the reduction is sudden and the interactions among the variables cannot adjust to this unanticipated change. For the sixth simulation, the same basic results are generated. Again, this confirms that the crucial policy variable is government expenditure and that inventory policies of the public sector are less crucial.

11.6 Conclusion

The major conclusion drawn from the set of simulations conducted here is that government intervention in agriculture has indeed mattered. The dynamics of the variable path responses to unanticipated shocks significantly change under altered values of the government expenditure variable. Having not had government intervention in the eight years from 1981 to 1988, or having had it gradually reduced to zero since 1981, would have pushed the farm sector in a cost-price squeeze in the long run and would have made it more vulnerable to money and exchange rate shocks.

A gradual decrease in government expenditure makes the long-run effect of monetary shocks on prices more pronounced and unexpected price increases less effective. The feedbacks from money to prices are stronger and last longer, while those from the exchange rate are unaltered. On the other hand, the feedbacks from prices to money do not change, while the feedbacks from prices to the exchange rate appear to be strengthened.

If the decrease in government expenditure is coupled with a reduction in total farm inventories, the effect of money is more persistent and the farm sector is pushed toward a cost-price squeeze in the long run. Monetary shocks have a more pronounced effect and farm prices react faster in the short run. On the other hand, while all the feedbacks from farm prices to money are basically unaltered, the feedback to the exchange rate would have been stronger, accentuating the instability of the foreign exchange market.

Two other conclusions can be drawn from the set of simulations and, more generally, from the theoretical and empirical analysis. First, not only has government intervention in agriculture mattered but it also has had some positive effects. Government support partially reduces the impact of unanticipated monetary shocks and prevents the farm sector from being pushed into cost-price squeezes, although it tends to make the feedbacks from farm prices to the exchange rate more pronounced. Secondly, by explicitly recognizing the existence of the forward and backward feedbacks among money, the exchange rate, and prices, we are able to account for the effects that government expenditure reductions can have on the monetary farm-sector linkages as well as farm price shock, money, and exchange rate linkages.

All past claims that reductions in government support of the agricultural sector would have made the sector more vulnerable to the negative impacts of monetary and exchange rate shocks have been based on forward linkages from money and the exchange rate to prices, while neglecting any backward linkages. In this study we have shown that feedbacks are significant in both directions and have also shown that money and exchange rate shocks affect prices. Thus, any reduction in government expenditure in agriculture affects the path by which price shocks feedback on money and the exchange rate. From a policy perspective, this is very important, since it implies that any change in government support of the farm sector should be evaluated from an integrated market point of view. This more integrated or global perspective is needed because expenditures and budget deficits, monetary, exchange rate, and farm policies are significantly related and their interactions far too strong to be neglected.

References

Ardeni PG, Rausser GC (1990) Interactions among money, exchange rates, and commodity prices. Working Paper. Department of Agricultural and Resource Economics, University of California, Berkeley

Barnett RC, Bessler DA, Thompson RL (1983) The money supply and nominal agricultural prices. Am J Agric Econ 65:303–307

Barro R, Grossman H (1976) Money, employment, and inflation. Cambridge University Press, New York

Bessler DA (1984) Relative prices and money: a vector autoregression on Brazilian data. Am J Agric Econ 66:25–30

Bordo M (1980) The effects of a monetary change on relative commodity prices and the role of long-term contracts, J Polit Econ 88:1088–1109

Cairnes JE (1874) Some leading principles of political ecoomy newly expounded. Harper, New York

Chambers RG (1981) Interrelationships between monetary instruments and agricultural commodity trade. Am J Agric Econ 63:934–941

Chambers RG (1984) Agricultural and financial market interdependence in the short run. Am J Agric Econ 66:12–24

Chambers RG, Just RE (1979) A critique of exchange rate treatment in agricultural trade models. Am J Agric Econ 61:249–257

Chambers RG, Just RE (1982) An investigation of the effect of monetary factors on agriculture. J Monetary Econ 9:235–247

Cukierman A (1979) The relationship between relative prices and the general price level: a suggested interpretation. Am Econ Rev 69:444–447

Cukierman A, Wachtel P (1979) Differential inflationary expectations and the variability of the rate of inflation: theory and evidence. Am Econ Rev 69:595–609

Cukierman A, Wachtel P (1982) Relative price variability and nonuniform inflationary expectations, J Polit Econ 90:146–157

Devadoss S, Meyers WH, Johnson SR (1986) Exchange rates, trade deficits, and US prices. Working Paper 86-WP-10, Center for Agricultural and Rural Development, Iowa State University, Ames

Dornbusch R (1976) Expectations and exchange rate dynamic. J Polit Econ 84:1161–1176

Engle RE, Hendry DF, Richard JF (1983) Exogeneity, Econometrica 51:277–304

Fischer S (1982) Relative price variability and inflation in the United States and Germany, Eur Econ Rev 18:171–(196

Frankel J (1986) Expectations and commodity price dynamics: the overshoooting model. Am J Agric Econ 68:344–348

Gardner B (1981) On the power of macroeconomic linkages to explain events in US agriculture. Am J Agric Econ 63:871–878
Hercowitz Z (1981) Money and the dispersion of relative prices. J Polit Econ 89:328–356
Hicks Sir JH (1974) The crisis in Keynesian economics. Blackwell, Oxford
Mussa M (1982) A model of exchange rate dynamic. J Polit Econ 90:74–104
Okun AM (1975) Inflation: its mechanics and welfare costs. Brookings Paper on economic activity vol. 2:351–401
Orden D (1986) Money and agriculture: the dynamics of money-financial market-agricultural trade linkages. Agric Econ Res 38:14–28
Parks RW (1978) Inflation and price variabililty. J Polit Econ 86:79–95
Rausser GC (1985) Agriculture, trade, and macroeconomics. Working Paper No 407, Department of Agricultural and Resource Economics, University of California, Berkeley
Rausser GC, Chalfant JA, Love HA, Stamoulis KG (1986) Macroeconomic linkages, taxes, and subsidies in the US agricultural sector, Am J Agric Econ 68:399–412
Rotemberg JJ (1982) Sticky prices in the United States. J Polit Econ 90:1187–1211
Schuh GE (1974) The exchange rate and US agriculture. Am J Agric Econ 56:1–13
Schuh GE (1976) The new macroeconomics of agriculture. Am J Agric Econ 58:802–811
Sims CA (1980a) Macroeconomics and reality. Econometrica 8:1–48
Sims CA (1980b) Comparison of interwar and postwar cycles: monetarism reconsidered. Am Econ Rev 70:250–257
Stockton DJ (1988) Relative price dispersion, aggregate price movement, and the natural rate of unemployment. Econ Inquiry 26:1–22
Tweeten L (1980) An economic investigation of inflation pass-through to the farm sector. West J Agric Econ 5:89–106
Vellianitis-Fidal A (1976) The impact of devaluation on US agricultural exports. Agric Econ Res 28:107–116
Vining DR, Elwertowski TC (1976) The relationship between relative prices and the general price level. Am Econ Rev 66:699–708